contents

INTRODUCTION V

ACKNOWLEDGEMENTS IX

1 ABOUT TIME 1

2 THE FUTURE IN HISTORY 13

3 VARIETIES OF FUTUROLOGY 36

4 THE RECENT FUTURE 59

5 SCIENCE FUTURES 78

6 POPULATION 95

7 ENERGY AND THE CLIMATE CRUNCH 116

8 ENERGY: A NEW REGIME? 134

9 WATER 157

10 FOOD SUPPLY 171

11 HUMANS AND OTHER SPECIES 194

12 FUTURE HEALTH 214

13 FUTURE WAR 238
14 DEALING WITH DISASTER 257
15 LIFE, SOCIETY AND VALUES 277
16 THE WORLD OF INFORMATION 296
17 MOVING ON, MOVING UP 315
18 INFINITE IN ALL DIRECTIONS 333

PICTURE CREDITS 355
INDEX 356

THE ROUGH GUIDE to

the future

by

Jon Turney

WW ... n

For Danielle, naturally

Credits

The Rough Guide to The Future

Editing: Tracy Hopkins, Richie Unterberger
Layout: Ajay Verma
Diagrams: Katie Lloyd-Jones
Proofreading: Jason Freeman, Kate Berens
Production: Rebecca Short
Cover design: Tom Cabot/ketchup

Rough Guides Reference

Director: Andrew Lockett
Editors: Kate Berens, Peter Buckley,
Tom Cabot, Tracy Hopkins, Matthew Milton,
Joe Staines, Ruth Tidball

Publishing information

This first edition published November 2010 by
Rough Guides Ltd, 80 Strand, London WC2R 0RL
11, Community Centre, Panchsheel Park, New Delhi 110017, India
Email: mail@roughguides.com

Distributed by the Penguin Group:
Penguin Books Ltd, 80 Strand, London WC2R 0RL
Penguin Group (USA), 375 Hudson Street, New York 10014, USA
Penguin Group (Australia), 250 Camberwell Road, Camberwell,
Victoria 3124, Australia
Penguin Group (NZ), 67 Apollo Drive, Mairangi Bay,
Auckland 1310, New Zealand

This paperback edition published in Canada in 2010. Rough Guides is represented in Canada by Tourmaline Editions Inc., 662 King Street West, Suite 304, Toronto, Ontario, M5V 1M7

Printed in Singapore by Toppan Security Printing Pte. Ltd.

376 pages; includes index

A catalogue record for this book is available from the British Library

ISBN: 978-1-85828-781-2

1 3 5 7 9 8 6 4 2

introduction

For I dipt into the future, far as human eye could see,
Saw the vision of the world and all the wonder that would be.

Alfred Tennyson, "Locksley Hall" (1842)

Of course this examination of the future is a Rough Guide. What else could it be? But even if you reckon that the future is unknowable – and I'm bound to agree – we humans spend an awful lot of time and energy thinking and talking about it.

Lately, those efforts have expanded in lots of ways. As well as the usual motley band of seers, prophets and pundits, astrologers and readers of runes, we have scenario-writers in think-tanks, horizon scanners in corporate planning units, foresight teams in government departments, all armed with more or less helpful methods for trying to see how where we are now influences where we can go next.

There are also more complexities and confusions in our thinking about the future now than there were in the past. On the one hand, the future has expanded – to an almost unimaginable extent for cosmologists looking into the fate of the whole universe. But even for the earthbound, our contemplation of the future over decades, centuries and millennia takes place against the background of a planet that will be around for many millions of years yet – even if we are not. And our smaller, personal futures (if we're lucky) stretch further too. Living to be a hundred may soon be the norm, not the exception, in the developed world. In other words, if you're in your teens, there's a real chance you'll get to see the twenty-second century.

On the other hand, the future seems more constrained. This is partly because more of us know more about the state of the world as it is now, and what we know gives us a sharper sense of the big problems, such as food supply, energy, climate change, environmental stress and species loss. Add in already established anxieties like nuclear war and pandemics, and – while we're not quite sure what form they'll take – there is a widespread sense that the rest of the century will feature some pretty bad things.

At the same time, there are some apparently well-informed people who think that we'll not only survive these perils, but come out the other side ready to embrace our destiny as some kind of superhumans – through genetic enhancement, brain-boosting drugs or perhaps linking our "wetware" brains with ultra-intelligent computers. Then again, there are others for whom this is exactly the kind of future they fear.

If you look around the ever-expanding mediasphere (itself a fast-changing landscape that invites speculation about the future), you'll encounter an uneasy mix of attitudes – ones we all probably share at various times. We view the future fearfully, and are often in denial. We are a bit cynical, or at least disbelieving about people who say they know what to expect – not least because of the many past "forecasts" which have disappointed. We also seem nostalgic for futures past. We're still hoping for the jet pack and the flying car, the domed cities and undersea hotels. All pretty silly ideas, but we still want them! Finally, even if we maintain that there is nothing truly useful that can be said about the future, we still think about it, in some way, almost every day.

HOW MANY FUTURES?

The Rough Guide to the Future looks at all of this – the history of ideas about the future, different ways of thinking about it, current attitudes to it and, of course, what the future might hold for six and a half billion human beings (and counting) over the coming decades.

One reason why a guide to the future has to be rough is the scope of the subject. The future pervades our lives. Every tool and device, every design, every advertisement, every contract, treaty or statute, every racing tip and investment analysis, and every marriage ceremony embodies, in its own way, a story about possible futures. On top of these everyday encounters with the future, we're aware that more and more of the decisions we make – from protecting an endangered species to building a power station – have consequences that reach further and further down the timeline. So any single-volume treatment of "the future" will inevitably omit many things that will be part of someone's future.

Our interest in the future is part of human beings' habit of putting themselves into a narrative – if possible, one that encloses an individual life in a larger story. So a book which is looking forward ought to see how far into the distant future we can make the story go. But the vast majority of thinking about the future relates to the medium term, and that is where there's the most material to choose from. From that area, I have selected topics because a) they're the issues most people are concerned about, b) they'll have the most important effects (often, but not always, the same thing) and c) there is more to build on from knowledge of the past and the present. That means there's a bias in the book towards basic needs like energy and food, rather than the more complex features of cultures and ideas, and the approach is global rather than focusing on the details of life in any particular place.

THE PLAN

Whatever their attitude to the medium-term future, most people seem convinced that it'll be eventful. So a survey of what we might imagine happening in the rest of this century, or what people have said will happen, takes up much of the book. Before that, chapters 1 to 4 examine where the future came from, in the uniquely human perception of time – and mortality – and the changing idea of the future in history. There's also a review of some of the ways people have tried to make their thinking about the future a little more disciplined, if not necessarily more certain.

Chapters 5 and 6 address two topics that seem essential to underpin any discussion of our current prospects: science (including technology) and population. This is both because they're important, and because we know something about how where we are now affects where we will be in the next few decades. These are followed by chapters 7 to 14 on the global basics – energy and climate, water, food, biodiversity, health, war and potential disasters. Though nothing is certain, there are trends that point fairly clearly in particular directions. These chapters are more or less self-contained, but their topics are interconnected in many ways.

The book then moves into the still more uncertain waters of life and culture with chapters 15 and 16, including a brief foray into politics, society and values. In Chapter 17 it takes a more detailed look at the possibility that technology will move mankind to a place where all that has gone before is irrelevant, as human (or perhaps post-human or "transhuman") history enters a new era. Excitingly, there are people who believe that this will happen before the end of the century. I consider how seriously we can take such a view, and how the arguments supporting it square with the multiple crises that seem likely to crop up in relation to energy, water and food before such an era might exist.

Finally, in Chapter 18 the book takes a tour of the far future which, apparently paradoxically, is a less uncertain realm than the rest of this century. Apparently paradoxically because it can only really be discussed in very general terms, informed by the sciences of geology, physics and cosmology, which aren't too concerned with small details like the emergence – or disappearance – of human life on Earth.

A decent guidebook should steer you round the main territory, but also offer further exploration if you want it. I've done this by mentioning important reports, books, people and organizations in the text where relevant, giving just enough detail so you can find them easily yourself on the Internet. I've also ended each chapter with a small number of recommended books and websites for further reading on the main topics.

FACTS AND OPINIONS, HOPES AND FEARS

How can this be a book of fact, when there are no facts about the future? Well, there are facts here, of several kinds: facts about the past and the present, and facts in the journalistic sense of accurately reporting what other people have said on the subject. Much of this is, of course, opinion, speculation or simple assertion. On the whole I've favoured other people's (often expert) opinions over mine, but some of my own views are also present. These belong to the elusive class of, as it were, temporary facts. That is, I meant what I said when writing them, but whether I still agree with all my ideas now is another matter. The future, after all, is a moving boundary.

To help do justice to other people's opinions, the book also features a range of disparate current views about possible futures from fifty thoughtful futurologists, scientists and other experts. Many of those experts have offered complex overviews of their specialist topics, or of the whole picture, but I tried to get a fix on their opinions by asking them all just three simple questions.

- What is your highest hope for what will happen?
- What is your worst fear?
- What is your best bet for what will actually occur?

In each case, they chose the topic of most interest or concern to them, considered in what direction it might go over the next fifty years, and came up with a brief reply. Their answers are spread throughout the eighteen chapters of this book.

As you will see, they tend to confirm that our collective view of the future, as we move further into the twenty-first century, is rather mixed. I tend to agree. And to give you a better idea of the overall opinions that shape this guide's selection of future signposts, here are my own answers to those three questions:

> **Highest hope**: We navigate through the eye of the needle of the middle decades of the century well enough to allow the bottom billion a real chance of a humane life.

> **Worst fear**: The environmental calamity so many informed scientists predict gathers pace faster than our efforts to forestall it.

> **Best bet**: Crises, muddling through and continuing vast inequalities are the order of the day. In spite of that, it remains, technologically and culturally, the most fascinating of times to be alive.

If you want to discuss just how fascinating the near future will be, or to suggest updates or additions to anything covered in these pages, you can follow my blog at www.unreliablefutures.wordpress.com or via the Twitter account @RoughFutures.

Jon Turney, 2010

acknowledgements

Gathering information and ideas about something as hazily defined as the future has been immensely stimulating and, at times, challenging. Those who helped me cope with it – and tested out some of my notions about how it could all fit together in one book – have mounted up nearly as fast as the downloaded PDFs, notes from the impressive range of futurist blogs and websites, and older style reports, books and journal articles.

They include groups who responded to talks about bits of work in progress – at Bristol Science Café, the science festival Fest in Trieste, the Alta Scuola Politecnica 2009 Winter School, Bardonecchia (thanks to Massimiano Bucchi), and the UK Futures Analysts' Network (thanks to Brian Bader). Network meetings where I did not get to talk were also extremely useful. Others who supplied documents, ideas, tips, encouragement and useful conversation included Ehsan Masood, Oliver Morton, Richard Sandford, Jerry Ravetz, David Roden, Adam Nieman, Adam Gordon, Martin Ince, Peter Reiner, Peter D. Smith and Andy Sawyer.

The editorial team at Rough Guides – Andrew Lockett, Peter Buckley and Ruth Tidball – gave invaluable feedback in impressive quantities (their collated comments would add up to a longish chapter), and Tracy Hopkins then licked the whole thing into shape.

I am also immensely grateful to all those who read parts of the text in progress and commented on what was missing, or misguided. Again in no particular order they included: Peter Washer, Edward Wawrzynczak, Clare Dudman, Pat Spallone, David Roden, Jack Stilgoe, Peter Reiner, Richard Sandford, Felicity Mellor and Jim Baggott. Closer to home, various family members were attentive first readers, including Danielle, Mike, David and Richard Turney, and Eleanor Turney, whose keen editorial eye did her father proud.

Sincere thanks also to other essential facilitators – my agent Louise Greenberg, and the authors of two software tools which supported the project, Scrivener (offline) and Ubernote (online).

Lastly, I'd like to acknowledge Christopher Freeman and the late Marie Jahoda, editors of the landmark post-*Limits to Growth* volume *World Futures: The Great Debate*. That collection, published in 1978, first really persuaded me that there could be serious futures literature, and that I might want to go into it in depth one day. That itch took a long time to develop into this book, and in some ways it now looks like a very different world. But I'd have to say that anyone who lays their tome down alongside this one will see that many of their preoccupations are the same.

1

about time

about time

It is not until an act occurs within the landscape of the past and the future that it is a human act.

Ursula K. Le Guin, *The Dispossessed*

Thinking about the future means looking ahead in time, a trick humans have made their own. But how special is it, and how does it blend imagination with knowledge of the past and the present? Our sense of time, especially of how much of it there might be, has changed throughout history, and our sense of the future has expanded to match.

TALES OF THE FUTURE

Listen for a moment to the people in this trio of stories:

> I cannot convey the sense of abominable desolation that hung over the world. The red eastern sky, the northward blackness, the salt Dead Sea, the stony beach crawling with these foul, slow-stirring monsters, the uniform poisonous-looking green of the lichenous plants, the thin air that hurts one's lungs: all contributed to an appalling effect.
>
> H.G. Wells, *The Time Machine*

They had lunch at one of the five-star Rejoov restaurants, on an air-conditioned pseudobalcony overlooking the main Compound organic-botanics greenhouse. Crake had the kanga-lamb, a new Australian splice that combined the placid character and high protein yield of the sheep with the kangaroo's resistance to disease and absence of methane-producing, ozone-destroying flatulence.

Margaret Atwood, *Oryx and Crake*

As for human beings making a comeback, of starting to use tools and build houses and play musical instruments and so on again: they would have to do it with their beaks this time. Their arms have become flippers ... Each flipper is studded with five purely ornamental nubbins, attractive to members of the opposite sex at mating time. These are in fact the tips of four suppressed fingers and a thumb. Those parts of people's brains which used to control their hands, moreover, simply don't exist any more, and human skulls are now much more streamlined on that account. The more streamlined the skull, the more successful the fisher person.

Kurt Vonnegut, *Galapagos*

These people are all in the future. But they are also all creations of people in the past, yours and mine. H.G. Wells showed us the world winding down in the far future. Margaret Atwood, a hundred years later, depicted a near future of organisms engineered for our convenience at the table. And the late, great Kurt Vonnegut portrayed an alternate far future in which the world carries on quite happily without recognizable humans, as mankind have evolved into happy, semi-aquatic beasts with no recollection of civilization.

Of course, whatever evolution, real or artificial, may hold, we cannot hear actual voices from the future. Yet we all imagine futures every day – though perhaps not ones as dramatic as these. Even mundane future-imaginings like planning to stop on the way home to shop for the evening meal are a uniquely human thing to do. There is a small amount of evidence suggesting that chimpanzees, our closest relatives among the primates, can plan – zoo-bound chimps have been observed hoarding stones to fling when visiting time comes around. But even chimps generally appear to be trapped in the present.

Other creatures which seem to realize complex projects are following an instinctual programme, rather than anticipating their future needs. Bees and beavers are both builders, but neither are architects, as someone with his own strong vision of the future, Karl Marx, pointed out in *Capital, Volume 1*: "What distinguishes the worst architect from the best of

PREDICTING... OR "NEXTING"?

American psychologist Daniel Gilbert, in his entertaining dip into the psychology of satisfaction *Stumbling on Happiness* (2006) agrees that human beings are the only animals that think about the future. But he points out a confusion. We use one word – prediction – to mean two very different kinds of things.

One is common to all creatures, and involves assessing what is going to happen in the next few moments, in "the immediate, local, personal future". All living things do this somehow. It does not require conscious thought. Watch a sea-slug that has been trained to avoid an electric shock or a human ball-player calculating the trajectory of a catch on the run: both are acting on an expectation of the future course of events worked up by their nervous system. This automatic or unconscious prediction also applies to things that are more distinctively human – readers, for instance, continually predict how a sentence will turn out, but do not notice unless the writer has done something unusual.

The other kind of conscious prediction is uniquely human. Predicting anything from interest rates and population growth to next week's Top Ten demands some level of considered thought, prior knowledge or shrewd deduction. But, more importantly, it requires a brain that can handle the concept of later, whether that is later today, later this year or later this millennium. It is an exercise of the imagination.

Gilbert says it would clarify things to use a new term for the first kind of unconscious future-mapping, and he suggests calling it "nexting". Though his argument seems logical, it seems unlikely to catch on somehow. But we will not be needing the term again here anyway, because this book is about predicting, not nexting. Exactly what it means to make a prediction, however, is another matter.

bees is this, that the architect raises his structure in imagination before he erects it in reality."

Invoking Marx, the prophet of revolution, underlines another feature of human thought about the future: imagining it does not make it so. A human architect trumps a beaver because he or she can design something that becomes a real building unlike any ever built before. Or it may remain a fantasy, a palace of the mind. Most will lie somewhere in between. Something gets built, but falls short of the splendid structure the architect really wanted. The budget is tight, the materials not quite up to it, the workforce cut corners, and the people who finally move in have different ideas about how to use the space. Predictions which extend beyond the immediate future nearly always fail. As the old line has it, nothing is certain except death and taxes (though the super-rich who keep their money in offshore accounts have largely dealt with one of those, and may soon have ideas about tackling the other).

THE LIVES TO COME

Prediction may be dodgy but it is addictive. Perhaps that's because the kind of highly developed brain that can hold a conscious image of the future also confers a great but unwelcome insight: death. We all know that there will be a great discontinuity in our futures – our own deaths. Here other creatures are again radically different from humans. Despite the myth of elephant graveyards, animals don't plan for their deaths, and they either ignore it when one of their number dies or appear mildly puzzled. Chimps sometimes cling onto a dead baby, but this looks instinctual. Lions may tidy up their dead by eating them, and elephants show some signs of recognizing elephant remains, but only humans bury their dead. Animals tend to ignore them.

In its broadest sense, the various afterlives that different cultures hope and plan for are part of the future. Archaeologists have found evidence of human burials going back a hundred thousand years or more. Tombs in Neolithic Britain were designed to receive burials from successive generations, suggesting that their builders were looking into the future of their tribe, as well as preparing for the time after death. In some cases, life after death seems to have been taken rather literally. Burial of "grave goods" perhaps reached a peak with the Vikings and Anglo-Saxons. A thousand years ago, a Viking warrior's voyage to Valhalla was helped along by interring him in his ship, complete with his weapons, armour, food and drink, live animals, birds and even slaves.

This ritual effort to ensure a smooth transition to the afterlife by provisioning for the future was overtaken when the Vikings took up Christianity, which commits the body to God. Other religions, of course, have their own routes to salvation. Some, such as Hinduism, involve repeated reincarnation, with the possibility of ascending to a more enlightened state. But this guide is not going to venture further into that religious or philosophical territory. Our tour is restricted to the possible futures which people may experience while they are alive, and how those futures have been imagined.

The facts of mortality do, however, undoubtedly affect our capacity to think about the future. For some, admitting the reality of death is the end of real hope, and leads to a feeling that the future is not worth caring about – they ask, along with Groucho Marx, what's posterity ever done for me? If you can retain an interest in the future, imagining the stuff that will happen after you've gone calls for an extra effort. It's also subject to a slight contradiction. On both small and large scales, our sense of the future is bound up with our sense of history.

REMEMBERING THE FUTURE

When you contemplate the future, the past is all you really have to go on, which is why futurists often start out as historians. The same is true for all of us in our personal lives. In fact, the way we build memories may be specifically tailored to allow us to efficiently anticipate situations yet to come.

Our memories seem to work by storing individual pieces of past experience separately, as part of a complicated, interconnected web, rather than as a chronological list or ordered library of discrete records. Our brains then assemble recollections of past episodes by adding together bits of information that seem to be related. This makes memory unreliable in crucial ways. For example, you may recall seeing two people in a particular place at the same time, when you actually noticed them there on different days. So why can't our brains have a memory system that is more like a videotape, which can simply be replayed to recapture the scene correctly?

Psychologists Daniel Schachter and Donna Addis of Harvard University suggest that the videotape would be less useful for imagining the future. When we think about the future, we are trying to anticipate events which will not exactly repeat the past, but will be something like it. Once, perhaps, people needed to think through the possibilities for a hunt, a trek to new hunting grounds or a fight. Today it might be anything from a board meeting, to a date… to a fight. A memory system which allows you to review sketches of past events, and recombine them to imagine new ones is a more flexible guide to possible futures, even though it makes mistakes in pure recollection.

According to Schachter and Addis, this idea is backed up by several lines of research. Amnesiacs sometimes lose their sense of the future as well as their past. Some deeply depressed people have little specific grip on the past or the future. And neuroimaging shows activity in the same regions of the brain whether the people being scanned are remembering past episodes or imagining future ones. Other supporting evidence comes from studies suggesting that older people who find it difficult to recall past events also have trouble imagining new ones. Further specific experiments to test the idea are under way, but the results lie, well, in the future.

PREDICTIONS FILE

Anne Skare Nielsen

Chief futurist and partner, Future Navigator, Denmark

Highest hope: That the majority of the world's inhabitants will come to the sensible conclusion that if we keep on asking others to change, nothing grand will ever happen. That we – as Buddhists say – have to be the change we want to see in other people. We should stop instructing and start constructing. I hope that we can let go of our need to control, learn to "listen louder" and co-create better solutions that will bring out the best in people.

Worst fear: That it is too late. That we have pushed the limits of the planet too far, and created an imbalance with nature so deep that our only hope is that aliens are watching ready to save us from the brink of disaster.

Best bet: That the world's corporations and sensible decision makers will see that nurturing a world of happy, healthy, sustainable and surplus-creating people is just good business.

This intricately carved Aztec stone in Mexico City's National Anthropological Museum is thought to be a fifteenth-century calendar.

In any case, the more history – the more experience – you have had personally, the easier it is for you to think in the longer term. Younger people, with long lives still ahead of them, often seem least able to think their way into the future, even though they will be around for more of it than their elders. This is one reason why teenagers' outlooks may differ from those of people in their sixties or seventies (though there may be others!).

DOWN THE GENERATIONS

Our sense of past and future is also strongly geared to the length of a human life, even though that is not an immediately obvious span. Being conscious of the passage of time does not necessarily lead to an image of time as we tend to think of it now – something measurable in agreed units of hours, days, weeks or years, something divisible and always passing. Very early societies probably saw time as marked by recurrent patterns – time's cycle, not time's arrow. The most obvious cycles were ecological, and seasonal. Someone living off the land, attuned to the changes of the seasons, would know which of their neighbours had been around for more summers or winters, but beyond those comparisons precision was hard to come by.

Astronomical events were more regular than ecological ones, so those ancient people who watched the skies closely had a basis for a more formal analysis of time. The Mayan civilization of the Americas, for example, knew that the year had 365 days from long observation of the phases of the moon and the movements of Venus. This was the basis for a calendar regulated by a caste of astronomer priests, who set the dates of key ceremonies followed by all. Their time, though, like their cultivation of maize, was still cyclical.

Across all cultures, the units for reckoning time past and time future can be years, or centuries – but generations have a stronger grip on our collective imagination. It is, surely, hard to really get your head around the idea of people living more than a few tens of generations away from us in either direction? True, the ideas of reincarnation or rebirth which figure in some of the major religions, and the notion of an afterlife which features in others, allow the belief that an individual's history might extend into eternity. Yet specific images of future lives are more likely to be restricted to hundreds or thousands of years rather than millions.

In the West, this restricted future used to be a mirror image of an abbreviated past. The scholarly seventeenth-century Irish archbishop James Ussher

(1581–1656), poring over the genealogies in the Bible, filling in a few gaps and linking some of the people named to historic events, came up with his famous calculation that the world was created in 4004 BC. It seemed eminently reasonable at the time – not much more than a hundred lifetimes since Adam.

Since Ussher's day, the past has been colonized by science, and has expanded enormously, if not immeasurably. The Victorians were gripped by the discovery of geological time. We in the West credit its announcement to the pioneering Scottish geologist James Hutton (1726–97), though he was partly anticipated in Chinese writing of the eleventh century. Even in the West few came across Hutton's *Theory of the Earth* (1795), as it was a notably turgid read until you reached his striking conclusion about the vastness of geological time. He explained that the rock formations showed "no vestige of a beginning, no prospect of an end". In other words, you can think about the Earth's sights as evidence of the continuous processes that form it (such as erosion and deposition), and these processes have been going on for far longer than anyone had contemplated before (see box overleaf).

By the early nineteenth century, Victorian intellectuals were digesting Charles Lyell's much more readable *Principles of Geology* (1830–33) with its vistas of mountain-building and erosion spanning hundreds of thousands, millions, and even hundreds of millions of years. And, as Hutton had implied, if there was that much history, why shouldn't the future be equally huge?

A new sense of scale

We contemplate a lifetime, but mainly think of shorter intervals. "Days are where we live", according to death-obsessed poet Philip Larkin. Getting a sense of just how much more time there is, beyond the days or years we can see passing by, is hard enough today. Imagine how much harder it was for the Victorians, who were the first in the history of Western culture to face up to the sheer extent of the past, and possibly the future.

Sigmund Freud, a product of mid-nineteenth-century Vienna, saw part of the human predicament in two great blows to our self-esteem. First, Copernicus's news that the Earth was not the centre of the universe set minds reeling. Then, four hundred years later, Darwin showed that human beings were descended from animals, that we were not separate, privileged or specially created in any way. Palaeontologist Stephen Jay Gould (1941–2002), contemplator of evolution and Earth history, reckoned Freud missed a third blow, between these two: the replacement of a young Earth, ruled by humans almost since it began, with an immensely old planet, almost all of whose history preceded ours.

When the numbers get so large, words fail. Douglas Adams' *The Hitchhikers' Guide to the Galaxy* (1979) informed us that "Space is big – really big – you just won't believe how vastly, hugely mind-bogglingly big it is. You may think it's a long way down the road to the chemist, but that's just peanuts to space." Adams makes fun of people who have tried to come up with metaphors for all that immensity, and challenges them to do better than him.

With time it is harder still because time itself remains mysterious. This book is not about that mystery – which fills volumes on physics and metaphysics – but it's worth noting that there are only a few ways we tend to describe time. It's like a river flowing from past to future. Or it's something all around us which we pass through. Either way, when you think about it, we are really talking about time as a kind of space.

JAMES HUTTON

In the 1770s, an Edinburgh farmer and intellectual thought long and hard about weathering, erosion, the origins of soil and the kinds of rock found in his native Scotland. Then he brought it all together in one grand theory in which molten rock welled up from the Earth's fiery insides. This led to new, solid rock on the surface, which was very slowly weathered away, making soil and undersea sediments. The sediment would sink down, be swallowed up by the Earth and eventually emerge again, reheated, as new molten rock. This theory of planetary recycling was inspired by ideas about the circulation of the blood. But it had a vital extra ingredient – lots and lots of time.

How much time? Hutton demonstrated that the gradual processes at work – deposition and erosion – would take far longer than James Ussher's biblical timescale to create landscapes. But he was not in the business of making estimates. In his mind, these processes were, in effect, eternal. Hutton's successors in Victorian England, who developed more detailed theories of geology based on his uniformitarian principles, were soon discussing the Earth in terms of millions, or even tens of millions, of years.

www.james-hutton.org.uk

That fits with the everyday words which get attached to time, too (long, short, passing slowly or too fast). It is certainly a challenge to find another way to think about time. The term coined by American writer John McPhee to describe geological time seems hard to improve on. What did the Victorians grapple with?, he asks. *Deep* time.

And it is, of course, really, really deep. McPhee offers a vivid and oft-quoted analogy for understanding the temporal insignificance of human history. It does not involve calendars or clocks, or human beings appearing in the final minutes of the last hour of the year. It is spatial *and* related to the body. Think of Earth's history as the old English yard – which was the distance from the king's nose to the tip of his outstretched hand. One pass of a file across the nail on his index finger removes all of human history.

McPhee also quotes a geologist on the benefits of working with deep time, every day, and adjusting your time sense accordingly: "If you free yourself from the conventional reaction to a quantity like a million years, you free yourself a bit from the boundaries of human time. And then in a way you do not live at all, but in another way you live forever."

COSMIC TIME

Forever is a long time for geologists, but cosmologists go further still. Since Lyell and Hutton, the scientific story has expanded even further, but lost its symmetry. The most common view departs from James Hutton: there *was* a beginning, of sorts. Life began with the "Big Bang", which inaugurated the cosmos, or at least this cosmos. But the end, if there is one at all, is a lot further away. The universe is thought to be around fourteen billion (that's fourteen thousand million) years old. Rather a lot of lifetimes. But mainstream opinion holds that the whole shebang will go on for much, much longer. It may get old, cold and boring, but there's likely to be a universe of sorts out there for, oh, at least ten trillion more years.

It's a thrill as well as a challenge to confront timescales like this. They make us feel small, but they can still be thought about. So, in that sense, they are spanned by our minds, or at least our mathematics. And while cosmologists speculate happily about endless futures, a few heroic academics now practise what they call "big history" – in which they try to integrate what we know about the past, over a complete range of timescales, into one vast narrative. David Christian's *Maps of Time: An Introduction to Big History* (2004) does the job in a mere five hundred pages. But how easily can we get away from our old habit of thinking on the scale of a single generation, or a few lifetimes in each direction, even if a lifetime is a bit longer on average than it used to be?

From our current, science-based understanding, thinking about the future – or the past – calls for serious recalibration of our sense of time. It is not just that there is far, far more of it than previously supposed. It's also that the sheer extent also takes in processes which work on many different timescales, most of them outside our normal intuitive grasp. When the poet Andrew Marvell speculated about matters "vaster than empires and more slow" in the seventeenth century, he did not know how vast or how slow they might really be. Nowadays we have a grasp of the evolution of the universe, the drift of continents, the change of species, over and above the more immediate sense of the movements of human life and culture. (There are also many processes and events invisible to previous science which take vanishingly small amounts of time, but they can be ignored here.) At the higher levels, the biggest picture of how things work includes a range of timescales that look something like this:

- Tens of years: the normal range of the personal (a career, a marriage, raising children)
- One hundred years: a lifetime (with luck)
- One thousand years: historical time (the rise and fall of a civilization is generally less than this)
- Ten thousand years: archaeological time (agriculture and large human settlements go back about this far)
- One hundred thousand years: anthropological time (culture emerged somewhere in this range)
- One million–ten million years: primatology (apes turning into hominids, the Grand Canyon formed)
- One hundred million years: recent evolutionary past (reptiles' reign ended, mammals rule)
- One billion years: geological time and deep evolution (continents shift significantly, development from simple to complex creatures)
- Five billion years: planetary time (age of the Earth: 4.5 billion)
- Ten billion years: cosmological time (age of the universe: 13 billion)

Not all of these scales are relevant to human concerns about the medium-term future. But the fact that we know, for instance, that it takes hundreds of millions of years to turn plants into oil, but only about a century to use half of it up, does suggest that the larger scales still relate to human decision-making. And while democracy is a fine thing, it does seem to fix politicians' attention to the next election, and perhaps the one after that, not to anything more than a few years away.

FROM DEEP TIME TO THE LONG NOW

Getting around our fixation with the short-term to instil a real awareness of the possibilities of a vast future calls for some creative thinking, and some

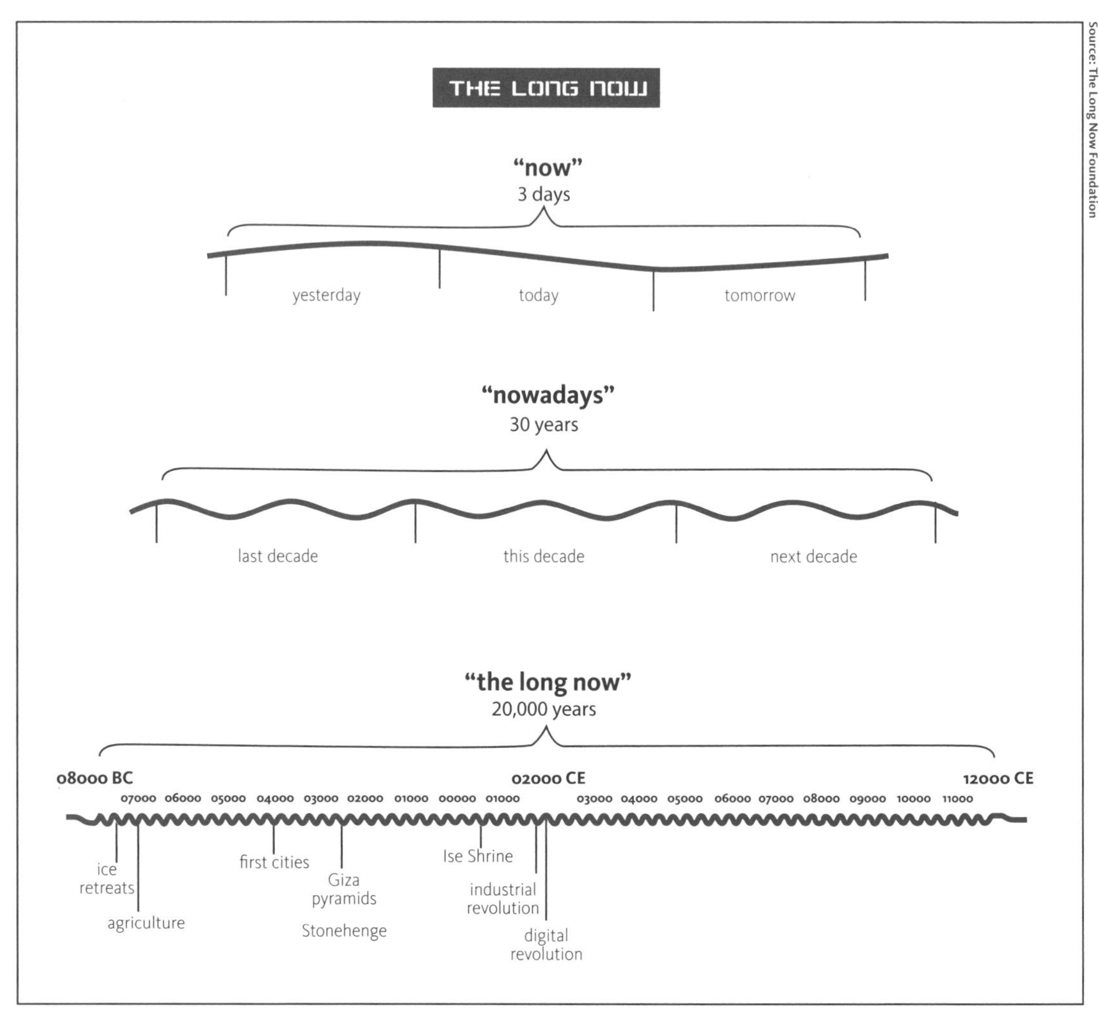

new cultural projects. And that's the idea behind The Long Now Foundation (www.longnow.org), set up at the instigation of computer scientist Danny Hillis and cultural entrepreneurs Stewart Brand and Brian Eno. Their mission statement says that "Civilization is revving itself into a pathologically short attention span. … Some sort of balancing corrective to the short-sightedness is needed – some mechanism or myth which encourages the long view and the taking of long-term responsibility, where 'long-term' is measured at least in centuries."

The foundation's best-known scheme is the planned Clock of the Long Now, an ultra-reliable timepiece which will last ten thousand years, roughly the same amount of time that human civilization has been developing since the end of the last Ice Age. It will operate really, really slowly. It will "tick" once a year, and sound out once a century. The first prototype has already been built, and the foundation has bought a site in Nevada for the final version. Building the actual device will be a long-term project, but it has already been realized in fictional form in Neal Stephenson's monumental science-fiction novel *Anathem* (2008).

The transformation in perception that the group hopes to bring about makes sense in the context of the extremely extended timescales we now consider when we look at the evolution of life, or the universe. As the foundation's diagram (opposite) tries to indicate, from this point of view the "long now" might be tens of thousands of years. The diagram also highlights another set of questions. How can we imagine what milestones might occur in the blank stretch to the right of the present, to match the key events noted in the past? And how can we think about what is to come? The next chapter begins to look at that in more detail by reviewing some of what people have thought about the future in the past.

FURTHER EXPLORATION

Stewart Brand The Clock of the Long Now: Time and Responsibility (2000) Brand explains the forward-thinking manifesto of The Long Now Foundation, alongside details about designing the clock itself.

Encouraging a long-term view: the prototype for the Clock of the Long Now, due to chime once every hundred years.

FORWARD INTO THE FUTURE?

"But Irish had an old soul, you might say. He was a man with a great future behind him, already."

Angela Carter, *Wise Children*

Carter's line pins down a particular kind of disappointment, and journalists like to use it as a neat put-down. But do we have to move *forward* into the future? It feels natural to English-users and, it seems, to speakers of most of the rest of the world's languages. But not quite all of them. A study reported around the world in 2006 showed that there is at least one group who take a very different view. For the Aymara Amerindians of Bolivia, Peru and Chile, the past is in front of them, the future behind.

This seems strange at first sight, though it is not quite as radical as it appears – the Aymara are still imagining time in terms of space. But the inversion of what we regard as the normal perception of the timeline is unusual. The Aymara word for past, transcribed as *nayra*, actually means eye, sight or front. Their word for future is *q'ipa*, which also means behind or the back. Cognitive scientist Rafael Núñez of the University of California, San Diego, and his colleagues filmed conversations about things which had already happened and things that were yet to happen, and confirmed that Aymara speakers really do imagine the future behind them. The recordings show that their gestures, as well as their words, point forward for the past and back for the future.

Aymara speakers, especially older people who have not learned grammatical Spanish, point one thumb over their shoulder to indicate the future, or wave their hand behind them. When speaking of the past they sweep their hands or arms forwards, and gesture further out for longer ago. That is, they do just what you or I might do, but in reverse.

Some anthropologists criticized the scientists' paper, and certainly the way it was reported in the media, for exaggerating the unexpectedness of this about-face. Others had already noted the spoken orientation of the Aymara, and the closely related Andean language Quechua, but not the gestures, and there are other languages, such as Maori, which invite talk of the past as in front. However, the words and gestures of this South American group – whose language is spoken by around a million people in the Andes – are the best documented example of a different approach to the future. It *is* exceptional, though, so here we can safely stick to assuming that the future lies ahead of us.

J.T. Fraser Time the Familiar Stranger (1989) Physicist Fraser spent thirty years puzzling over the human meanings of time, and this is his best effort at capturing them in one book. Out of print, but well worth ferreting out. His later volume *Time, Conflict and Human Values* is nearly as good and easier to get hold of.

Stefan Klein Time: A User's Guide (2008) A popular science writer's reflections on the human sense of time, time management and how we sense where our lives are going.

G.J. Whitrow What is Time? (2003) Latest edition of the classic overview of time by the late British cosmologist and science historian.

2 the future in history

the future in history

History since the French Revolution has changed its role. Once it was the guardian of the past: now it has become the midwife of the future. It no longer speaks of the changeless but, rather, of the laws of change which spare nothing. Everywhere history is seen as progress...

John Berger

Aside from mirroring the ever-lengthening past, and getting much, much longer, what other changes have come over the future? That is a complex history itself. But, still concentrating on the Western culture which now dominates the planet, those changes in outlook can be summarized simply. The biggest shift was the idea that the future will be different. And with that came the notion that we might have some influence over whether it will be better or worse.

FROM PROPHECY TO PREDICTION

Older cultures which viewed the future as essentially static still experienced plenty of small-scale variation. The normal hazards of life preoccupied them in ways we still recognize. We do not know exactly when humans, or perhaps their proto-human ancestors, first became conscious that there was a future. But the wish to see beyond the present must have been born almost immediately. The future is always uncertain, and ways of reducing uncertainty – even illusory ones – help make life bearable. When the next meal depends on the success of the hunt, or surviving the next season depends on the weather at harvest time, the idea that we might have some foreknowledge of what is to come offers reassurance. And, subsistence aside, the hazards of love and war increase the incentive to find clues about how things will turn out.

Throughout history, most cultures have had special rituals, often performed by special people, for reading the signs and getting hints about the likely outcome of their actions. Even if the local belief system did not include choice – holding instead that everything was in the hands of fate or the gods – there was still the question of what they held in store. And soothsayers, priests, astrologers and oracles offered answers.

Old practices to aid prediction which survive today, such as astrology, tarot reading or palmistry, testify to the perennial attraction of such guidance. Like the trances of the ancient oracles, their results are typically loosely worded and open to interpretation. Some, such as the ancient Chinese system of consulting the *I Ching*, or *Book of Changes*, moved on from divination to offer advice on what actions the asker might take. A great text from around 1100 BC, the *I Ching* evolved from being a guide to the meaning of the patterns made by scattered yarrow stalks into a mainstay of Chinese culture and conduct.

One thing all these traditional prophetic aids have in common is that they are used within a framework that assumes a basic continuity between past, present and future. They are geared to the medium term, and don't typically try to look very far ahead. Historian Robert Heilbroner, in *Visions of the Future* (1995), reckons that until just a few hundred years ago, one thing united all of humanity's diverse cultures and past societies: none conceived of the material conditions of life changing appreciably. Real change – good or bad – came only in the afterlife. Future-fiction

A TOUCH OF THE VAPOURS

The best-known oracle of the ancient world was the Oracle of Delphi in Greece. The Pythia, a priestess trained in the disciplines of the Delphic temple, voiced the gods' answers to questions posed by visitors from far and wide. The gods varied over the many centuries the tradition endured – from about 1400 BC to the fourth century AD when a newly Christian Rome took exception to it – but the longest association was with Apollo, the powerful son of Zeus.

The Pythia's cryptic answers were delivered in a trance, and had to be interpreted by the priests. Even then, they were reliably ambiguous. Anyone consulting the Oracle was first told "know thyself" – an early recognition of the power of wishful thinking that continues in modern astrology to this day, where success often relies on letting clients hear what they want to hear. The questions posed ranged from everyday concerns about planting crops to important matters of state. The most often cited example of Delphi leading someone astray is King Croesus of Lydia, who consulted the Oracle in around 455 BC on whether or not to go to war with his threatening neighbour, Cyrus the Great of Persia. After the temple had banked his gold, he was told that war would result in "the destruction of a great empire". Following his defeat and capture by Cyrus, Croesus queried the prediction. Ah, they said at Delphi, did you think we meant the other empire?

The whole tradition reveals much about the Ancient Greeks' attitudes to what can be known about the future, and how it can be known. They held that the Pythian trance was induced by vapours from deep underground. Early twentieth-century geologists debunked this theory, finding no evidence of a chasm beneath the site and no trace of volcanic gas. However, more recent investigations have revived the idea. Two fault lines appear to intersect beneath the old temple, and there are traces of the narcotic gas ethylene in nearby springs. So even after three millennia, it seems, the Oracle is still able to surprise.

scholar I.F. Clarke (1918–2009) agrees. In his comprehensive review of tales of the future, *The Pattern of Expectation* (1979), he writes: "for most of human history the image of the future has been a blank".

In fact, compared with the way we see it now, the future in the past was in a sense fundamentally uninteresting. Most early cultures had an oral tradition, steeped in myth, and they focused on the past or the

THE PROPHECIES OF NOSTRADAMUS

It seems as though the best way to become famous as an oracle is to offer prophecies – lots and lots of prophecies – which are worded vaguely enough that the faithful can later reinterpret them as successful predictions of things that have actually happened. This has the advantage that clever, if misguided, followers of your writings do all the actual work. It has the disadvantage that your main reputation comes after you're dead.

Undisputed champion of this strategy was French apothecary-turned-prophet Michel de Nostradame (1503–66), better known as Nostradamus. His most famous work, *The Prophecies* (c. 1555), combined reviews of past prophets with his own astrological efforts to offer hints of what was to come. The scope for interpretation was increased by the fact that the thousand-odd quatrains (four-line segments) were not in chronological order, were published in a mixture of French, Latin, Greek and Italian, and included many deliberately obscure references and allusions. Nostradamus's most enthusiastic followers appeared in the twentieth century, and a vast number of books and websites catalogue his supposed predictive hits, including the rise of Hitler, the use of atomic weapons and the death of Princess Diana. After the September 11 terrorist attacks, several passages of apparently relevant predictions whizzed around the Internet, but they were exposed as fakes, at worst, or composites, at best.

The original text of *The Prophecies* can be used to reassure, as well as to warn. The impressively titled Director of the Nostradamus Society of America insists on his website that there is no truth to the widely discussed Mayan "prophecy" that the world will end in 2012. No, no, it won't happen for another 1785 years, for the old man said everything would be over in 3797, when burning stones will rain down upon the Earth. You have been warned.

present. In this outlook, life as it is now is basically a recapitulation of life as it was before, and as it will continue to be. If there is significant change, it will be in getting closer to the long-lost perfection of a distant, arcadian past. Therefore, the probing of oracles did not extend to questions about the far distant future. It was restricted to the next day, the next year or, at most, the next generation. As all generations would live in the same kind of world, prediction was about the everyday details. Would one enjoy good fortune, win the battle, be cured of an illness or get the girl? Otherwise, nothing new under the sun. Even extensive works of prophecy, like the compendium put together in the sixteenth century by Nostradamus, operated at this level (see box opposite). His world had a fiery end, but until then life went on much as before.

This picture of stasis with small variations contrasts starkly with our modern view. We *expect* changes over time. We are continually moving away from a known past, and approaching an unknown future, which holds more change. Whether we are moving towards something, or simply away from the state we were in before, we have direction. The future lies down a road to some new destination.

Sticking with this broadbrush approach to history, the big change in Western culture came in two stages. Simply put, the classical world, like previous cultures, saw history as cyclical, a pattern of endless repeats. The first shift came with the rise of Christianity, which brought a dramatic beginning and a big finish. The drama at the foundation of Christian thought tells of the Fall, and ultimate redemption.

Of course, this particular forward-looking story has a special end. Significant change on Earth is not necessarily in the picture, but history leads inexorably toward judgement day. Over the centuries, different religious denominations have upheld many variations of the details, but the basic framework – which is similar to the other Abrahamic religions, Judaism and Islam – is Final Judgement preceded by widespread destruction on Earth and resurrection of the dead. With that expectation, striving for improvements in human life is not an obvious priority. The point of life is right conduct to ensure that one is on the path to heaven rather than hell in the eternal hereafter.

That view still shapes the lives of millions of people today (see box overleaf). But the idea that history has a forward motion has taken on another, more modern aspect. The second shift from the classical world is the now familiar notion that human history can include change for the better – in a word, progress.

ENTERING THE AGE OF PROGRESS

The idea that the future will be different – and perhaps better – was one result of a whole raft of changes in ideas and outlook which overtook the Western world a few hundred years ago. These linked cultural and intellectual transformations – the Renaissance, the Enlightenment, the scientific revolution – had complex causes and consequences, but a change in attitude to the future, and in who was best qualified to talk about it, was one of the most important results.

Studies of fiction help us nail down when attitudes to the range of possible futures changed. Critic Adam Roberts' *The History of Science Fiction* (2006) locates the crucial period as the seventeenth century. He sees the key stimulus as modern science's new view of the universe, especially the astronomical observations

RAPTURE READY

Modern views of future change are a clear departure from those of the past, but not a clean break. The satisfyingly apocalyptic biblical prophecy of the Book of Revelation still has its adherents, and they have no doubt about what is to come: the end of life as we know it on Earth, with only the faithful being saved and taken up to heaven. Many believe it will happen soon, and devote themselves to recording the signs. It's hard to say how many people believe in the imminent Rapture, though it appears to have a much stronger cultural presence in the US than in other predominantly Christian countries. The *Left Behind* novel series (1995–2007) by Tim LaHaye and Jerry Jenkins, which fictionalizes the Rapture, has sold in the millions – but then so has Dan Brown.

Those less devoted to scripture, or who interpret it differently, may find this devotion a tad scary. When Ronald Reagan was president of the US, there was concern that his attitude to nuclear war was influenced by his belief in armageddon, even though his commitment to the "Star Wars" defence system indicated that he was quite interested in preserving life on Earth for the time being. Environmentalists have also frequently quoted secretary of the interior James Watt who told a congressional committee in 1981, "I do not know how many future generations we can count on before the Lord returns". However, Watt was not seriously proposing abandoning the Earth to its fate either. After this pious reference to uncertainty about the Lord's timetable, he went on, "whatever it is we have to manage with a skill to leave the resources needed for future generations". Phew.

For a less reassuring reading of the state of the world, check out the Rapture Index. It presents a compilation of factors – ranging from unemployment levels and interest rates to false Christs and events involving witchcraft – that can be used as an indicator of how near we are to the end of life on Earth. The Index's creator provides a useful analogy: "You could say the Rapture Index is a Dow Jones Industrial Average of end-time activity, but I think it would be better if you viewed it as a prophetic speedometer. The higher the number, the faster we're moving towards the occurrence of pre-tribulation rapture."

The handy guide continues with the numbers:

- Rapture Index of 100 and below: Slow prophetic activity
- Rapture Index of 100 to 130: Moderate prophetic activity
- Rapture Index of 130 to 160: Heavy prophetic activity
- Rapture Index above 160: Fasten your seat belts

At the time of writing, the index stood at 167, but the publisher decided to go ahead with this book anyway.

www.raptureready.com/rap2.html

and theories of Kepler, Copernicus and Galileo. The change began with the theory of an Earth-centred cosmos, with the stars and planets circling it fixed to perfect crystal spheres. That was displaced by the dizzying new spectacle of planets going round the sun, which became merely a star seen up close. Then, thanks to Isaac Newton's mathematical description of gravity, the whole shebang could be described in terms of matter in motion, with every trajectory conforming to known laws. God might have been the prime mover, but the mechanism ran without any further divine intervention.

Equally boggling, the idea that all the planets were, in fact, other worlds invited speculation about what kinds of worlds they might be, and whether other stars might have them too. The notorious martyrdom of Italian philosopher Giordano Bruno (1548–1600), burnt at the stake by the Roman Inquisition, was not just because of his advocacy of a sun-centred solar system, but because he imagined an endless universe, an infinity of other worlds. Such distant worlds would be a particular problem for the Catholic Church because if Jesus did not get round to visiting them all, any inhabitants there would remain unredeemed.

These were only one part of Bruno's heretical beliefs, which conflicted with orthodoxy on many more strictly religious matters as well. But his expansive view of the universe symbolizes the size of the shift in thought that was under way at that time. And for all the conflict this shift produced, it was also a liberation of the imagination. As Roberts puts it:

> "It is because of the opening up of a secular idiom for cosmological speculation in the seventeenth century that a new sub-genre of specifically secular futurology can come into being ... Just as the seventeenth century was the time that provided some writers with a newly materialist physical universe to explore, so the same period enabled the imagining of futures not determined by the Revelation of St John."

It took a while for this to sink in, evidently. Roberts, following a study by Paul Alkon in the *Origins of Futuristic Fiction* (1987), notes just two harbingers of the new sub-genre actually produced in the seventeenth century. The first was Jacques Guttin's *Epigone, histoire du siècle futur* (1659), but it used a visit to the future to relate a traditional epic reminiscent of the *Odyssey*. The next tale which was actually "set in a consistently imagined future", didn't appear for another three quarters of a century.

By then, the future was getting attention for reasons closer to home than the far-flung theories of astronomers. The scientific revolution was succeeded by the industrial revolution. Social and technological change sped up. Suddenly, people could notice significant change in their own lifetimes. That personal experience also opened the way for imagining futures which were fundamentally different from the past or present. Millennial prophecy had to compete with predictions of what the British philosopher and statesman Francis Bacon (1561–1626) called the "relief of man's estate" in his treatise *The Advancement of Learning* (1605).

Bacon was an influential early theorist of science, advancing the cause of reason and experiment over deference to ancient wisdom. He also had a towering vision of what science might achieve. His *New Atlantis* (1627) was poised between old and new literary traditions. It was a fantasy voyage of discovery, rather than a trip through time, but it seems fair to call his island kingdom of Bensalem futuristic. Bacon sketched an ideal society, in the tradition of many other utopias before and since, and he depicted a wealth of new inventions. But most critics agree that the crucial feature of his strange new world was that he showed a glimpse of a world improved by humans. The leaders of his new society turned away from a quest for a heavenly salvation to concentrate on making things materially better on Earth. Some would say this secular dream is pretty much what most of us have been pursuing ever since.

A RIOT OF INVENTION

Francis Bacon's *New Atlantis* (1627) is not a modern text; it is still set in a religious framework, and much of it is a commentary on the affairs of the court of James I. But it is impossible to resist reading parts of it in modern terms, particularly the sections related to Salomon's House, Bacon's fictional scientific institute where researchers gather knowledge to inspire new inventions and benefit mankind.

Four centuries on, it is striking how many areas of life Bacon thought science would infiltrate. The second half of his short book is basically a long list of cool stuff, as told to the narrator by one of the "fathers" of Salomon's House, the priestly figures who preside over his temple of knowledge. Compared with previous visions of the future, Bacon's book covers just about the entire range of human activity, and most later lists of future invention can be placed in the Baconian scheme.

The institute's inventions include, among other things, super foods and drinks which confer longevity and superhuman strength as desired, human-like robots and mechanical animals, flying machines, powerful explosives and incendiaries, and new breeding techniques which allow other species to be modified at will.

Each subject the father mentions is worked through systematically and in surprising detail:

> "We have also sound-houses, where we practise and demonstrate all sounds and their generation ... Divers instruments of music likewise to you unknown, some sweeter than any you have; with bells and rings that are dainty and sweet. We represent small sounds as great and deep, likewise great sounds extenuate and sharp ... We have certain helps which, set to the ear, do further the hearing greatly; we have also divers strange and artificial echoes, reflecting the voice many times, and, as it were, tossing it; and some that give back the voice louder than it came, some shriller and some deeper ... We have all means to convey sounds in trunks and pipes, in strange lines and distances."

So recording studios, mixers, synthesizers, hearing aids and telephones, then. The developments in other areas are even more impressive:

> "We have also engine-houses, where are prepared engines and instruments for all sorts of motions. There we imitate and practise to make swifter motions than any you have, either out of your muskets or any engine that you have; and to make them and multiply them more easily and with small force, by wheels and other means, and to make them stronger and more violent than yours are, exceeding your greatest cannons and basilisks. We represent also ordnance and instruments of war and engines of all kinds ... We imitate also flights of birds; we have some degrees of flying in the air. We have ships and boats for going under water and brooking of seas, also swimming-girdles and supporters."

Bacon cannot explain exactly how it is all done – that would have to wait for later eras, and later writers – but what is new is the scale of the ambition, and the idea that all of this is possible because invention has been properly organized. Salomon's House sends people out in secret to gather intelligence from other lands, and augments this with extensive consultation and experimentation of its own. For Bacon, the future is there to be made.

THE POLITICS OF PROGRESS

Most of Bacon's inventions in *New Atlantis* were thinly imagined speculation, but over the next few hundred years it seemed as though some of them were coming true. The idea of the future itself, as something to be invented, as a project, gripped more and more people. It was not just a matter of applying Baconian discipline to experimenting with technology. The idea of future change was intensely political.

This comes through clearly in a landmark future fiction, Louis-Sébastien Mercier's *L'An 2440* (1771), in which the narrator travels through time simply by taking a prodigiously long snooze. The book's young, idealistic author gave his readers a beguiling vision of a well-ordered, Parisian society seven hundred years in the future. According to I.F. Clarke, Mercier's "detailed account of a world of peaceful nations, constitutional monarchs, universal education and technological advances was an extension of the scheme Bacon had presented." Unlike *New Atlantis*, Mercier's book was both banned by the state (in Spain) and became a bestseller (it went through eleven editions in pre-revolutionary France). But it also played well with fans of technology. The second edition of 1786 had a new chapter on global air travel, inspired by the first balloon flights just a few years earlier. The future was catching up with the present – or was it the other way round?

Either way, the first flush of Enlightenment optimism was marked by a conviction that progress was indivisible: improvement would move ahead in all areas of life, material, political, moral, even spiritual. If the technological advancements have been more impressive, then and ever since, this was at first assumed to be only temporary. Revolutionary writers, such as French philosopher Condorcet (1743–94), outlined sweeping visions of future societies which were enlightened in every respect (see box overleaf).

As technology became a more visible force, however, authors responded with a burst of diverse futuristic visions of new societies, new inventions and, occasionally, new catastrophes. On the whole, though, the stories of the future remained optimistic, fuelled by a hope that the wonders of science really would permit "the effecting of all things possible", as Bacon grandly put it. By the early decades of the nineteenth century there were so many future fictions in print that they had their first historian, another Frenchman, Félix Bodin (1795–1837), whose *Le roman de l'avenir* (1834) charted the rise of the idea of progress. He even struck a modern note by reviewing some past fictional predictions and pointing out their mistakes, but he never questioned the belief that things were generally going to get better.

THE DISMAL SCIENCE ANSWERS BACK

This fiesta of optimism did not include everyone, of course. There were also stories about the end of the world, as compelling an idea in Mary Shelley's *The Last Man* (1826) as it is now. There were laments about the soulless, materialistic world that the new industrial technology would create. And, in a more direct confrontation with the prophets of reason and progress, there were arguments about why the idea of an ever-upward curve was doomed to disappoint.

That confrontation was stark in 1798. That year, British doctor Edward Jenner (1749–1823) published his research on vaccination, just about the best news

THE HEIGHT OF PERFECTION?

The peak of optimism from the days when the idea of progress first took hold was probably the Marquis de Condorcet's *Sketch for a Historical Picture of the Progress of the Human Mind* (written 1793–94). Mathematician, philosopher and secretary of the French Academy of Sciences, Condorcet (1743–94) became caught up in the liberatory fervour of the French Revolution, and was one of the secretaries of its first legislative assembly in Paris. He set down his ideas on the future in a work that featured a new, progressive reading of the past.

Condorcet saw the whole human race as following a path which resembled that of a maturing child, becoming better educated, more skilled and, well, more enlightened. History so far had covered nine stages of human development, from simple hunting societies through the invention of agriculture, and on to a more complex division of labour and forms of government. The subsequent history of society spanned further advances from the Ancient Greeks to the revolution of his own day. And after that? In Condorcet's eyes, there were no limits. He foresaw a rational, atheistic future which would bring the political reform he craved, improved education and, finally, the "abolition of inequality between nations, the progress of equality within each nation, and the true perfection of mankind".

In this first flush of enthusiasm for the idea of humanity taking charge of history, the future was to bring as much advance in morality and politics as in science and technology – and each would feed off the other. Knowledge would increase, and become accessible to all. Agriculture and manufacture would become ever more productive and efficient. The higher standard of living that ensued would help bring out the best in everyone. Complete equality for women, for instance, would come as a matter of course, because "this inequality has its origin solely in an abuse of strength, and all the later sophistical attempts that have been made to excuse it are in vain." The newly equal, rational citizens would lead long, disease-free lives.

Condorcet's vision of the limitless perfectibility of human life was an Enlightenment landmark, but sadly the author fell foul of the French Revolution. Hounded by the ultra-radical Jacobins, he wrote the *Sketch* in hiding, and died soon after in prison.

anyone had ever heard. The advent of his smallpox vaccine just two hundred-odd years ago, which saved innumerable lives, is a powerful reminder to scientists to deliver on the promise of technology. Here was a deadly, highly infectious disease, estimated to have killed four hundred thousand Europeans a year during the eighteenth century, that most people got a dose of early in life. If it didn't kill you, it left you pockmarked, at best, or blind. Then you waited to see if it would kill your children. No wonder Jenner's work was hailed by a committee of British parliamentarians as "unquestionably the greatest discovery ever made for the preservation of the human species".

Ah, but preservation of the human species would only bring trouble, according to another work published that same year. In 1798, Reverend Thomas Malthus (1766–1834) looked into the future, and saw more people, all starving. Increase population, he warned, and increase hunger. Demand would inevitably grow faster than supply. This was not a fictional speculation, or an inspired prophecy. It was a mathematical prediction. More, it was a natural law.

Malthus's classic work, *An Essay on the Principle of Population* (1798), was a clear statement of an idea which futurists still argue over today. It was also an early example of combining impeccable logic with dodgy conclusions. Here's the logic. Any population – whether mice or men – tends to increase geometrically, or as we now more often say, exponentially thus: 1, 2, 4, 8, 16, 32… This does not require every breeding pair to have four children: even a tiny fraction above two produces exponential growth. The doubling time will increase compared with the more dedicated reproducers, but if you plot the size of the population against time, it will still curve upwards ever more steeply. This is why getting compound interest on your savings account is better than stashing the cash in an old sock.

But how will we feed all these people, you ask? Well, says Malthus, perhaps more land and better farmers could provide meat and potatoes for one doubling of the population. But there's much less chance that we'll be able to double food supply again in the time it takes the number of mouths to go up another twofold. And it is vanishingly unlikely it can rise in line with population growth after that. Indeed, if food supply can be boosted at all, the basic tendency will be for the increase to be arithmetic, or as we now more often say, linear (2, 4, 6, 8, 10…).

The result, according to Malthus, is that any breeding population must outgrow the available food supply. So if nothing else keeps population in check, starvation will eventually do the job. Or, in his own words:

> "The germs of existence contained in this spot of Earth, with ample food, and ample room to expand in, would fill millions of worlds in the course of a few thousand years. Necessity, that imperious all pervading law of nature, restrains them within the prescribed bounds. The race of plants and the race of animals shrink under this great restrictive law. And the race of man cannot, by any efforts of reason, escape from it."

Malthus discussed various ways of evading the great hunger, including war, emigration, delayed marriage and "vice" (which he took to include contraception.) But failing that, the outcome was clear: "gigantic inevitable famine stalks in the rear".

At one level, Malthus's logic works. A modern equivalent is that any process which grows exponentially in a closed system will eventually reach the limits of that system, and come to a crashing halt. Malthus argued that this general truth was ignored by Enlightenment thinkers who believed in the perfectibility of man. Regrettably, Malthus reported, a general improvement in living standards was a fantasy. Oh, and those Poor Laws for the relief of poverty-stricken families in English parishes merely made things worse – they encouraged the poor to breed.

This is where logic joined seamlessly with politics, and the results went beyond economic theory. Malthus's baleful influence was visible in Britain's policy towards famine in its richest imperial possession, India. Let the brown people starve, went the official thinking: it's nature's way of keeping the population in check. Similar arguments were raised against helping the Irish during the Great Potato Famine of the 1840s, even though the immediate cause was a sudden crop blight. Let the enterprising ones emigrate; the rest could go hungry! But this conclusion is not contained in the logic of the Malthusian principle. It was simply invoked as an excuse for the consequences of human action – or inaction. Neither the starving Indian provinces nor blighted Ireland were necessarily closed systems, and treating them as such was a decision, not a law of nature.

Malthus's landmark forecasting essay attracted as many critics as it did admirers, but his influence was undoubtedly strong, and it can be seen in Charles Darwin and Alfred Russel Wallace's theories of evolution by natural selection. However, the continued economic and technological expansion of the nineteenth century helped ensure that most images of the future remained upbeat. Filthy factories, child labour and urban squalor all loomed large in the present – but most believed that the future could be better.

THE FUTURE HAS ARRIVED

Visible new technology helps convince people that they ought to think about a different future, and new technology was everywhere in the last quarter of the nineteenth century. Crucial innovations came together to underpin a world which was, in many ways, recognizably modern.

Discussions of future trends today often claim that the pace of technological advance and social change has got intolerably fast, but the Victorians saw some pretty startling changes too. From the 1860s, things definitely seemed to speed up. The dynamo enabled commercial electricity generation, and electric motors and electric trains soon followed. The telephone and the typewriter became standard office equipment. The two-stroke, and then the four-stroke, engine found its way into the horseless carriage – and car engines have changed little since.

Next came moving pictures and the phonograph, steel-framed skyscrapers, x-rays and wireless telegraphy, the first plastics, the commercial production of aluminium, and half-tone printing for newspaper photographs. This mix of new processes, new wonders and new gadgets gave a distinctive new flavour to history. The impressive thing was not how different the future would be from the present, but how different the present already was from the past. The future had, in fact, arrived.

The idea is captured in this squib from the magazine *Science Siftings* in 1894:

> "What did George III know [he died in 1820]?
>
> He never saw a match
He never rode a bicycle
He never saw an oil stove
He never saw an ironclad
He never saw a steamboat
He never saw a gas engine
He never saw a type-writer
He never saw a phonograph
He never saw a steel plough
He never took laughing gas
He never rode on a tram car
He never saw a fountain pen
He never saw a railway train
He never knew of evolution
He never saw a postage stamp
He never saw a pneumatic tube
He never saw an electric railway
He never saw a reaping machine
He never saw a set of artificial teeth
He never saw a telegraph instrument
He never heard the roar of a Krupp gun
He never saw a threshing machine…"

Even so, with any view of the recent past so densely packed with stunning innovations, it was easy to believe that the future would hold countless more. Many of the details of invention that Francis Bacon had been unable to describe in *New Atlantis* were now available for inspection. The "futuristic" novels of the hugely popular French author Jules Verne (1808–1925), for example, were rather methodically derived from his reading of contemporary science news, with some imaginative licence but little genuine novelty. Like innumerable science-fiction works since, "Les voyages extraordinaires" series he began in 1863 (and which included *A Journey to the Centre of the Earth* and *Twenty Thousand Leagues Under the Sea*) reflected his readers'

Technology took a huge leap forward in the Victorian era, with new innovations including Henry Ford's first commercial car – the 1896 "Quadricycle".

world back at them in only slightly altered form. Verne's novels weren't the only place the Victorian public could get their fill of contemporary innovation. In 1851, Britain held the first World's Fair (or the Great Exhibition of the Works of Industry of All Continents as it was properly, and rather laboriously, titled) to showcase the technological accomplishments of the era. The tradition of the World's Fair as a glimpse of the future continued through the twentieth century, but like the riot of invention it celebrated, it was during the second half of the nineteenth century that it really flourished.

PREDICTIONS FILE

Bruce Sterling

Science-fiction writer, futurist, author of *Tomorrow Now: Envisioning the Next Fifty Years* (2002) and blogger at www.wired.com/beyond_the_beyond

Highest hope: People will come to recognize that the future is history that hasn't happened yet. We've already had plenty of "highest hopes"; we are the highest hopes. Hope shouldn't be a mystically empty optimism but an engaged effort to make sense of things.

Worst fear: A collapse of civilization, a new dark age. I'm a science-fiction writer, so I'd describe that as my worst plausible fear. It's childishly simple to invent amazing, horrible fears.

Best bet: The next fifty years will be rather like the last two hundred years, except dirtier, more crowded and somewhat more technically advanced within certain areas. A lot of the things we thought we'd learned will be cheerfully forgotten, and the same mistakes committed with some different labels attached. "A comedy for those who think, a tragedy for those who feel": that's a pretty safe description of the great historical parade in any epoch.

THE CHICAGO WORLD'S FAIR OF 1893

Held in the glass-walled marvel of Crystal Palace under the watchful eye of Queen Victoria, London's Great Exhibition of 1851 was a landmark event that set the precedent for the numerous international fairs that followed (see also p.63). From the Paris Exposition Universelle of 1889, which was overlooked by the newly built Eiffel Tower, to the 1915 San Francisco Exposition celebrating the completion of the Panama Canal, each event sought to outdo the previous one. But the 1893 World's Columbian Exposition in Chicago had no trouble competing with its predecessors, as visitors were duly wowed by a panoply of the latest inventions.

As well as experiencing electric boats, steamships and a tethered hot-air balloon and marvelling at neon lighting, they could ride on the first elevated electric railway, try a moving pavement – bravely stepping between a three-miles-per-hour strip and an adjoining one moving at a dizzying six miles per hour – or watch moving pictures on Thomas Edison's kinetograph. Other diversions included electrical gadgets such as Gray's Telautograph, which copied handwriting at a distance, and the opportunity to have your first ever taste of Shredded Wheat, Juicy Fruit gum and diet soda. Towering above all this, George Ferris's 250-foot wheel was a huge, and highly profitable, sensation. "The Eiffel Tower involved no new engineering principle, and when finished was a thing dead and lifeless. The wheel, on the other hand, has movement, grace, the indescribable charm possessed by a vast body in action", gushed one journalist.

Towering technological accomplishments: George Ferris's 250-foot wheel turns on the World's Columbian Exposition fairgrounds, Chicago, Illinois, 1893.

More than 27 million people – well over a third of the US population in 1890 – visited the fair. And as well as seeing the future in action, they were enthused by newspaper articles like the American Press Association series that asked 74 commentators to predict what life would be like 100 years in the future. On the whole, 1993 seen from 1893 looked pretty good. The detailed forecasts were, of course, mostly wrong – the mail, for example, would still be delivered by stagecoach – but the idea that life would be better pervaded nearly all of them. The healthy, leisured citizens of 1993 would work a three-hour day, live to the age of 120 or so and reside in the new suburbs, where there would be no unemployment or poverty and little crime.

There could hardly have been a starker contrast with the reality experienced by millions of people living in the teeming industrial cities of 1893. For despite these illustrious visions and all the technological wizardry on show, the lives of the labouring classes in the US seemed little improved. Perhaps that's why the nineteenth century's most widely read vision of an improved future – Edward Bellamy's utopian novel *Looking Backward* (1888) – was more concerned with a transformed social order than scientific innovation (see box overleaf).

SOCIALISM IN AMERICA? SURELY NOT

Massachusetts author Edward Bellamy (1850–97) is notable in the history of utopian writing for his vision of a classless America, complete with nationalized industry and equal distribution of wealth. His novel *Looking Backward: 2000–1887* (1888) depicted a new commonwealth as seen by a protagonist who, after the traditional long nap, wakes up in twenty-first century Boston.

Bellamy took it for granted that improved technology would allow his new society to provide material comfort for all, and he folded in the technological optimism prevalent at the time. His hero, Mr West, awakens to a fully mechanized, electrical utopia, in which people work three days a week for an equal wage, and retire at 45. It is a world of skyscrapers, streets sheltered by transparent awnings which roll out automatically when it rains, and broadcast (well, piped) entertainment. The citizens of the future buy stuff in elegant shopping malls using state-issued debit cards and find it already delivered by pneumatic tube when they get home. Socialism could coexist with a consumer paradise.

Wanting to inspire others with a convincing prospect of what America could become, Bellamy said *Looking Backward* was "a forecast ... of the next stage in the industrial and social development of humanity". His book galvanized socialist politicians in Britain, as well as several experimental communities in the US. It also appealed to people in places where industrialization was far less advanced; Tolstoy, no less, produced the first Russian translation, and the book was banned from libraries by the czar.

Looking Backward sold hundreds of thousands of copies worldwide, far outstripping the host of other idealized visions of the future published in its wake. Its success probably stemmed from its combined portrayal of the beneficial application of technology that was already becoming familiar with social reforms. Bellamy's readers presumably shared his desire to see technology harnessed without the inequality, urban squalor and industrial conflict they saw around them.

THE REALLY LONG VIEW

Jules Verne, and the legion of authors of pleasantly appointed techno-futures who entertained the Victorian reading public, were still thinking in old terms in one respect: their futures were a few decades or a few centuries away at most. They normally observed the old imaginative limit of just a few generations. But as well as enjoying the fruits of invention, the nineteenth century saw vast changes in science, including the discovery of deep time (see p.8). And the deep future harboured more disconcerting prospects than electric trains and skyscrapers. Two of the great scientific insights of the century had a particular impact on the hope of progress. The first, which came in physics, was Lord Kelvin's new law of thermodynamics and the realization that the universe was gradually running down; the second, in biology, was Charles Darwin's theory of evolution by natural selection.

LORD KELVIN AND THE END OF THE UNIVERSE

In 1852, Scottish physicist Sir William Thomson (later Lord Kelvin, 1824–1907) proclaimed a new law of thermodynamics. It was, and is, a tough bit of physics to get hold of, but the simplest non-mathematical way of putting it is that energy cannot flow from cold to hot objects. For the engineers of the industrial revolution, that also meant that no engine could convert heat to work at a hundred percent efficiency. In more technical terms, Thomson's law said that the entropy of a closed system always tends to increase. Entropy is a measure of disorder. Its irreversible growth means that you can't unscramble an egg or put Humpty Dumpty back together again. It also means that the whole universe is going to run down. The sun will go out, the stars will fade away, and there will be nothing left except cold stones circling slowly, endlessly, in the dark.

In fact, you can actually see the universe running down as you watch the ice cubes melt in your drink. This was a stark challenge to Victorian self-confidence, and to visions of unlimited advance for endless ages. God had apparently ordained that energy could only be created or destroyed by him (the law of conservation). But he had also arranged things so that useful energy would eventually run out, inevitably, incontrovertibly. So much for progress. The "heat death" of the universe (a phrase coined by fellow physicist Hermann von Helmholtz, 1821–94, to describe the implication of Thomson's mathematics) might be millions of years in the future, but it was still a bleak prospect.

This conclusion had added impact because Thomson's argument was quickly recognized as such an important piece of physics. Along with the existence of atoms, the second law of thermodynamics is often touted as the most important single idea in science. It was a landmark in the physics of energy, a great nineteenth-century preoccupation. It has been reworked over the years, and reformulated by a number of other physicists, but never contradicted.

There are a host of literary, artistic and philosophical reactions to this notion, but atheist philosopher Bertrand Russell (1872–1970) put it most memorably in his 1903 summary of what the nineteenth century had taught him:

> "That all the labours of the ages, all the devotion, all the inspiration, all the noonday brightness of human genius, are destined to extinction in the vast death of the solar system, and that the whole temple of Man's achievement must inevitably be buried beneath the debris of a universe in ruins – all these things, if not quite beyond dispute, are yet so nearly certain that no philosophy which rejects them can hope to stand."

If human beings' consciousness of the future means we are always grappling with our own deaths, this was now firmly linked to broader ideas about ultimate destruction.

The origin – and extinction – of species

The uneventful end of the universe also haunted English naturalist Charles Darwin (1809–82), who saw deep time as the backdrop for the evolution of life. His evolutionary theory was the second insight of the era to have an impact on the perception of progress, though its implications were ambiguous. The scheme by which new forms of life appeared on Earth in gradual succession, as formulated by Darwin, could be depicted as progressive. After all, it had given rise to human beings, who now understood the outline of the whole story. But, on the other hand, the agent of "natural selection" was death. And the theory drew attention not just to the death of individuals, but also to the death of species: extinction.

The Darwinian picture thus allowed for both wildly optimistic imagined futures and radically

THE EVOLUTION OF MACHINES

During a brief spell working as a sheep farmer in New Zealand, disillusioned English cleric Samuel Butler (1835–1902) wrote a letter to a Christchurch newspaper addressing the idea that machines could evolve beyond human beings.

> "We are ourselves creating our own successors; we are daily adding to the beauty and delicacy of their physical organisation; we are daily giving them greater power and supplying by all sorts of ingenious contrivances that self-regulating, self-acting power which will be to them what intellect has been to the human race. In the course of ages we shall find ourselves the inferior race."

These words, which were published under the heading "Darwin among the Machines", were written in 1863, just four years after Darwin's theory of evolution in the living world was unveiled. The ideas expressed formed part of Butler's notes for his utopian novel *Erewhon*, published anonymously in 1872, some years after his return home. In three of the twenty-nine chapters – "The Book of the Machines" – he developed the idea of technological evolution, through a kind of artificial selection, leading to mankind being dominated by mechanical creations that develop their own consciousness.

Although the book was widely read, this notion was taken seriously by few readers, as it was generally regarded as a way of mocking Darwin. Butler protested that he was merely following the logic of the future of technology as he saw it. Machines would evolve, and they would do so faster than natural organisms. He had lit the trail leading to all the robot wars of twentieth-century fiction. As a character in *Erewhon* puts it, "I fear none of the existing machines; what I fear is the extraordinary rapidity with which they are becoming something very different to what they are at present."

pessimistic ones. Species change and, in some sense, improve. Perhaps humans will do the same? But species also disappear. Could that be the fate of *Homo sapiens* too? And, if we were to disappear, who or what might replace us as the dominant life form on Earth? The increasingly visible impact of technology in the nineteenth century meant that the idea that machines might also evolve, and eventually outstrip their creators, appeared almost as soon as Darwin's *Origin of Species* made its sensational appearance in 1859. This notion formed the centre of Samuel Butler's classic utopian satire, *Erewhon*, in 1872 (see box).

THE MAN WHO SAW THE FUTURE: H.G. WELLS

By the late nineteenth century the main ideas about how the future might unfold were fully formed, though contradictory. But whichever way things were destined to go, the evolutionary story, especially when extended to include the development of the whole universe, was much longer than history as previously understood. The future, like the past, might be much, much larger than people had previously imagined. And, since it seemed possible to probe the past scientifically – after all, this was how deep time was uncovered by geologists – perhaps one might take a more scientific approach to the future too.

These two thoughts came together in the mind of probably the most influential writer on the future in modern times, H.G. Wells. Although he is now best remembered as a science-fiction novelist, Wells also trained as a scientist and wrote the earliest systematic attempt at actually forecasting the future, rather than just imagining it.

With its shining brass frame and sumptuous red velvet chair, the titular object in the film version of *The Time Machine* (1960) seems firmly rooted in the nineteenth century. Rod Taylor plays George, the inventor embarking on his greatest adventure.

Herbert George Wells (1866–1946) did not own a time machine, he just wrote about one. But his influence as one who could perceive *The Shape of Things to Come* – the title of one of his later predictive forays (1933) – was so great that perhaps some people believed that he did. He made his reputation with his earliest novels, the ones that are still read most often today: *The Time Machine* (1895), *The Island of Dr Moreau* (1896) and *The War of the Worlds* (1898). But alongside these "scientific romances", he also turned out reams of essays on science, culture and world affairs, and his nonfiction became enormously influential.

Escaping the genteel poverty of the lower middle class by force of intellect, the young Wells learned all about biology and the new vistas of evolution first hand from Darwin's charismatic disciple, biologist Thomas Huxley (1825–95). Wells's tales of alien invasion, the manipulation of species, the degeneration of humankind and the ultimate extinction of life on Earth were the fruits of a giant imagination which was gripped by the idea of deep time, past and future, and the possibilities of progress far beyond the present. His imagination was shaped by Thomson's theories about the heat death of the universe and the ambiguous implications of a Darwinian view of life.

Wells laid out his thinking about the future in a lecture at the Royal Institution, London in 1902. In that lecture, entitled "The Discovery of the Future", he highlighted recent discoveries about the past brought to light by Victorian archaeologists looking at the evidence all around them in new ways, and argued that there might be equally effective ways of probing the future. Of course, we can never know our personal futures in the same way we can know our personal pasts, but Wells believed we could use observations about the past to infer, or induce, general propositions about the future – giving us, what he called, an "inductive knowledge" of the future. He declared: "I believe that the time is drawing near when it will be possible to suggest a systematic exploration of the future."

And what a future it would be. Although he sometimes offered much darker views in his novels – the degeneration of the ape-like Morlocks in *The Time Machine* or the monstrous creations of Dr Moreau – at the Royal Institution, Wells waxed lyrical about the possibilities of human evolution. The future he proposed to explore scientifically was solidly in the tradition of progress, and he closed his lecture with a ringing declaration of evolutionary faith:

> "Everything seems pointing to the belief that we are entering upon a progress that will go on, with an ever-widening and more confident stride, forever." In time, it would usher in a successor race of beings who "shall stand upon this Earth as one stands upon a footstool, and shall laugh and reach out their hands amid the stars."

This optimistic take on the future, which characterized much of his early nonfiction, was outlined at length in his 1901 book *Anticipations* (or *Anticipations of the Reaction of Mechanical and Scientific Progress upon Human Life and Thought*, to give it its full title). Here, Wells worked more methodically to develop his programme of genuine forecasting, rather than evolutionary sermonizing, though he retained a weakness for climbing into his secular pulpit. *Anticipations* was not always original. Much of the book, which was first serialized in *The Fortnightly Review* as one of many efforts to peer into the new century, drew upon the assorted visions of the future that had poured from the presses in the previous thirty years. But Wells synthesized them more grandly and expressed them more powerfully.

He began with a fairly conventional look at "locomotion" in the twentieth century, which he predicted would take advantage of a soon-to-be-perfected "light, powerful engine" to switch from rail to a network of roads carrying trucks, cars and buses. He also reckoned it not "at all probable that aeronautics will ever come into play as a serious modification of transport and communication". Moving on to citie, he forecast not suburbs exactly, but widely diffused populations due to increased commuting speeds.

Turning to social issues, Wells sketched a vast expansion of the middle classes and the technically educated, and addressed the condition of individual households, the future of democracy and the coming world state. The final section, Darwinian in the extreme, discussed the ethics of his "New Republic", and sanctioned extreme eugenic measures to rid the world of the "unfit" and prevent the "inferior races" from procreating. But it's not Wells's destination that's important, or the details of his many speculations along the way (such as how houses will be built or food will be cooked); what's important is his stepping aside from fiction to produce a considered series of connected forecasts of the way life will change in the coming century – and beyond.

As he aged, Wells's view of the future darkened, and the bleak undercurrents that had always played a role in his fiction became more apparent in his academic writing. In *Anticipations*, he had managed to put a positive spin on tanks, chemical warfare and even atomic weapons, arguing that if such technologies were perfected and used, they would pave the way for a benign world government that would preside over a war-free world and really would know what was best for everyone. But two world wars destroyed his confident anticipation of this New World Order.

Like millions of others, Wells was appalled by the devastation wrought by World War I, the first great war to be shaped by modern technology. And although he was the man who dubbed it "the war to end war", he spent the following awful decades fearing the coming of the next. His advocacy of the League of Nations during the interwar years quickly turned to disillusionment when it became clear that it wouldn't live up to his expectations, and his view of the future tipped into terminal pessimism with the outbreak of World War II. One can only imagine his despair when real atomic weapons were used at the war's close, just a year before his death in August 1946.

PREDICTIONS FILE

Tom Lombardo

Center for Future Consciousness,
Scottsdale, Arizona

Highest hope: We will pursue and realize our highest ideals and goals; we will have settlements on Mars and the asteroid belt; we will teach wisdom and enlightenment in higher education; we will develop a symbiotic and highly empowering relationship with robots and computer minds; we will revel in the miracle of existence and the evolutionary journey we are on; we will create new forms of life and enhance our present forms – we will beautify nature; we will transcend, psychologically and biologically, our present human capacities – cognitively, emotionally, aesthetically; we will extend the normal human life span to double its present average, hopefully more; we will evolve ethically; we will love learning, knowledge, intelligence, wisdom, ethics and enlightenment more than we presently seem to love power, money, technogadgets, food and sports. We will have a vision that inspires us, in which we see and appreciate our place in the cosmos.

Worst fear: We will wallow in our present mindset of mediocrity, superficiality, presentism (the here and now), surface appearances, speed, glitz, egocentricity, spoiled children (in the West), gluttony, anti-intellectualism, all-our-heroes-are-movie-stars-and-athletes, the-newest-technogadgets-make-the-world-go-round, big vehicles and mindless fundamentalism ruling a third of the world.

Best bet: We will be significantly, if not totally surprised. The world of fifty years from now would probably frighten us from our present perspective – too strange and unanticipated.

Wells's final word came in the updated edition of his mega-selling *A Short History of the World* (1922), published a month after his death. The first time round, he acknowledged the new doubts besetting the twentieth century, but he kept the evolutionary optimism and faith in science that he had shown in "The Discovery of the Future" back in 1902.

> "If the dangers, confusions and disasters that crowd upon man in these days are enormous beyond any experience of the past, it is because science has brought him such powers as he never had before. And the scientific method … gives him also the hope of controlling these powers …
>
> Can we doubt that presently our race will more than realize our boldest imaginations, that it will achieve unity and peace, that it will live, the children of our blood and lives will live, in a world made more splendid and lovely than any palace or garden that we know, going on from strength to strength in an ever widening circle of adventure and achievement?"

Yet, just 25 years later, his verdict was far bleaker:

> "A tremendous series of events has forced upon the intelligent observer the realization that the human story has already come to an end and that *Homo sapiens*, as he has been pleased to call himself, is in his present form played out."

The apparent disparity between these two positions (mankind's eternal progression or terminal decline) suggests a gradual slide from optimism to

A bleak future beckoned at the end of World War II. A tremendous mushroom cloud erupting in the Marshall Islands during the Able nuclear tests, 30 June 1946.

pessimism as Wells aged, but it wasn't that simple. For although he spent a great deal of energy promoting progressive politics and an extraordinary future for humanity, both positive and negative outlooks had always coexisted in his writing. The balance may have shifted as the twentieth century took its toll, but Wells the Darwinian had been cautionary about the future as long ago as 1893:

> "Man's complacent assumption of the future is too confident … Even now, for all we can tell, the coming terror may be crouching for its spring and the fall of humanity be at hand. In the case of every other predominant animal the world has ever seen … the hour of its complete ascendancy has been the eve of its entire overthrow."

This extract is from an essay published in the *Pall Mall Gazette* under the title "The Man of the Year Million". It underscores the fact that the modern view of the future is nearly always marked by extremes of thought. We might achieve paradise on Earth, or we might destroy ourselves. We might transcend our humanity, or be replaced by some superior creature. In the century since *Anticipations* was written, the full range of opinion about the future has continued to develop: utopian, dystopian, progressive and pleasurable, degenerate and dreadful. The new developments in thinking about the future were set in the newly expanded – larger, older and longer – universe that science had just allowed Wells and his contemporaries to begin to explore. But while a great deal of future fiction was still to come, the shift to nonfiction that Wells inspired led to a new generation of commentators finding fresh ways to explore what lies ahead.

FURTHER EXPLORATION

Henry Adams The Education of Henry Adams (1918) This classic by the American historian and critic is a reflection of the changes that came upon the world during the nineteenth century, from learning to ride a bicycle at fifty to the overwhelming impact of the World's Fairs in 1893 and 1900.

William Butcher Jules Verne: The Definitive Biography (2006) The best way to approach Jules Verne is to read his novels (they're widely available online), but this modestly subtitled account of his life by translator and scholar Butcher is also worth a look.

Christophe Canto History of the Future: Images of the 21st Century (2001) A review of nineteenth- and twentieth-century images of the future; it's been translated from the French, so it has plenty of commentary on Jules Verne.

I.F. Clarke The Pattern of Expectation, 1644–2001 (1979) A long out-of-print classic by a pioneer scholar of science fiction, this comprehensive review of tales of the future is well worth hunting for.

Patrick Parrinder Shadows of the Future: H.G. Wells, Science Fiction and Prophecy (1995) An English professor puts Wells's career in context, and explains that his imaginative visions of the future were the guiding thread in his writing.

Oona Strathern A Brief History of the Future (2007) A breezy guide to futures past, by a journalist turned futurist.

www.hgwellsusa.50megs.com The H.G. Wells Society, founded in 1960.

3 Varieties of futurology

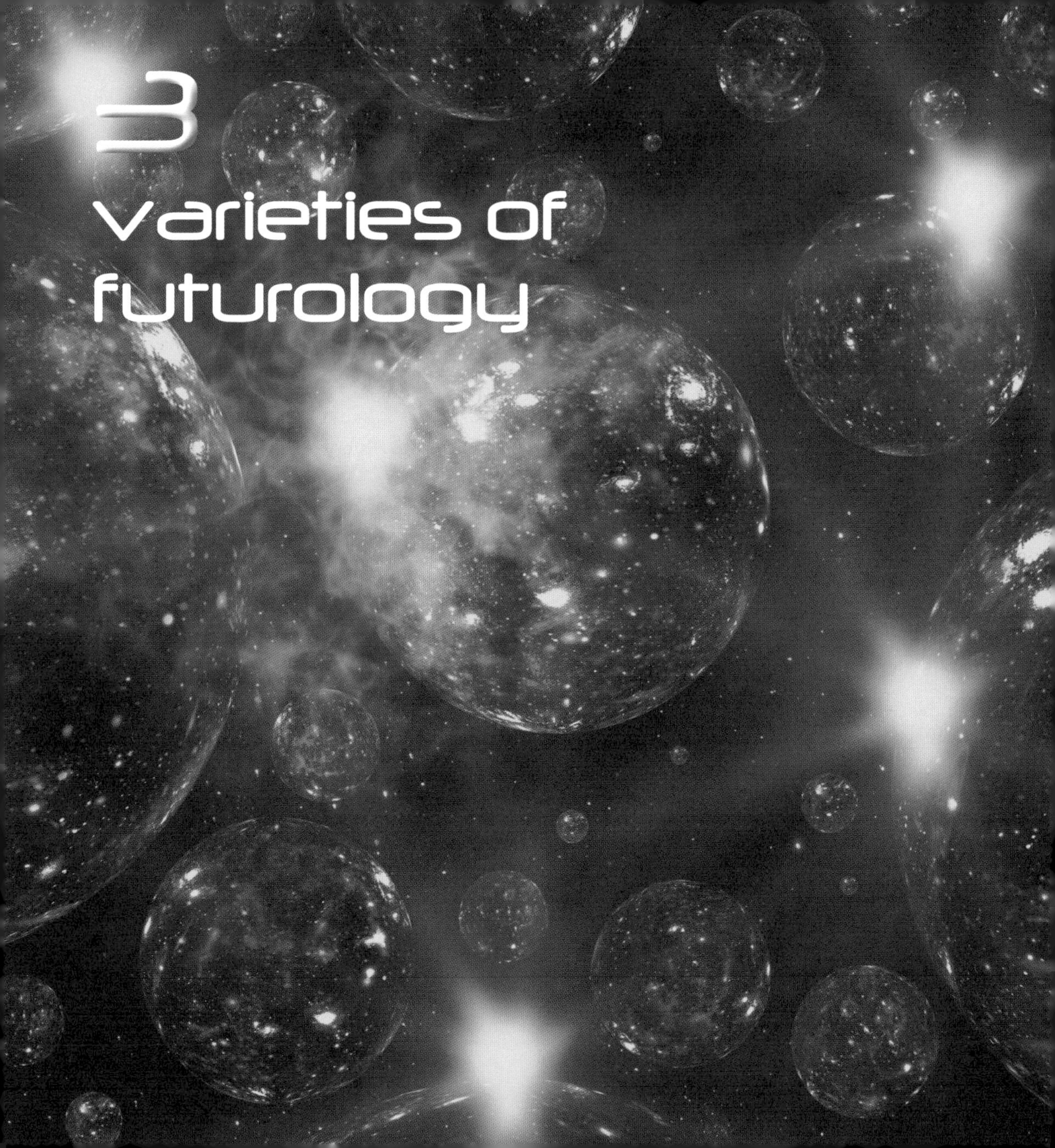

Varieties of futurology

A writer may tell me that he thinks man will ultimately become an ostrich. I cannot properly contradict him. But before he can expect to bring any reasonable person over to his opinion, he ought to show that the necks of mankind have been gradually elongating, that the lips have grown harder and more prominent, that the legs and feet are daily altering their shape, and that the hair is beginning to change into stubs of feathers.

Thomas Malthus, *An Essay on the Principle of Population*

Once the future became a more or less respectable subject, efforts emerged to study it using formal methods. It would be hard to claim that those methods have improved much (science writer Nigel Calder was at least half serious when he suggested that the best tool for thinking about the future was a bunch of sentences strung together), but they have certainly changed over the years, as futurism developed from a fringe interest to the work of a few pioneering academic centres to a modern community of think-tanks, government "foresight" outfits and corporate scenario sketchers.

Real prediction, or dumb luck?

The urge to predict is old. So what is new in the art of forecasting? A first glance suggests not much. Old-style prediction goes on, but no one takes it very seriously, or at least admits that they do. Ronald Reagan was widely derided in the US press when it emerged that his wife's enthusiasm for astrology meant that much of his presidential schedule was adjusted to accord with his horoscope. Besides, close scrutiny quickly shows that the near-term predictions you can read in the paper every day are worth very little.

Study the astrology column, or the tips suggesting which horse to back or which shares to buy by all means, but don't expect their advice to give you a better chance of making good decisions. If racing tipsters got it right, the bookies would have gone out of business a long time ago. But tipsters of all kinds survive because the alternative is to use our own unaided judgement or simply to choose at random. This is usually equally effective, but too scary for most people.

Bigger predictions, over longer periods, presumably have even less chance of being useful. Does that mean that all forecasting is pointless? No, but not because some things foretold do come true. Make enough predictions and some are bound to be right (see box).

DECISIONS, DECISIONS

Adverts asking if you've got any spare cash to invest often declare that this or that fund manager has outperformed the market or the average fund in, say, four of the last five years. Impressed? You shouldn't be, as maths guru John Allen Paulos explained in his illuminating book *Innumeracy* (1988).

Suppose a stock market commentator sends out letters to 32,000 prospective investors. Half the letters predict a particular group of stocks will rise, half that they will fall. A week later, 16,000 people get a letter highlighting the fact that he made the correct "prediction". Again, half these letters make a further prediction of a rise, half of a fall. Another week passes, and 8000 people get a letter reminding them that they have now seen two correct predictions, 4000 predict a rise the following week, 4000 a fall. Carry on for another three rounds and 500 astounded recipients have had six spot-on predictions in a row. They're asked to pay highly for the next one. Who could refuse?

Doing this is actually illegal, but the temptation to believe that this kind of prescience is possible is behind the essentially similar pitch of the fund managers. The trick is that the information they offer is always incomplete. At the very least, a fund manager's performance should be compared with the results of buying and selling at random. The advice Paulos offers was first popularized in finance professor Burton Malkiel's classic book *A Random Walk Down Wall Street* (1973), but it's still widely ignored by ever-hopeful investors.

CAN SCIENCE HELP?

The ideas H.G. Wells outlined in "The Discovery of the Future" in 1902 (see p.32) marked a change in thinking about times to come, and science lay at the heart of that change. In fact, Wells's notion of how science could be used to get a clearer picture of the future derived from the great changes that occurred in scientific theory in the seventeenth century. Useful scientific theories predict what will happen under some defined set of conditions, and the new theories at the centre of the scientific revolution were impressive at predicting. Isaac Newton's mathematics modelled a "clockwork universe", in which matter moved in an orderly motion under the influence of gravity. Get the equations right, and a man of science could predict the motions of the planets, and even the comets – long taken as portents by earlier seers.

Newton's success led to an intoxicating vision, not of control but of knowledge. His laws of motion were universal: as in the heavens, so on Earth. They described what mathematicians now call "dynamical systems", which have one fascinating property. If you know the state of the system at time t, its state an instant later, t+1, is directly related to the state at time t, and the equations of the system tell you how. The same goes for t+2, and so on indefinitely. Just keep plugging the numbers back into the equations and reading off the result to get a picture of the system as far ahead as you'd like to go.

This suggested a rather bold hypothesis, most famously summed up by French mathematician Pierre-Simon Laplace (1749–1827), of a superior intellect being able to analyse matter in motion and "see" the future:

> "An intellect which at any given moment knew all the forces that animate nature and the mutual positions of the beings that comprise it, if this intellect were vast enough to submit its data to analysis, it could condense into a single formula the movement of the greatest bodies in the universe and that of the lightest atom; for such an intellect nothing could be uncertain, and the future just like the past would be present to its eyes."
>
> *Philosophical Essay on Probabilities* (1825)

But Laplace's vision is unattainable. For one thing, no one ever knows enough to do such calculations. His all-seeing intelligence would need complete information, and the data simply isn't there. In fact, no one *can* know enough, as more recent science sets limits to what can be recorded about matter in motion. Secondly, even in today's technologically advanced age, the necessary calculations would require a computer of unimaginable size and speed. And finally, the equations used may just be wrong.

The newer science of chaos and complexity, which grew from work on the limits of weather forecasting, shows that some things are, in principle, unpredictable. Chaos theory describes how small changes in the initial conditions of a system can produce vast differences in outcome as the consequences of those changes work through. That is why, to take the most familiar daily example, the weather forecast for the next six hours is pretty good, but the five-day forecast is less useful, and the long-range forecast for, say, three weeks ahead says something vague like "temperatures will again be at least above average if not well above average, both by day and night." And re-running the forecast with a larger or faster computer will not make it any better.

A TV weather forecast in 1954. Chaos theory suggests even today's powerful computers can't make reliable long-range forecasts.

Laplace's fine words are often taken as hubris, but that's unfair. His vision of plotting the path of everything from stars and planets to atoms is certainly a strong statement of a universe where everything is determined by a few simple laws. Causes and effects work themselves out inexorably once someone (God, presumably) kick-starts the great cosmic machine. But Laplace does emphasize that this is a statement of principle. We can't actually match the intellect he is imagining, and the feeble human mind will always remain "infinitely removed" from the real thing.

However, none of that stops the mechanical approach working well much of the time. The Apollo astronauts could trust that NASA had aimed their craft at the position the moon would be in when they got there because Newton's equations are more or less correct (with a small correction for effects which Einstein knew about but Newton did not). But the best answers are still only an approximation; they only work for systems which behave like the one Laplace thought he lived in.

People, for example, are not like atoms. Predict their behaviour, and the prediction becomes part of

PLEASE SHOW YOUR WORKING

Weighing up the future state of the world means assessing probabilities. Any honest forecast will cover a range of outcomes, but with a bit more thought we might be able to narrow that range, or at least increase the likelihood that what actually happens falls within it. Population is a good example because, if not completely uncontentious, it is at least well-studied.

The probability that there will be one hundred billion people on the planet in 2050 is as close to zero as makes no difference. Likewise the probability that there will be just one or two. A professional demographer will agree, but will also say that the most likely range actually centres on 9.4 billion. But any figure like that should come with some additional information about its pedigree.

My broad projection of between a single person (or perhaps a breeding pair) and one hundred billion people could be based on little more than the intuition that we will muddle through the first half of this century without calamity – but that hundred billion would mean an awful lot of babies in an implausibly short time. The demographer's forecast will be better not just because it is more precise, but because it includes decent estimates of the existing population, and of current trends in birth and death rates. A professional estimate will also offer lots of detail about how these estimates are made. The Population Division of the US Census Bureau, for example, which builds up a global population projection from more than two hundred individual country calculations, offers its best numbers with a three-thousand-word explanation of how they're worked out and continually re-evaluated.

As this suggests, the details get quite complicated quite quickly. But you can get a sense of whether this, or any, forecast is a potentially helpful guide to thinking about the future, or if it's sheer speculation, by asking four simple questions:

- Do the authors reveal their original data?
- Do they say where the data comes from, and how good it is?
- Do they spell out their assumptions for working up this data into a forecast?
- Do they show their working?

If the answer to all four of these questions is yes, as it is here, that still doesn't mean you can put complete faith in the results. It does mean, however, that low-quality data, sloppy working or questionable assumptions can be challenged, or even improved, by others. If any of the questions cannot be answered with a yes, suspect the authors of trying to avoid such challenges.

the system. It also turns out that atoms aren't much like atoms either. Planets are big enough to follow deterministic laws, but on very small scales, effects described by quantum mechanics are much more important. The uncertainty principle summarized by Werner Heisenberg (1901–76) says that you cannot accurately measure both the position and the momentum of a particle at the same time. When it comes to matter in motion, if you know where something is, you don't know where it's going, and vice versa. That spells eternal frustration for a vast intelligence hoping that "the future, just like the past, would be present to its eyes".

A further limitation on prediction, even with well-understood systems, is the quality of the data about past and present. In principle, some uncertainties can be reduced by looking at large numbers of cases, which smoothes out irregularities to give a larger, clearer picture. But how clear depends on how good the information is. Take, for example, a twenty-year-old chosen at random, and it would be foolhardy to say how many children or grandchildren they might be inviting to their seventieth birthday party – or even if they'll live to see that happy event. Take a country with a million people, and you can say far more about the likely population size fifty years on.

Consider the entire planet and the prospect for decent prediction improves a bit more (see box opposite). There will still be uncertainties, but it's a good bet that there will be at least a few billion more people in 2050 than the present 6.5 billion, but not 5 billion more. That sounds like a wide range, but it is at least, well, a rough guide. Anyone supporting an estimate of future population outside that range would have to explain why they disagree with virtually the entire profession of demography.

Mapping possibility

Formally, then, predicting the future is not on the agenda, and modern futurists routinely emphasize that they are not making predictions. That is partly on principle, and partly because there are now so many past predictions on record that it's easy to choose almost any subject and find confident assertions that have proven wide of the mark. If there are some that score better with hindsight, does that mean their authors really knew what was what, or just that they got lucky?

Still, accepting in principle that prediction is a mug's game, and that uncertainty rules, does not stop people thinking about the future, or trying to find better ways of doing so. Both boil down to two basic questions. How can we get a feel for what the future might be like? And how can we tell if some futures are more likely than others? That second question also usually implies a third: what can we do about it?

The oldest method for peeking beyond the present is storytelling. Old-time oracles told stories, modern-day prophets still do. Some come labelled as fiction, others have a more ambiguous foothold in reality. But even teams of professors with mountains of data, elaborate systems for analysing it and superfast computers still flesh out their prognostications with stories. The coherence which comes from wrapping a forecast in a narrative makes it easier to grasp, and more convincing. This is both good and bad. Stories are great fun, and an unbeatable aid to communication. But the better they are, the harder it is to spot what has been left out. A good working approach to any forecast is to take it as a story and ask: why should we believe this particular tale of the future?

Storytelling has remained a mainstay of contemplation about the future, as the rise of science fiction

DISSOLVING DISBELIEF

Future stories are often written as warnings. But how do you get your warning heeded before the unthinkable happens? The all-time prizewinner here is Rachel Carson, who did it with a story just two pages long in her landmark environmentalist book *Silent Spring* (1962). "A Fable for Tomorrow" depicts a mythical small American town, once a picture of harmony and prosperity, overcome by a mysterious blight. Cattle are stricken. Children who play outdoors fall gravely ill. The birds disappear. Roadside vegetation shrivels and dies. Everything is briefly, but vividly described. Then comes the punch line: "This imagined tragedy may easily become a stark reality we all shall know."

Why should we believe her? Because she provided another sixteen chapters of reasons, the evidence gathered from years of news reports, scientific papers and interviews with researchers, naturalists and officials, all documented in a long list of sources. The culprit in her imagined future, as in many smaller disasters that had already happened, was the use and abuse of chemical pesticides.

Carson's short story was striking and it hit close to home for many people, though it was dismissed by her critics in the chemical industry as mere science fiction. The much longer, fact-filled portion of the book helped convince the public and policy-makers this was a future to avoid. The two together were remarkably effective. The worst offender among the pesticides, DDT, was banned for most uses in the US in 1972.

The ban is still controversial, with some claiming that anti-malaria efforts were hampered by the failure to use DDT and similar chemicals against mosquitoes. However, the agents were already losing their effectiveness in many areas because the insects were developing a resistance to them (a topic also dealt with in the book).

in the twentieth century confirms. But the urge to turn a tale of the future into nonfiction of some kind was also a hallmark of the century. As futures studies have developed, the storytelling has been beefed up by a range of techniques aimed at getting a better feel for what comes next, and reducing – or at least defining – some of the uncertainty.

THE USES OF HISTORY

The past is the obvious source of clues about the future. From the eighteenth century to the early twentieth, ambitious historians tried to establish patterns and trends which they could use to predict future paths on the largest scale. The period saw intellectual, scientific and industrial innovation combined with social and political change. The French Revolution became the archetype of upheaval.

The early futurist writers like Condorcet (see p.22) dreamed of social transformation, and the founders of social science began to read the lessons of history in a new and progressive way. Social theorists like Henri de Saint-Simon (1760–1825) and Auguste Comte (1798–1857) were men with a mission. As historian Frank Manuel put it in *The Prophets of Paris* (1962):

> "They were intoxicated with the future; they looked into what was about to be and they found it good. The past was a mere prologue and the present a spiritual and moral, even a physical, burden which at times was well-nigh unendurable. They would destroy the present as fast as possible in order to usher in the longed-for future, to hasten the end."

Their planned destruction was nonviolent, though; they relied on the development of mankind's better nature, along with scientific achievement, to realize their visions. Their influential successor, Karl Marx (1818–83), had a different view, seeing growing contradictions in society as setting up tensions which would only be released by revolutionary action.

Marx's theory of history and society envisaged a progression of stages of human social organization, linked to advances in technology and culminating in a classless, communist society. The aftermath of the various supposedly Marxist revolutions of the twentieth century left a hefty dent in his predictions. But even so, they were notable for being predictions made to inspire political action to make them come true, in line with his famous dictum that philosophers have only interpreted the world, but the point is to change it.

More recent large-scale readings of history have been less action-oriented, as indicated by the title of political scientist Francis Fukuyama's *The End of History and the Last Man* (1992). Fukuyama also saw a long-term process of social evolution, but he reckoned it has already reached its last and greatest stage: contemporary liberal democracy. All that remains is for the nations that have not yet embraced this enlightened regime, with its parliamentary apparatus and free markets, to fall in line and, he said, the important part of history will be over. This conclusion was controversial: history normally ends with either utopia or apocalypse. Fukuyama's rather bland vision of carrying on more or less as we are was nicely pitched to annoy devotees of both of those extremes.

Although grand schemes are generally out of fashion with historians, futurecasts still invariably make use of historical data. They also commonly refer to interpretations of the past as a guide to the kind of things which may lie ahead. This can be simple-minded (Hitler was a bad man. Saddam Hussein is a bad man. Therefore, Saddam Hussein is like Hitler and must be stopped before he tries to take over the world). More promising is the search for processes, which are likely to continue, that can be studied in the past, or for analogies between contemporary happenings and similar past events. As H.G. Wells observed, this is how sciences like geology operate (see p.32), and perhaps the same techniques can be applied to society and technology too. For instance, if the history of print technology tells us that revolutions in communications have wide-ranging and long-lasting effects, we might expect the development of the Internet to have similarly epoch-making consequences.

Occasionally, this search for analogies creeps back towards grand generalization, but whether this is useful is a matter of opinion. For example, some who project a major shift in society, technology and culture in the twenty-first century support their cause by suggesting that everything tends to describe an "S-shaped" curve. They observe that population graphs expanding exponentially and then levelling off have a characteristic shape, and that they look a bit like graphs for other things which can move from one relatively stable state to another, through an unstable mid-phase called a "phase transition". The Global Scenarios Group, a coalition of environmentalists who produced a weighty 2002 report on the state of the world, claim that this is a universal pattern, covering everything from the emergence of matter in the wake of the Big Bang to the growth and evolution of life. They want this to reinforce belief in a new "Great Transition" in society which they hope will become a self-fulfilling prophecy.

It's true that there are plenty of things that fit this picture, if only because (as discussed in Chapter 2) anything which enters a phase of exponential growth has to level off eventually. Human population is

TWENTY PREDICTIONS

Of all the predictions uttered in the past, here are a score which seem, looking back, to be on the extremes of either spectacularly wrong or commendably (or sometimes eerily) accurate. But would you have known at the time which ones were which?

"My principles in brief, are, that Jesus Christ will come again to this Earth, cleanse, purify, and take possession of the same, with all the saints, sometime between March 21, 1843 and March 21, 1844."

William Miller, 1834

"By the influence of the increasing percentage of carbonic acid [i.e. CO2] in the atmosphere, we may hope to enjoy ages with more equable and better climates."

Svante Arrhenius, 1896

"No place is safe – no place is at peace. There is no place where a woman and her daughter can hide and be at peace. The war comes through the air, bombs drop in the night. Quiet people go out in the morning, and see air-fleets passing overhead – dripping death – dripping death!"

H.G. Wells, *The War in the Air*, 1908

"We do not consider that aeroplanes will be of any possible use for war purposes."

Richard Haldane, British Secretary of State for War, 1910

"You will be home before the leaves have fallen from the trees."

Kaiser Wilhelm, to German troops, August 1914

"British rule in India will endure. By 2030, whatever means of self-government India has achieved, she will still remain a loyal and integral part of the British Empire."

Lord Birkenhead, *The World in 2030 AD*, 1930

"We shall escape the absurdity of growing a whole chicken in order to eat the breast or wing by growing these parts separately under a suitable medium."

Winston Churchill,
"Fifty Years Hence", *Popular Mechanics*, 1930

"The energy produced by the breaking down of the atom is a very poor kind of thing. Anyone who expects a source of power from the transformation of these atoms is talking moonshine."

Ernest Rutherford, 1933

"This new phenomenon [a nuclear chain reaction] would also lead to the construction of bombs, and it is conceivable – though much less certain – that extremely powerful bombs of a new type may thus be constructed. A single bomb of this type, carried by boat and exploded in a port, might very well destroy the whole port together with some of the surrounding territory. However, such bombs might very well prove to be too heavy for transportation by air."

Albert Einstein, letter to President Roosevelt, 1939

"Automobiles will start to decline as soon as the last shot is fired in World War II. Instead of a car in every garage, there will be a helicopter."

Harry Bruno, aviation publicist, 1943

"Consider a future device for individual use, which is a sort of mechanized private file and library ... a device in which an individual stores all his books, records and communications, and which is mechanized so that it may be consulted with exceeding speed and flexibility. It is an enlarged intimate supplement to his memory.

It consists of a desk, and while it can presumably be operated from a distance, it is primarily the piece of furniture at which he works. On the top are slanting translucent screens, on which material can be projected for convenient reading. There is a keyboard, and sets of buttons and levers. Otherwise it looks like an ordinary desk."

Vannevar Bush,
"As We May Think", *Atlantic Monthly*, July 1945

"Where a calculator on the ENIAC [an early electronic computer] is equipped with 18,000 vacuum tubes and weighs 30 tons, computers in the future may have only 1000 vacuum tubes and weigh only 1.5 tons."

Popular Mechanics, 1949

[In response to reporters' questions about the prospects for interplanetary travel] "It's utter bilge. I don't think anybody will ever put up enough money to do such a thing ... What good would it do us? If we spent the same amount of money on preparing first-class astronomical equipment we would learn much more about the universe ... It is all rather rot."

Richard van der Riet Woolley, British Astronomer Royal, 1956

"Our descendants may well consider it a barbarous extravagance to propel a man in two tons of steel, with 200 brake horsepower, used only when the traffic lights change, and in the rare moments when there is a free stretch of road in front."

Denis Gabor, 1963

"[By 1985], machines will be capable of doing any work Man can do."

Herbert A. Simon, 1965

"The battle to feed all of humanity is over. In the 1970s the world will undergo famines – hundreds of millions of people are going to starve to death in spite of any crash programs embarked upon now."

Paul Ehrlich, *The Population Bomb*, 1968

"There are good prospects for what the Europeans would call *la belle époque*, or, if you will, a good era, similar to that experienced between the turn of the century and World War I – a worldwide period of growth, trade, peace and prosperity on the whole, and a time, generally speaking, of optimism about the future."

Herman Kahn, *The Year 2000*, 1969

"There is no reason anyone would want a computer in their home."

Ken Olson, president of Digital Equipment Corp, 1977

"I do believe ... that if concentrations of CO2, and perhaps of aerosols, continue to increase, demonstrable climatic changes could occur by the end of this century."

Steven Schneider, 1976

"The Ice Age is coming, within five to seven years of 1988."

Dr Larry Ephron, 1988

"We are on the edge of the greatest die-off humanity has ever seen. We will be lucky if twenty percent of us survive what is coming. We should be scared stiff."

James Lovelock, *The Times*, 2007

currently moving into the more level part of the curve after a very steep increase. But applying this analogy to society as a whole leaves vague exactly what is being measured to draw the curve. It's more of a poetic flourish informed by wishful thinking than a law of history.

DELPHIC DELIBERATION

Non-historians have had their own ideas about how to develop ideas about the future. One step beyond making up your own story is to ask other people what they think. Ask a lot of people and you can assemble a composite of their views, in the hope that many contributions produce a wiser result.

A slightly more elaborate way of doing this, the Delphi technique, was developed in the 1950s by the RAND Corporation (see box on p.49), a US think-tank which then mainly served the military. The technique, which uses a feedback system to stimulate thought and try to reach a consensus, works like this: collect a group of experts on a particular topic (the future of technology in the earliest efforts) and ask each of them for their individual opinion. Then tell them what everyone else said, and what their reasons were (but without telling them who said what, to prevent anyone deferring to predictions made by someone presumed to be smarter or more powerful). Then ask each person if they'd now like to modify their forecast.

This cycle can be repeated, if the budget allows, until everyone's views stabilize. In the best case, the method produces a consensus, but in most cases it tends to narrow the range of opinion. At the very least, it means that someone who sticks with a prediction that's wildly different from the middle ground has to be sure they have a good reason.

Maintaining anonymity during the exercise helps to minimize the "bandwagon effect" and allows the experts to express their opinions without the pressures of group dynamics. However, an obvious criticism of Delphi-style exercises is that they tend to induce "groupthink" and weed out wacky ideas which might yield real insight into future possibilities. More importantly, even if all the participants are experts in a particular field, they are not experts on the future and they can easily get it wrong (see box opposite).

Delphi-style opinion-taking is still in use, though it's less fashionable than it once was. The National Institute for Science and Technology Policy in Japan, for example, carries out a large-scale Delphi survey every few years, alongside numerous other technical forecasts using different methods (www.nistep.go.jp). This is one of a clutch of government-funded and private outfits dedicated to "foresight", a term popularized by a pioneering effort in the UK in the 1990s. Foresight is an aspiration rather than a method, and it uses a variety of techniques, but the core of most programmes remains talking to experts, or getting them to talk amongst themselves, and trying to discern the factors likely to influence a particular area. The British government's Foresight outfit say that they try to "use the best evidence from science and other areas to provide visions of the future". This, they suggest, can help support policy-makers by identifying potential risks and opportunities, and offering strategies for dealing with them.

This is a good summary of the general mission of such units. In 2008, the Millennium Project's *State of the Future* report (see p.72), found 28 future strategy units around the world, from Argentina to Venezuela. The UK unit began back in 1994 and is one of

IMPROVING ON THE ORIGINAL?

Does the modern, systematic polling of experts under the banner of Delphi improve on the trance-induced prophecies of the ancient oracles (see p.15)? Some Delphi studies are old enough to judge, and the results don't look good. Take the "Report on a Long-Range Forecasting Study" submitted by two RAND Corporation researchers in 1964, which contained summaries of fifty-year forecasts covering six areas of analysis (science, population, automation, space progress, probability of war and future weapon systems).

The forecasters were not foolhardy enough to claim that their questionnaire, feedback and resubmission method would yield reliable predictions, only that they might be less unreliable than other methods. And they expressed a hope that their findings might "represent a beginning in the process of sifting the likely from the unlikely among the contingencies of the future". That's just as well, as their science and technology forecasts indicated that all the experts expected automatic language translators, limited weather control, primitive artificial life and ocean-floor mining before the year 2000.

To be fair, they also foresaw effective oral contraceptives (not too hard in 1964), "operation of a central data storage facility with wide access for general or specialized information retrieval", which could be seen as a version of the World Wide Web, and new organs through transplants or prosthesis. They were also correct in consigning man–machine symbiosis, the control of ageing, breeding intelligent animals for low-grade labour and two-way communication with extraterrestrials to the more distant future. Most also reckoned, correctly, that world population growth would begin to level off before the end of the twentieth century, though apparently the teeming billions of the future would be well fed because the "commercially efficient production of synthetic food may be expected within forty years". So a mixed result, at best.

The accuracy in more restricted spheres like computing, automation and space technology was rather worse, and we're still waiting for many of the "highly likely" developments they predicted. For example, a Mars landing and a moonbase seem further away now than they did in 1964. The major social problems the panels feared were massive technological unemployment and "social upheaval" induced by widespread automation and the introduction of robots into industry and services.

the most developed. It has run a series of studies on specific topics, which typically begin with a review of existing science and then get a group of experts to assess where the science and technology might be heading in the next ten to fifteen years. This is known as "horizon-scanning" (which sounds much better than well-informed guesswork), and it is combined with analysis of social and economic trends.

The UK Foresight programme has also produced a handy online guide to tools for "strategic futures thinking", which is one of the best of its kind. These tools have lots of different names, but the majority are workshop-style discussions, supported by carefully compiled data, which are designed to elicit structured opinions about what might be possible, and preferable. If you fancy having a go at futures work yourself you can access this toolkit at www.foresight.gov.uk/microsites/hsctoolkit. The rest of this book draws repeatedly on exercises of the kind described there.

WORLD-MODELLING

These more or less respectable foresight efforts generally have tightly focused topics and modest aims. But is their emphasis on expert opinion-sampling, trend analysis and carefully structured conversation as far as the art of future-gazing has got? Yes and no. There are more ambitious efforts to inform discussion of the future, but they don't necessarily produce more convincing results. They tend to depend on beefing up the analysis and synthesis which humans can manage with computer treatments of various kinds. This leads to attempts at a global computer simulation.

The best guesses made by experts in different fields – in a Delphi-style exercise, for example – need to be brought together in a way that allows things happening in one area to affect trends in another. This stage of future-building, known in the trade as cross-impact analysis, quickly becomes difficult to keep track of. When the effects are well-understood and the numbers are known, you can plug everything into a super spreadsheet, and watch as one variable is altered and the others move accordingly.

It's not quite so simple when you're looking at the big picture, but there have been attempts to summarize the behaviour of the planet with equations which capture the interactions and feedbacks between different processes. This has become familiar from the ever more detailed attempts to model the future of the Earth's climate. The global interactions between sunlight, atmosphere, clouds, oceans, rocks and living things are incredibly complex, and even a simplified version of all the relationships involved, expressed in some set of mathematical equations, is not going to be any use without help doing the calculations.

Nowadays, that help comes from superfast computers. (We will look at how helpful they are in climate prediction in Chapter 7, but for now remember the limitations of Laplace, see p.39.) In more general future-oriented work, the features of the world that are modelled are usually aspects of population, economy, environment and technology. The pioneers that produced the computer model used to underpin the analysis of the landmark study *The Limits to Growth* (1972) had a simple model, little data and a computer incredibly slow by today's standards. So now we have better data and faster machines, the models will have got better too, right? Well, only up to a point.

In simplifying the wonderful, messy profusion of the real world, models often depend on projecting past trends into the future, based on the overarching assumption that key chains of cause and effect will stay the same as they are now. As the critics of *The Limits to Growth* (see p.66) showed, that assumption can determine the result of the forecast, and faster calculation by a supercomputer does not alter that.

NOT ONE FUTURE, BUT MANY FUTURES

The broader point, though, is not that the assumptions are *necessarily* wrong, but that they are so variable, no one knows which ones will turn out to be correct. In short, there are many possible futures. This is another way of putting the discouraging fact that the future is inherently unpredictable, and in ways that are unlikely to be overcome (the one safe prediction, perhaps).

The standard response to this among futurists is to acknowledge that possible futures are infinitely

THE RAND CORPORATION'S WAR GAMES

Like much else in the modern futures analyst's toolkit, scenario-building has its origins in military planning. At the end of World War II, the US military were so impressed by the power of science and technology – the atom bomb had that effect on people – that General Hap Arnold, commander of the Army Air Force, commissioned an émigré Hungarian aeronautical engineer to convene a group of experts to forecast the future of aircraft technology. Arnold was so pleased with the results that he set up the research and development group, Project RAND under the management of the Douglas Aircraft Company.

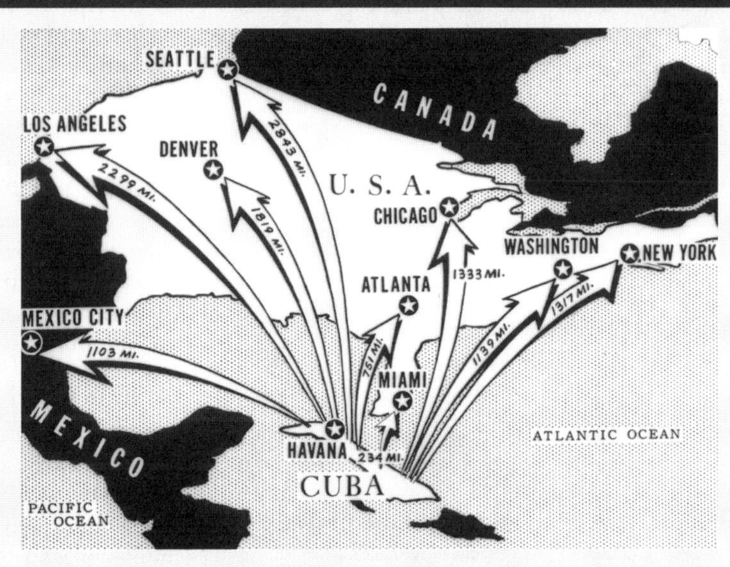

War scenarios: a newspaper from 1962 illustrates the distances from Cuba to major US cities during the Cuban Missile Crisis.

The project became independent in 1948 as the nonprofit RAND Corporation, and initially focused on futures and design studies for the military. During the Cold War, RAND were preoccupied with understanding what might happen during a hypothetical nuclear exchange between America and Russia, a speedier event than military planners were probably used to – ballistic missiles could cross continents in hours.

One response was the development of the "war game". Such games tried to simulate future conflicts, and assigned the players roles in the resulting combat. The idea was not to make predictions, but to set up possibilities, and to try to understand where they might lead. In the words of RAND researcher Herman Kahn, they allowed strategists to "think the unthinkable". Crucially, they let people rehearse decisions that would have to be taken instantly in a real war, and to reflect on the results (Hmmm... just lost two-thirds of the population in a counterstrike. Better try that one again...).

Less dramatically, the same approach has informed scenario exercises which do not lead to the end of the world, or at least not so quickly. In fact, scenario-building has become a common tool in future-oriented projects of all kinds. It's less common to use it as a basis for "war game" style exercises, but if budgets and time permit, games do allow people to explore scenarios from the inside in ways that are hard to match just by reading about them.

www.rand.org

varied, and to sketch that variability by drawing key factors together into a set of coherent possibilities, or scenarios. Exercises like this can incorporate many techniques, but often conclude with stories of how different futures may unfold. These stories can be written in deliberately evocative and powerful ways to try to prevent the realization of some scenarios, or to promote others (as Rachel Carson's "A Fable for Tomorrow" proved, see box on p.42).

This type of scenario-sketching seems to meet the needs of lots of different organizations, and allows the results of other kinds of forecasting to be brought together. Again, its roots lie in the work of the RAND forecasters (see box on previous page). RAND strategist Herman Kahn refined scenarios as an aid to business planning in the 1960s. He also became a famous – and famously optimistic – futurist, from his new home at the Hudson Institute, the prototype private sector think-tank. In the 1970s, scenarios went respectable in business, with notable efforts being made at the oil company, Shell.

PREDICTIONS FILE

Athena Andreadis

Neurobiologist, University of Massachusetts Medical School, author of *To Seek Out New Life: The Biology of Star Trek* (1998), blogger at www.starshipnivan.com/blog

Highest hopes: In decreasing order of likelihood, that we will conquer – or at least tame – dementia, which will make the increase in average life expectancy individually worthwhile and collectively feasible; that we will pick up an unambiguous SETI signal (search for extraterrestrial intelligence); and that we will decipher the ancient script Linear A and find out that the Minoans were indeed enlightened, if not matriarchal.

Worst fears: Most of our activities will devolve into inward navel-gazing ("social" Internet, virtual reality) rather than outward exploration, and our politics (broadly defined) will force all research into applied/profit mode, doomed to produce results and reagents that will make the long-term survival of the planet and all its species increasingly problematic.

Best bet: Barring a natural or human-created catastrophe, we'll muddle along just as before and run out of resources and *lebensraum* before we're able to establish either a sustainable terrestrial footprint or expand beyond Earth.

CASE STUDY 1: SHELL

Success in producing scenarios that helped explain the possible causes and consequences of rising oil prices convinced Shell to develop techniques for corporate planning during the late 1970s and 80s. Shell employees who moved on, notably Peter Schwarz of the futures consultancy Global Business Network, spread the word, and scenarios soon became a common part of company strategies. Corporate scenarios are, of course, geared towards helping executives figure out how their organizations can make money, and continue making money. But they can also generate stories that incorporate various visions of the future.

In 2008, Shell published scenarios that explored the interactions and trade-offs between three overarching goals that governments may pursue up to 2025 – efficiency, social justice and security. Sophisticated though their analysis may be, the "signposts" the Shell team cite as indicators of current trends are basically a collection of items from newspapers, government agencies and private think-tanks. In other words, the scenario-builders' stories are not based on secrets about the world that are only available to colossal companies like Shell, but on collating and selecting information that anyone can access. Like other futurists they are trying to read the signs around them; so the usefulness of their stories must come from the way they're put together.

The scenario habit has now spread much more widely, and is often used by public sector organizations and pressure groups to highlight issues they may have to deal with in the future. Their goals and interests differ, so the way they present the future varies as well: futures have to be sold.

CASE STUDY 2: THE BRITISH HEALTH AND SAFETY EXECUTIVE (HSE)

The British Health and Safety Executive (HSE), the outfit that tries to reduce hazards in UK workplaces, published a set of scenarios in 2006 about, you guessed it, the future of health and safety. These were based on "hot topics" gathered by their Horizon Scanning team and "critical issues" identified by a series of interviews with stakeholders. Their description of their methods is a good example of the kind of procedure common to many scenario-building efforts. They took all the hot topics and critical issues, covering an array of social, scientific, economic, environmental and demographic factors, and boiled them down to two crucial questions:

- Are public attitudes towards risk those of personal responsibility, or of the "blame culture"?
- Will the UK increase its competitiveness in the global economy?

The first question has just two simple answers (either/or). Assuming that the second question also has just two answers (yes/no), bring all the alternatives together and, hey presto, you've got four possible scenarios.

However, that was not the end of the process. The point of the scenarios was to get people to think, so the HSE used them in internal conferences as "incasting" and "wind-tunnelling" exercises. The jargon here cloaks something fairly straightforward. Incasting involves imagining what issues might arise in each of those future scenarios, and wind-tunnelling involves testing existing policies against those scenarios.

It's tempting to have fun with the consultant speak that infests foresight projects and scenario exercises, but organizations and governments will be better prepared for the future if they try to think about it systematically and to recognize which uncertainties are the most important. It's at least got to be better than just assuming things will carry on as they are: muddling through isn't much of a contingency plan.

The questions posed in the original HSE interviews to identify the key health and safety issues were relatively simple, and, most interestingly, would probably prove helpful in all sorts of forecasting contexts:

- Clairvoyant: if you could spend some time with someone who knew the outcome, a clairvoyant or an oracle, what would you want to know? (i.e. what are the critical issues?)
- An optimistic outcome: what would be a good outcome and what would be the signs?
- A pessimistic outcome: how could the environment change to make things more difficult? What could go wrong?
- The internal situation: what needs to change if your optimistic outcome is to be realized?

- Looking back: over the past ten years, what successes can we build on and what failures can we learn from?
- Looking forward: what decisions need to be made in the near-term to achieve the desired long-term outcome?
- The epitaph: if you had a mandate, without constraints, what more would you need to do?

The list reads like it could have been taken from a self-help book on how to get what you want out of life. But self-help books exist because people need help – and organizations do too. This kind of forethought isn't even close to rocket science. It's simply a way for organizations to lift their heads from last week's balance sheets or next week's bureaucratic edicts to try to think seriously about the medium term and how their decisions can affect it.

The trendier bits of technique that are added to the basic list may obscure this simplicity, or add something of genuine value. Which way it goes probably depends on what those techniques are, and how skilfully they're applied. In this case, the HSE report says they added "confirming data" and references to other organizations' scenarios, and that they considered feedback loops and how the "resulting changes affect not only actions and infrastructures, but also deep structures like worldview, values and identity."

That all sounds pretty comprehensive. The four scenarios the study focused on were much less so, as they were the result of just the two crucial questions mentioned above. Perhaps that rather narrow definition of the relevant key features of possible futures was right for an organization with the HSE's responsibilities, but it does seem a bit parochial. Their report certainly shows how scenarios can take on the preoccupations of the customers, or authors.

Case study 3: the Centre for Responsible Nanotechnology

That same point applies to a final example: the Centre for Responsible Nanotechnology (CRN), a nonprofit think-tank concerned with the social and environmental implications of nanotechnology. The CRN created a series of scenarios which were designed to promote public discussion about the possible future of "molecular manufacturing" – the yet-to-be-achieved process of building objects atom by atom (see p.86). Their scenario-building project had interesting similarities and differences with the exercises of the smart commercial thinkers at Shell and the forward-looking bureaucrats at the HSE.

For one thing, as a small, nonprofit "virtual think-tank", the CRN proves that almost anyone can play the scenario game, with a bit of effort and organization. They brought together more than fifty people, from diverse backgrounds and specialties, across six continents, to conduct "a unique series of 'virtual workshops', using a combination of teleconferencing, Internet chat and online shared documents". Over several months, their Scenario Development Project came up with eight very different scenarios.

Look closely at the results (www.crnano.org/ctf-scenarios.htm), however, and you'll find that the final scenarios are not very well written. Most of the ideas seem like suggestions for derivative science-fiction stories, and they're not worked up convincingly. Further proof, perhaps, that scenario exercises, especially highly speculative ones like this one, are heavily dependent on the quality of the storytelling.

IMMORTALITY OR EXTINCTION: TAKING SCENARIOS INTO THE FOURTH MILLENNIUM

For the novice futurist, the results of the scenario boom can be a little overwhelming. A single set of scenarios, well drawn, can be helpful in highlighting key issues and assessing what factors influence how those issues will develop over time. But there are now so many scenarios out there, it's hard to know which ones to choose, or how they relate to each other.

The Millennium Project, a United Nations-inspired effort, has been offering scenarios of its own since the mid-1990s, and trying to keep track of everyone else's. By 2007, their index of semi-fictions about possible worlds had over seven hundred entries, and was growing by about fifty a year. But, as many of these are multiple scenario sets – no one here is claiming to offer stories about *the* future – the total is even higher. Of course, many of the scenarios have common features and speculate along similar lines, and the number of ideas about politics, technology, economics, the environment, or whatever that can be fed in is limited. But the details vary widely, often because a particular sponsor has asked for thoughts about a very specific topic of interest – whether it's life on Mars for NASA, or diabetes and obesity for healthcare company Novo Nordisk.

The Millennium Project itself has joined in the scenario game enthusiastically, offering a menu that includes global scenarios and more detailed considerations of big topics, such as energy, terrorism and Middle East politics. If that is not enough, they have even stretched their imaginations with six very long-range scenarios designed to show what might affect human life in the year 3000 – well, it is called the "millennium" project.

And the results of that exercise? By the year 3000, humans will apparently either be "functionally immortal" at the turn of the fourth millennium (scenario number one), or be extinct, but succeeded by "a system of robots, computers and networks" (number two). Then again, humans could survive and command technologies which allow time travel (three); learn to live rather uneasily alongside enhanced cyborg creatures and completely artificial beings who are following their own path (four); or be happily enjoying the endless possibilities of life as super-powerful human-machine hybrids (five). Finally, there is a scenario in which recognizable humans, disappointed that they still haven't made contact with life elsewhere in the galaxy, are continuing to try to develop space travel, in order to ensure that the species survives the eventual end of life on Earth – a very long-term project indeed.

Once again, these scenarios are essentially science fiction. Whatever the use of scenarios to inform discussion of the medium term, the longer horizon is probably best left to the novelists.

www.millennium-project.org/millennium/scenarios.html

TWENTY-TWENTY HINDSIGHT IS THE WINNER

So, where does this tour of future-gazing techniques leave us? It's pretty clear that the future remains radically uncertain, and there's really not much we can do about it. At the least, anyone who actually makes "predictions" invites scepticism. Oracles aside, all we can really hope for is to collect the best information from the most reputable sources, and then rely on imaginative, informed, carefully considered opinion, checked against a spread of other views.

Nevertheless, some of the information available is getting better (especially about the past), as is detailed in the following chapters. Futurists are becoming more adept at gathering those informed opinions systematically, and helping to eliminate, or at least make explicit, their various prejudices and presuppositions. And despite – or perhaps because of – our fears and uncertainties about the future, there are a growing number of efforts to think about it, and even to try to plan for it.

Besides, uncertainty about the future is itself a good thing: imagine how it would feel if the future was certain. Entrepreneurial types (and insurance underwriters) regard uncertainty as opportunity. That may not be much comfort to a subsistence farmer wondering whether this year's crop is going to be hit by drought, but may still be felicitous for people looking for new adventures and opportunities outside of the money-driven marketplace.

The most useful advice for aspiring futurists (which is easier to offer than to follow consistently) is to review as many forecasts and scenarios as possible and weigh them against each other. But keep in mind that you won't really know which ones are correct until you can apply the only truly effective technique: twenty-twenty hindsight. (Not that hindsight settles all arguments, though, as a glance at any discussion of the recent history of Iraq will instantly remind you.)

TOP TIPS FOR ASPIRING FUTURISTS

If sampling a wide range of forecasts or scenarios is the best way to plot possible futures, how are we to judge those forecasts? With so many predictions, promises and warnings now on offer, are there any rules to indicate which ones are more plausible, and based on sound thinking and good data (about the past and the present), and which are more likely, or more achievable?

Italian critic and novelist Umberto Eco is not a futurist, but he has one piece of sage advice for would-be predictors: "Never fall in love with your own airship". He is referring to the way futurists in the early twentieth century were beguiled by the airship or zeppelin – the wonder of the age – and incorporated fleets of lighter than air craft in their visions of the days to come. But a few decades, and a few disasters, later the airship was a dead duck, and heavier than air craft were the technology of choice. Though perhaps one should add to this technological parable not to forget the airship, either. A shift away from fossil fuels could yet see the economics of (safer) airships making them a feasible substitute for gas-guzzling aeroplanes.

Eco's point is about reading trends too simply. Trends from the past may be all we have to go on, but be prepared for discontinuity. Past performance is no guarantee of future results, as it says in the investment guides' small print. At the most basic level, it's reasonable to assume that the sun will rise tomorrow, but since Scottish philosopher David Hume (1711–66) expressed scepticism about inductive reasoning (see p.32), futurists have accepted that, however psychologically compelling that conclusion may be, it's not backed by logic. After all, the sun could go nova, be sucked into a black hole or serve as a snack for a giant space dragon before you wake up tomorrow. It just isn't very likely.

So much for the problem of spotting when change could overcome apparent stability. But what about mapping change that is already under way? Bjørn

"Never fall in love with your own airship". Once a symbol of the future, the airship was abandoned following the 1937 *Hindenburg* Disaster, but may yet make a comeback.

Lomborg, author of the controversial, figure-laden tome *The Skeptical Environmentalist* (2007), criticizes many reports and forecasts that claim to portray the state of the world now and in the future, saying that they often neglect the fundamentals.

His first principle is that useful assessments are based on comparisons. "The world is in a terrible state." Ah, yes. Compared to what? The most relevant comparison, he suggests, is with how the world was before, but such trends are best summarized using

PREDICTIONS FILE

Austin Williams

Architect, director of the Future Cities Project

Highest hope: The production of housing and buildings will become industrialized, as new technologies such as rapid prototyping are applied to the design, manufacture and prefabrication of dwellings.

Worst fear: The philosophy of sustainability will entrench the contemporary, accusatory mantra that "mankind is a problem", resulting in the mainstreaming of misanthropic Malthusianism whereby resources are privileged over humanity.

Best bet: Humanity's aspiration for material (as well as social and political) improvement will triumph over restraint. The historic role of humanity is to argue against "limits" in order to play an active, progressive role in the future. Whether these limits are socially constructed or naturally imposed, allowing our mindset to be governed by what is deemed to be "environmentally responsible" is leading to a survivalist mentality. Only by maintaining a human-centred challenge to the risk-averse culture of limits, will we retain a 360-degree vision of the future.

global data. There will be much local variation, and thus a temptation to pick regions that support a particular view. Only global figures, he says, can indicate "whether there have been more good stories to tell and fewer bad ones over the years or vice versa". This argument isn't entirely persuasive, as averaging on the largest scales can also obscure important details – though this might mean the global trend is being read using the wrong indicator (more on this in Chapter 4). Lomborg also suggests that comparing various global problems can help you determine how important they really are.

Lomborg's second recommendation is to look at long-term trends, as in decades or centuries, rather than years. In particular, he advises watching out for people cherry-picking a span of years during which a longer term trend was temporarily halted or reversed. Finally, he says that people's needs and desires are the most important features of any assessment of the state of the world. Even without a perfect method of discovering what those needs and desires are, it's good to be reminded that futurism is political. As Leo Howe wrote in his *Predicting the Future* (1993):

> "Prediction and planning are not … neutral processes but ideological ones. Prediction is rarely done for its own sake; it is almost always prediction undertaken for some specific reason, and carried out by middle-class pundits, academics and politicians. …The issues those in power want forecasts about are not very likely to correspond to what the disadvantaged are most concerned with."

Lomborg's final principle is an ethical stance, not a recommendation about method. His other ideas seem like good advice, though opinions differ over whether he follows them himself – the main charges levelled against him being that he makes selective use of the evidence, and misinterprets trends. Pots and kettles, anyone?

SIX RULES FOR EFFECTIVE FORECASTING

Clearly an awareness of political or personal agenda is essential when evaluating whether a particular prediction is up to scratch, but knowing what makes a good forecast is equally important. Paul Saffo, a veteran technology forecaster, gave the readers of the *Harvard Business Review* six rules for effective forecasting in 2007. Here's a summary:

▶ **Define a cone of uncertainty:** Uncertainty is cone-shaped because it gets larger as you move away from now. "Mapping the cone" is Saffo's way of saying the forecaster is not predicting, but finding the limits, or range, of what might happen.

▶ **Look for the S-curve:** This echoes the popular assumption that lots of natural, social and technological developments follow a trajectory that starts off exponential, but then levels off – generating the "S" shape which the Global Scenarios Group saw as ubiquitous (see p.43). The pattern summarizes many forecasts, from demographers' projections of future population to Moore's law charting increases in computing power (see p.82). The trick is gauging how long the exponential phase will last.

More important, though, according to technology specialist Saffo, is spotting when it is about to start; when something that is growing slowly suddenly speeds up. The lesson for forecasters, he says, is not to allow this kind of apparently sudden take-off to cause them to overestimate short-term change, but underestimate long-term change. As he puts it, "never mistake a clear view for a short distance".

▶ **Embrace the things that don't fit:** Anomalies, interesting failures and oddities are hard to spot because they do not mesh with our habitual ways of thinking, but they may be indicators of impending change. This sounds helpful, but seems less so when you realize that there are always innumerable odd things going on at any one time. The trick is choosing which ones to notice.

▶ **Hold strong opinions weakly:** This is Saffo's way of saying forecasters should look for evidence that contradicts their hypothesis – to counter the tendency to seek only that which reinforces what you already think. You have to have strong opinions to come to any conclusion at all, but they should still be discarded easily. "If you must forecast, then forecast often, and be the first to prove yourself wrong."

It's also useful advice to bear in mind when reading criticisms of forecasts. Finding out that a particular forecaster has been wrong in the past might be to their credit, not a reason to ignore their latest prediction. It all depends on what they got wrong, and whether they owned up and could explain why.

▶ **Look back twice as far as you look forward:** Looking back further than the recent past makes it easier to find effective historical parallels. Again, this can be applied to new technology, with the early days of the Internet being compared to the early days of broadcasting. But remember that the parallels are never exact: "History doesn't repeat itself, but sometimes it rhymes." When evaluating a forecast, ask if a historical analogy offers a point-by-point comparison, and recognizes differences as well as similarities. Saddam Hussein is like Hitler, you say? How, exactly?

▶ **Know when not to make a forecast:** As a professional forecaster, Saffo put it a bit more gently, but sometimes it's pretty obvious that no one has a clue what is going to happen. When the Berlin Wall was pulled down in 1989, for example – part of a sequence of events no one had foreseen – the shape of the post-Cold War world was ill-defined, and instant interpretations were highly suspect. Better, says Saffo, to wait and ponder – looking for signs missed before the change and signs emerging in the ensuing turmoil.

Saffo concludes by admitting that "forecasting is nothing more (nor less) than the systematic and disciplined application of common sense". Evaluating a forecast using these six rules assumes that it is one where the forecasters have actually shown their working – spelled out what their prediction is based on, and explained the logic that led to their conclusions. If this is not the case, then what you have is the work of a prophet, a diviner or a seer, not a forecaster. Then you are on your own. You can have faith, or not. It's up to you. But don't expect everyone else to join you if you choose to sign up.

FURTHER EXPLORATION

Alex Abella Soldiers of Reason: The RAND Corporation and the Rise of the American Empire (2008) Abella's book demonstrates the tight coupling between the futuristic think-tank and US government policy – especially military policy – since the 1940s.

Adam Gordon Future Savvy: Identifying Trends to Make Better Decisions, Manage Uncertainty and Profit from Change (2009) An admirably clear guide to reading forecasts, and filtering out the flaky, biased or just plain fanciful ones. Written for business, but useful for everyone else.

Paul Halpern The Pursuit of Destiny: A History of Prediction (2000) Physicist and popular science writer Halpern looks at our addiction to prediction.

Nicholas Rescher Predicting the Future: An Introduction to the Theory of Forecasting (1998) The author is a philosopher by trade, so this is a systematic and fairly rigorous review, but it's still readable. He concludes that complete predictability in life would be as dispiriting as complete unpredictability, but fortunately we live somewhere between the two.

Peter Schwarz The Art of the Long View (1991) The classic text on the corporate scenario-building approach pioneered at Shell, and now often used by the Global Business Network (www.gbn.com).

www.foresight.gov.uk The UK's "official" futures think-tank has a handy guide to tools for futures work and how best to use them (look for "Toolkit").

www.millennium-project.org This UN-backed futurology clearing house publishes a wealth of resources, including a "Futures Research Methodology" handbook that runs to 1300 pages on CD.

www.oecd.org The Organisation for Economic Co-operation and Development has a whole department devoted to its International Futures Programme, which maintains quite a comprehensive list of other futures organizations and websites (find it under "By Department").

4 the recent future

the recent future

When I was a child, people used to talk about what would happen by the year 2000. Now, thirty years later, they still talk about what will happen by the year 2000. The future has been shrinking by one year per year for my entire life.

Danny Hillis, in 1993

Take your eyes off the future for a little while and it turns into the past. It always has done. But as twenty-first-century people, we seem to be acutely conscious that we are living in what used to be the future. This is adding new layers of complexity to our attitudes to the futures that have yet to crystallize as history.

THE NEW, EXPANDED FUTURE

The twentieth century, people tend to agree, was pretty eventful, giving rise to changes that fed the futurist imagination as never before. In the wake of H.G. Wells's turn-of-the-century writing (see Chapter 2) came a wondrous profusion of images of the future, and discussions of what it might bring. Some came from the development of forecasting techniques (as described in Chapter 3), others from the flowering of science fiction in print and on film. Yet more came from the efforts of propagandists, corporate image builders and campaigns to accelerate or decelerate the various trends of a turbulent era. Trying to round up all of these visions would be impossible, so here is a sketch of the most pertinent ones, painted with the broadest of brushes.

Even when tempered by the mechanized slaughter of World War I, optimism about the future seemed limitless in the West at the start of the century. Sometimes, this sense of unbounded possibility came attached to real scientific discoveries. For instance, when atomic scientist Frederick Soddy (1877–1956) pondered the future in *The Interpretation of Radium* (1909), he foresaw harnessing the newly discovered energy of radioactivity as bringing about some kind of golden age. If people did that, he wrote, they could "transform a desert continent, thaw the frozen poles, and make the whole world one smiling Garden of

Eden." Wells duly incorporated atomic power into his 1913 novel *The World Set Free*. His vision included a strikingly prescient description of atomic war, but it truly was a "war to end war": atomic destruction was followed by a nuclear-powered utopia ushered in by a new, rational world government.

Other visionaries went even further in the following decades. In his 1924 essay *Daedalus, or Science and The Future*, biologist J.B.S. Haldane (1892–1964) forecast the end of fossil fuels and their substitution by solar and wind power.

> "Four hundred years hence the power question in England may be solved somewhat as follows: The country will be covered with rows of metallic windmills working electric motors which in their turn

OLAF STAPLEDON: A ONE-MAN FUTURE FACTORY

The writer who did the most to expand the horizon of the literature of the future after H.G. Wells was fellow Brit Olaf Stapledon (1886–1950). As well as writing fiction and non-fiction, Stapledon drove ambulances in World War I, taught adult education and took a PhD in philosophy at the University of Liverpool. In his two great fictional epics, *Last and First Men* (1930) and *Star Maker* (1937), he tried to express a new vision of the human future, and beyond. However, he characterized his own writing not as prophecy, but as essays in "myth creation".

Last and First Men is a story told two billion years in the future by a member of the eighteenth species descended from present-day humans, as they face their own demise. This distant future-dweller can communicate with the author, and uses him to relate the rise and fall of successive human races across the Solar System. Although Wells sent his time-traveller into the far future in *The Time Machine* (1895), and pondered the great question of what is to come after people as we now know them, Stapledon went much further, unfurling a tale that is both cosmic and tragic. His great achievement was to imagine how a species' story could extend over aeons, as he ambitiously conjured up eighteen distinct versions of mankind, each with its own strengths and weaknesses.

Stapledon's aim seemed to be to provide some perspective by opening up a grander scale to measure current affairs against – as NASA's images of Earth as seen from space did in the 1960s, or as The Long Now Foundation are currently trying to do with their various projects (see p.10 and p.69). As he put it in the preface to *Last and First Men*, "To romance of the far future ... is to attempt to see the human race in its cosmic setting, and to mould our hearts to entertain new values."

Stapledon composed a further message from the same future in *Last Men in London* (1932), and created an even larger-scale narrative in *Star Maker* five years later. That book set the rise and fall of innumerable other species on other planets against a truly cosmic backdrop, and sowed the seeds for countless more conventionally novelistic science-fiction dramas in the 1940s and beyond.

Although Stapledon went on writing up until his death in 1950, and produced a good deal of non-fiction, his interests were mainly philosophical. His extraordinary imagination stands largely separate from the practical concerns of the forecasters and scenario-sketchers who created non-fictional literature about the future in the second half of the last century. Much of his work is now out of print, but it can be found online.

> supply current at a very high voltage to great electric mains. At suitable distances, there will be great power stations where during windy weather the surplus power will be used for the electrolytic decomposition of water into oxygen and hydrogen."

Haldane also foresaw instant communications, new drugs and stimulants, chemical foods and a host of advances in biology, though he recognized these would all be more problematic. This impression was reinforced by novelist Aldous Huxley (1894–1963) in his vision of a regimented, dystopian society, *Brave New World* (1932).

Marxist and crystallography pioneer J.D. Bernal (1901–1971) took future fantasies into a whole new realm – and mankind into space – with *The World, The Flesh and The Devil* (1929). Physics would control the forces of nature, biology would fix the human body, and psychology would improve the personality, but this was just the beginning. Eventually, a "man of the future" would evolve, with artificial replacement limbs and organs linked to the brain:

> "We badly need a small sense organ for detecting wireless frequencies, eyes for infra-red, ultra-violet, and x-rays, ears for supersonics, detectors for high and low temperatures, or electrical potential and current, and chemical organs of many kinds."

Even today, it's enough to make a person restricted to the normal human senses feel a bit inadequate. But even Bernal's speculations were put in the shade by a writer whose future extended further than anyone else had yet dreamed, another dazzling Englishman, Olaf Stapledon (see box on previous page).

Of course, optimism did not reign unchallenged – it never does. E.M. Forster's short story *The Machine Stops* (1909) provided a memorably chilling challenge to the general enthusiasm for technology. His far-future world is inhabited by a subterranean humanity who are better off than Wells's Morlocks (see p.32), but utterly dependent on their automated servants. They rarely leave their individual cells, and communicate with their fellows via an Internet-like network of video links, through which they exchange pallid "ideas" about stuff they have learned through the same channels. The idea that one might visit the dangerous, dead surface of the Earth is horrifying. All except one are content with their impoverished underground lot, but when their machine-god begins to run down, he is as helpless as the rest.

Although this warning about the rise of the machine maintained a cautionary tradition, it contrasted with the general view of a golden future

PREDICTIONS FILE

Melanie Swan

Futurist, MS Futures Group, Palo Alto, blogger at futurememes.blogspot.com

Highest hope: The world will have transitioned to a post-scarcity economy such that all material needs are met at zero to low cost and intelligent beings (whether physically or digitally embodied) are exclusively pursuing their own interests.

Worst fear: An existential risk, such as bioterror or bioplague, may cause a significant barrier to human progress.

Best bet: The pervasive adoption of one or more new technologies that are currently unforeseen and have the same impact as the Internet, but do not change society structurally since human morphology could likely remain fixed in this period.

made possible by advances in science and technology. As the years passed, the devastation wrought by two world wars, plus events such as the sinking of the *Titanic* and the destruction of the *Hindenburg* airship, continued to put futuristic optimism to the test, but for the first two-thirds of the century it survived more or less unchanged. In fact, two of the most severe setbacks (the Great Depression and the atomic bombing of Hiroshima) provided indirect boosts to the idea that everything might one day be (technologically) wonderful.

JAM TOMORROW

The economic depression which began with the Wall Street Crash in 1929 and stretched throughout the 1930s led to a search for signs of better times to come. As historian Joseph Corn put it, this was when many businesses discovered that allusions to the future in their advertising could boost sales. A futuristic gloss transferred to design as well, and during the Depression the US was wowed by visions of sexy, streamlined trains and cars, and state-of-the-art teardrop-shaped radios, desks, adding machines and even pencil sharpeners, many of which were unveiled to an eager public at the 1939 New York World's Fair (see box on p.65).

The second challenge to optimism about utopian techno-futures was

The future's looking great: stylish, streamlined designs promised better times to come in the 1930s, Depression-era US.

the nuclear obliteration of Hiroshima and Nagasaki in 1945, and the ensuing Cold War-driven arms race between the US and the Soviet Union. Whether there would actually be a future became a genuine concern, and a troubling topic for policymakers, rather than a question left to novelists or philosophers. But both governments – capitalist and Communist – remained committed to the idea that technology was the key to a prosperous and secure future, even while spending vast sums on ever-more-powerful weapons of mass destruction. And they tried to reassure their citizens that the most conspicuous new technology – atomic power – was a force for good.

This reassurance came in the form of a tide of propaganda about the nuclear wonders to come. Popular culture was abrim with contradictory messages. The fiction of the 1950s was written in the shadow of apocalypse, with thousands of novels, stories, films and comic books depicting mutants picking through the irradiated rubble. At the same time, popular magazines featured breathless, and often fanciful accounts of life with unlimited supplies of nuclear energy. Plans for nuclear-powered cars, ships, trains and planes were plentiful, and this was the age of confident predictions that electricity would be "too cheap to meter". That was not all. Nuclear technology would also, somehow, solve food shortages and fulfil Frederick Soddy's 1909 vision of warming the poles and greening the desert.

This hopeful mood was not confined to nuclear-enthusiasts, though. Even those sceptical about nuclear power shared the rose-tinted view of a technologically enabled life in the future. A lengthy article by the science editor of *The New York Times* published in *Popular Mechanics* in 1950 is as fine an example of pop-futurology as you'll find. His "Miracles You'll See in the Next Fifty Years" include the wide use of solar, not nuclear, power. They also allow suburban families to live in a luxurious but cheap house built of synthetic materials, fill it with washable furniture and disposable crockery, eat microwave meals, use videophones for teleconferencing and teleshopping, and travel by family helicopter for local trips or supersonic rocket plane if they can afford it. Cancer has not yet been cured, but medicine has kept seventy-year-olds looking like forty-year-olds. In short, life is good, and getting better.

That feeling was reinforced in the US by the extended consumer boom. The good life of the future was easy to imagine because you could see it arriving. As Bill Bryson recalled in *The Life and Times of the Thunderbolt Kid* (2007), about growing up in the 1950s:

> "People looked forward to the future in ways they never would again. Soon, according to every magazine, we were going to have underwater cities on every coast, space colonies inside giant spheres of gas, atomic trains and airliners, personal jetpacks, a gyrocopter in every driveway, cars that turned into boats or even submarines, moving sidewalks to whisk us effortlessly to schools and offices…"

Would people ever again look forward like that? The feeling certainly survived into the 1960s to some extent. The most visited exhibit at the 1964 New York World's Fair was, just as in 1939, a General Motors extravaganza called "Futurama". This time it visualized trips to the moon, undersea holidays, blooming deserts and cities with super-skyscrapers and moving sidewalks. Meanwhile in the jungle, as the blurb describes, visitors saw "trees felled by searing laser beams. A monster road-building machine follows,

PROMOTING THE WORLD OF TOMORROW

The Depression era's sales pitch for the future reached its peak at the 1939 World's Fair in New York, which was devised to lift the spirits of the American people and promote American businesses. Although it followed in the illustrious footsteps of Chicago's 1893 Fair (see p.26), the New York exhibition was more about flashy consumer gadgetry than scientific innovation. The fair's theme was "The World of Tomorrow", and visitors left with button badges proclaiming "I have seen the future".

A host of corporate-sponsored pavilions featured soaring cityscapes and homes of the future, but the biggest attraction was General Motors' "Futurama". Their exhibit was a ride through a 36,000-square-foot model of a future American landscape of super-cities linked by seven-lane highways. The pavilion was, in effect, a huge advertisement for the interstate highway system that the government built in the decades to come. But, more than that, it was a place to get a concentrated dose of optimism. As David Gelernter noted in his book *1939: The Lost World of the Fair* (1995), when the pavilion's designers turned their minds to the future, "What they saw was good. Technology in particular was good. The future in general was good."

An animatronic giant octopus frightens a visitor at General Motors' Futurama exhibit at the 1964 New York World's Fair.

leaving in its path an elevated superhighway". Driving roughshod through jungles seems today a problematic emblem of progress, one reason why this vision of the future is now such a symbol of the past.

NEW PROPHETS OF DOOM

By the late 1960s, the post-World War II optimism that endured in the West, despite fears of a nuclear holocaust, was sharply challenged. Increased discussion of two issues in particular – pollution and population – suggested a bleaker outlook for the world. More people, and more people seeking a higher standard of living, would probably mean more pollution, and global population was increasing at nearly two percent every year, the highest growth rate in history. That striking fact was highlighted by dramatic warnings of impending famines in Paul Ehrlich's *The Population Bomb* in 1968. Hundreds of millions would die, he said. Malthus's time had come.

The improved reporting of population statistics was one benefit to emerge from a massive effort by the League of Nations to gather better data about the global state of affairs after World War I. That effort was boosted after World War II by the foundation of new international organizations such as the United Nations (UN), the Organization for Economic Cooperation and Development (OECD), the World Bank and the International Monetary Fund (IMF). Their policies have often been controversial, but their huge archives of statistics have proved very useful, offering humanity a new and often clearer picture of the world.

At the same time, new methods for analysing data were being developed (see Chapter 3), and there was an increasing trend for developing systems thinking that highlighted global interdependencies. When all of this was pulled together with a little computer modelling, forward-thinkers were left with new ways of predicting what might happen in the future, and a think-tank came up with a theory that could be summed up in one phrase: "limits to growth".

THE LIMITS TO GROWTH

The 1972 book *The Limits to Growth* was not the only forecast of bad times to come, but it was certainly the most influential. It was presented by a group of scientists, businessmen and academics calling themselves the Club of Rome, and featured work done by economists and computer modellers at the Massachusetts Institute of Technology (MIT). The authors plotted the interactions between population rise, economic growth, food production, resource use and pollution, and concluded:

> "If the present growth trends continue … the limits to growth on this planet will be reached sometime within the next 100 years. The most probable result will be a rather sudden and uncontrolled decline in both population and industrial capacity."

The book was mostly read as a clear warning of an impending crash, and although it is more complex than that, the charge seems fair. The authors argued that the worst outcomes could be avoided if population growth was curbed, but there was absolutely no indication of how this could be achieved, short of famine or war. Some were convinced by its message, but it also drew strong criticism.

Critics pointed out that the model used to project future states of the world was too simple, based on unjustified assumptions and lumped too many things together (for example, all the figures were global, not regional, and "pollution" was treated as a single variable). Furthermore, despite the new global data-gathering efforts, most of the information that went into the models was poor quality, or used guesstimates that were off the mark. But the main charge was that the *choice* of variables assumed to change exponentially, such as population and pollution, and those held to linear growth, such as food and mineral production, meant that the projections of collapse were already built into the models. Picking up the computer programmers' slogan of the time, "garbage in, garbage out", the critics said all *The Limits to Growth* demonstrated was that you could write software to give you "Malthus in, Malthus out".

Unperturbed, the Club of Rome produced a second edition, *Beyond the Limits*, in 1992, and a thirty-year update in 2004. They still maintain that they are basically right. Just like the Malthusian principle (see p.22), their logic works if you accept their assumptions, but it doesn't indicate exactly when the limits will be reached. The fact that the global system rolls on does not refute their original ideas, which forecast business continuing as usual until perhaps the middle of this century. But they now say that we are in "overshoot" mode, living beyond the limits without realizing it.

PREDICTIONS FILE

David Fisk

Professor of Engineering for Sustainable Development, Imperial College, London

Highest hope: Societies will configure themselves to make people happy, not just make some of them rich. Problem pull will dominate science push. Economics departments will close at universities across the world through lack of interest. Business schools are full.

Worst fear: A pervasive social autism – much of it emanating from the propaganda of the "science" media industry – in a world then perceived as overcrowded, disabling social discourse and tipping the world into drab misery and dysfunction. *Blade Runner* without decent robots.

Best bet: History does not end – it gets reloaded. It's back to muddle, success and failure, with 2005 viewed as about as unrepresentative of the twenty-first century as 1920 was of the twentieth.

The debate launched by *The Limits to Growth* was a strange one – and it still is. It led to plenty of critical exchanges between forecasters, and fuelled discussion about whether continued growth was desirable, and how it should be measured. But the majority of governments went on pursuing growth for their own citizens. Leaders in less developed countries, especially those with rapidly rising populations, had overwhelmingly good reasons for doing so, but wealthier countries wanted more growth too, though they might link it with the new adjective "sustainable".

HERMAN KAHN AND THE CORNUCOPIANS

Some still held to the verdict of futurologist Herman Kahn, who published a series of detailed and upbeat forecasts just a few years before *The Limits to Growth* appeared. The former RAND strategist (see p.49) argued in *The Year 2000* (1967) that there was likely to be a smooth continuation of positive economic trends for the next thirty years and beyond, and that humanity's "capacities for and commitment to economic development and control over our external and internal environment are increasingly without foreseeable limit." No limits to growth there, then.

Challenged by the world-modellers' predictions of impending catastrophe, Kahn's team responded strongly. In 1976, they looked further ahead, in a book entitled *The Next 200 Years*, and essentially denied the constraints others perceived as built into the Earth's systems. They argued:

> "Any limits to growth are more likely to arise from psychological, cultural or social limits to demand, or from incompetency, bad luck and/or monopolistic practices interfering with supply, rather than from fundamental physical limits on available resources."

In other words, an end to growth would only happen through a kind of failure of will, or a failure to make capitalism run smoothly.

These two schools of thought are still very much alive in debates about possible futures that are directed at influencing policy and policy-makers. The new Malthusians are still pursuing their argument with their opponents – let's call them Cornucopians. The two sides dispute the effects of growth, the hidden costs of industrial production, the capacity of the planet to absorb the effects of human activity, and the prospects for technological improvement. As the following chapters will examine in detail, the disagreements have spread into some new areas of contention, but the fundamental difference of opinion stays the same.

HOW MANY DIFFERENT FUTURES ARE THERE?

The Malthusians and the Cornucopians tend to operate with different assumptions about how things will change in the future. And as these kinds of academic and think-tank forecasting efforts proliferated after the 1960s, it became clear that the picture you get of the future depends on how you see the world now. There have been periodic attempts to review all the forecasts vying for attention, and they show that while data sources and analysis techniques are important, the crucial distinguishing feature of future visions is the

LONG BETS

One way to sharpen futures thinking is to put predictions on the record, name a date and back up your forecast with cold, hard cash.

An early example featured in the "limits to growth" debate, when the Cornucopian economist Julian Simon bet ecologist Paul Ehrlich that the price of several key raw materials would fall between 1980 and 1990. Ehrlich, who was convinced they would go up, agreed to a price-tracking wager featuring five metals each worth $200. He would pay Simon the difference if the inflation-adjusted price of this $1000 basket fell – Simon would pay him if it rose. The result? All five of the prices fell, and Simon got a cheque from Ehrlich for more than $500 to settle the bet.

More recently, the Long Now Foundation (see p.10) has encouraged bets to develop arguments, promote long-term perspectives and encourage people to make accountable predictions, rather than to settle differences of opinion like this. You can register a prediction with Long Bets for any period over two years, but they prefer longer, and they promise that the Long Now Foundation will stick around to see – and help judge – the results.

The topic must be a sensible one (and preferably one "of societal or scientific importance"), and once it's posted, with a supporting argument and for a small fee, you wait to see if anyone wants to take you on. If someone does challenge you, they must explain their reasons for disagreeing. The point isn't to win, the cash stakes are low and the winner a nominated charity, but anyone registered on the site can discuss the arguments and vote on the likely outcome. The result is a kind of Internet-based Delphi dialogue. You can read what everyone else thinks about a particular argument, from the future price of electricity to whether the Large Hadron Collider will destroy the Earth, then send a comment supporting one side or the other, or pointing out something others may have missed.

At the time of writing, more than five hundred predictions had been posted, most not looking ahead more than ten or twenty years, and ranging over science, technology, politics and society. Only a couple of dozen had actually turned into bets, and five had already been decided, the rest are awaiting challengers.

www.longbets.org

Place your bets: will the Large Hadron Collider destroy the Earth?

world-view of the authors. Values and political convictions colour any effort to portray a possible future, and to judge whether it is viable, sustainable or desirable.

Jim Dator, a futurist at the University of Hawaii, developed a classification system in the 1970s that he has used ever since to order discussion about possible futures. He argues that there are just four main visions underlying attempts to outline possible or preferable futures. Here's an outline of his four "Generic Images of the Future":

PREDICTIONS FILE

Nick Bostrom

Director of the Future of Humanity Institute, Oxford University

Highest hope: Utopia. Utopia is not a location nor a form of social organization. Utopia is the hope that the scattered fragments of good that we come across from time to time in our lives can be put together, one day, to reveal the shape of a new kind of life. The kind of life that yours should have been.

Worst fear: There are existential risks that could cause the extinction of Earth-originating intelligent life or a permanent and drastic loss of its potential for desirable future development.

Best bet: The further the point in the future we consider, the more likely that the human condition will have ended, and our descendants will either have suffered an existential catastrophe or attained some kind of post-human condition.

▶ **Continued growth**: This is still the most common view, and certainly the "official" view of most political and academic discussion. Growth is desirable because it has made good things possible for some people already, and will bring more good things to more people in future. The idea that growth might falter is usually discussed only in terms of economic recession, and almost always assumed to be a "Bad Thing" – as a glance at any newspaper will confirm.

▶ **Collapse of economic structures**: This is the family of futures which descend from Malthus via *The Limits to Growth*. It has a popular constituency, who believe that the carrying capacity of the planet (see p.114) has already been exceeded, and that growth cannot be sustained much longer. The last straw may be climate change, oil depletion or a variety of other things, but the consequences are similar.

▶ **Disciplined, sustainable society**: This is the first addition to the simple Malthusian versus Cornucopian visions. It means trying to manage things to avoid the worst. The "third way" is outlined in many detailed plans for organizing a transition from the current social and economic system. The premise is that growth cannot go on forever, and avoiding collapse is overwhelmingly important. So these scenarios try to outline paths to a sustainable, steady state. What form the transition might take is controversial, partly because of the difficulty of designing a no-growth economy that works according to the currently dominant capitalist model and does not fall into depression.

▶ **Transformation**: These kinds of visions are about transformation, not transition, because they embrace a radical, usually technologically driven, alteration of the conditions of human life, and possibly of humanity itself. Under this heading are filed the future pictures which see the next stage of evolution as involving the immensely powerful development of, for example, artificial intelligence, robotics, genetic engineering or nanotechnology. These are "post-human" futures that perhaps include moving to off-Earth environments – one pretty convincing way of escaping from a closed system.

The first three images of the future cover the majority of current scenarios, which focus on the basic issues of supporting the growing population of the twenty-first century and beyond – food, fuel, pollution control and so on. The fourth set, "transformation", build on ideas that have a history in futurist writing, but have been less often worked up into formal scenarios or policies, perhaps because they are harder to describe in detail.

There are also low-tech variations of the fourth set, in which the transformation is spiritual rather than scientific. These often come from people who are deeply dissatisfied with present-day society for one reason or another, and think it needs to change radically. They tend to use arguments about the unsustainability of the current economic/technological system to support calls for a change in values. In other words, work now for a new, communally congenial ecotopia or prepare to meet Mad Max.

Dator reckons that "everyone's idea about the future falls into one of these four images". We'll see.

GROWTH FOR THE POOR?

The "limits to growth" debate, whether it featured two alternate paths or four, read differently in different parts of the globe. It still does. It's one thing to argue that limits are about to be exceeded and that we must curb consumption of resources in the developed world; it's quite another to say the same thing in the large parts of the globe where many people still live in poverty. This understanding sharpened the sense that discussion of the future is intensely political.

While the industrial revolution brought unprecedented prosperity, it was incredibly uneven. Whether you believe the world is heading for growth or collapse, or something else altogether, one thing is certain: the world as it is today is a remarkably unequal place. If life is getting better all the time, as some believe, *who* is it getting better for? The last two centuries of economic and technological development have allowed a minority of the vastly expanded global population to accrue enormous benefits. Some of the rest have also seen improvement in their living conditions, especially in recent decades, but they still trail far behind. Moreover, the energy-intensive way of life, based on the burning of fossil fuels, that defines prosperity in developed nations is not going to be viable for much longer. And there is already quite good evidence that it's beginning to make life worse for people in other parts of the world (see Chapter 7).

So it seems unlikely that the billions of people presently excluded from the good life can be brought inside the charmed circle anytime soon, especially if they're also relying on burning fossil fuels to power their way out of poverty. As this particular environmental problem starts to make "limits to growth" on the old model look more plausible, a key futures question arises: if we lucky, wealthy folks in the developed world expect to retain a high material standard of living in the future, can we help the rest of the world achieve one too? Aside from the moral imperative of tackling vast global inequalities, there's the prospect of political pressure increasing as such inequalities become more starkly visible. In the words of the UN Millennium Project's *State of the Future* report (2007):

> "The synergy between economic growth and technological innovation has been the most significant engine of change of the last 200 years. But unless we improve our economic, environmental and social behavior and close the gap between the rich and the poor, the next 200 years could be difficult."

What this understated remark means is that increasing inequality will eventually lead to increasing conflict. With that rather uneasy prediction in mind, much futures debate now turns on whether it's even possible to increase the chances of attaining a decent standard of living for the majority of humanity. People disagree about what might help achieve this, and what represents a "decent" standard of living, but the basic question is unavoidable. The rest of this book considers some of the answers to that question in more detail, but there is enough optimism around to hold on to the view that a positive answer is still possible, if far from inevitable. There is even some evidence that we're moving in the right direction.

THE MILLENNIUM PROJECT'S STATE OF THE FUTURE INDEX

The advent of the new millennium helped motivate a number of massive future-oriented projects which tried to take a global view on things. The largest big-picture effort was the Millennium Project, which was set up in the mid-1990s by the Smithsonian Institution and a bunch of UN-related NGOs. Its main concern is gathering information and "improving thinking about the future".

This global futures think-tank shouldn't be confused with the now-completed Millennium Project commissioned by the UN proper in 2002 to develop concrete action plans for achieving the UN's "Millennium Development Goals". These eight international goals focus on reversing "the grinding poverty, hunger and disease affecting billions of people" around the world, with non-profit organizations, world leaders and various experts working together to try to achieve their aims by 2015.

But getting back to the other, future-researching Millennium Project, one of their notable contributions is a kind of ultimate indicator about the future. One thing everyone always wants to know about the future is: will it be better or will it be worse? The extended network of futurists who contribute to the project decided to try to answer that question, and the result is the annual *State of the Future* report with its State of the Future Index (SOFI).

The index was created roughly as follows. Assemble as many indicators as you can think of about human wellbeing and the things that affect it. See which ones can be measured with reasonably reliable data, and review those for current values and any obvious trends. Then ask a large panel of experts to join in a Delphi-style exercise (see p.46), in which they estimate how each indicator will move over the next ten years. In regular Delphi fashion, they can adjust their best guesses in response to other people's guesses and their reasons behind them, until you reach some kind of consensus. Then all you have to do is decide how to add the results of all those indicators together to arrive at a graph of how well we will be doing in the next decade. The conclusions of the 2009 SOFI can be seen in the diagram opposite.

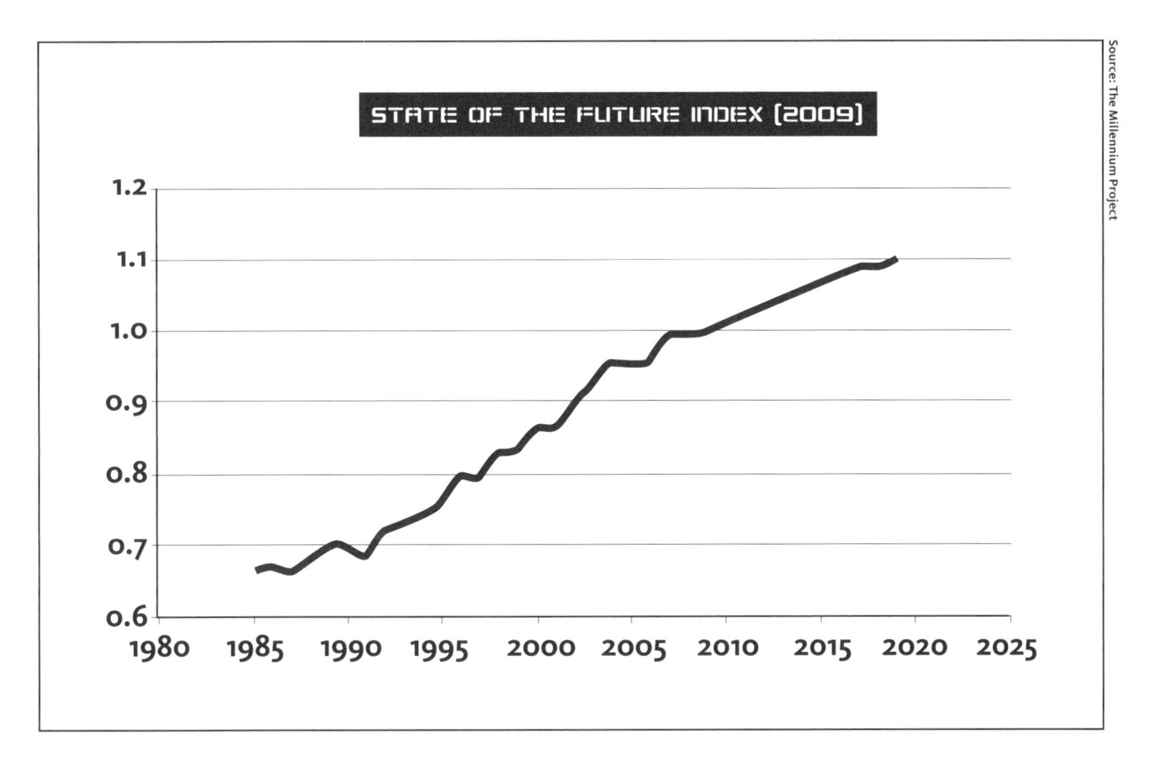

The State of the Future Index is a complex distillation of numerous experts' predictions, which suggests that enough things are getting better to improve life for most people.

But is this neat graph plotting such an abstract variable, however wondrously complex its creation, any more worthwhile than, say, the Rapture Index (see p.18)? Well, perhaps it is. The process itself is instructive. In 2007, they revisited the basic components of the index, in yet another extensive Delphi exercise, asking what variables should be included, and why. The list the compilers settled on, slightly longer than the one used in previous years, included 29 indicators, which were rated for their importance in measuring future quality of life. In descending order, with their units of measurement, they were:

- Population lacking access to clean water (percent)
- Literacy rate (percentage aged 15 and over who can read a fairly simple text)
- Levels of corruption (15 largest countries)
- Education (percentage in secondary school)

- Deep poverty (percentage of population living on $1 a day in low- and middle-income countries)
- Countries having or thought to have plans for nuclear weapons (number)
- CO_2 emissions (global, kilotonnes)
- Unemployment (percent of total labour force)
- GDP per unit of energy use (constant $2000 per kg of oil equivalent)
- Number of major armed conflicts (deaths greater than 1000)
- Population growth (annual percent)
- R&D expenditure (percent of national budget)
- People killed or injured in terrorist attacks (number)
- Energy produced from renewable (non-nuclear, non-fossil) sources (percent)
- Food availability (calories per person)
- Population in countries that are free (percentage of total global population)
- Global surface temperature anomalies
- GDP per capita (constant $2000)
- People voting in elections (percentage of population of voting age)
- Physicians (per 1000 people)
- Internet users (per 1000 people)
- Infant mortality (deaths per 1000 births)
- Forest lands (percentage of all land area)
- Life expectancy at birth (years)
- Women in parliaments (percentage of all members)
- Refugees (number per 100,000 total population)
- Total debt service (percentage of gross national income in low- and middle-income countries)
- Prevalence of HIV (percentage of population)
- Homicides, intentional (per 100,000 population)

The list is a bit of a hodge-podge. There are some obvious items, some less obvious ones, and plenty of room for discussion of their relevance. But it has at least three virtues. One is that it's so varied, which makes the point that life in the future will be influenced by many things, and that those things are interconnected. Secondly, the variables can be grouped into six areas – health, wealth, intellectual, moral, physical and security indicators – which can be separated out to create individual measures of "progress". The first three look quite good, the second three rather worse. But it's helpful to include the full mix. The list is a little light on environmental and ecosystem indicators, but perhaps that's a useful reminder that the full inventory of things to think about in the future is always in flux.

The list's third virtue is that it avoids the parochial, as it was reviewed by people from many countries. With a couple of exceptions, they are all global indicators. It's also impressive that they can all be estimated, though to be fully convinced of that you'll need to dig deeper into the project's extensive documentation about how the figures are assembled. Most of them come from agencies that are striving to keep track of things on a global scale – the World Health Organization, the UN Food and Agriculture

THE FUTURE AIN'T WHAT IT USED TO BE

Historian Robert Heilbroner (see p.15) saw the middle of the last century as marking a change in the way humanity saw the future. In contrast to the stasis that marked earlier cultures' ideas about time, and the notion of continued advancement that took hold around three hundred years ago, we now have a more complex view of possible futures. Change is still envisaged, but the hope of progress has grown fainter. In fact, many people, especially in more developed nations, are pretty pessimistic. They can foresee as many possible disasters as desirable futures, and tend to think the former are more likely.

It's certainly true that mankind's view of the future has got more complicated in recent decades. Contemporary views of the future are often contradictory, and in more ways than the simple contrast between Malthusians and Cornucopians suggests. Visions of technology-fuelled utopias now coexist with extended analyses of reasons for the collapse of civilizations. Even techno-enthusiasts often discuss the possibility that their favourite innovations might lead to disaster (see computer scientist Bill Joy's concerns about self-replicating machines in the box on p.93). Optimism and pessimism now also accommodate an overlay of apprehension.

Faith in progress has loosened its grip, and so too has faith in any kind of predictions. The apparent failure of many past forecasts induces scepticism about hopeful visions of the future, as well as a kind of affection for those past images. Books and websites cataloguing fantastic futures are easy to come by and invite nostalgia for futures past – but they also inhibit efforts to envision new ones. Technologists' predictions still attract breathless reporting, but futurists are equally likely to be derided as purveyors of future schlock.

Humorist P.J. O'Rourke noted in his 2008 article "Future Shlock" in *The Atlantic* that when Disney's Tomorrowland theme park was completely refurbished in 1998 (see picture) the corporation described their latest creation as "a classic future environment". It was built in a style O'Rourke drily dubbed the "Jules Vernacular". Tomorrowland's equivalent in Paris, Discoveryland, is actually a tribute to Verne, among other past futurists. And nowadays *Futurama* is a science-fiction cartoon devised by the inventor of *The Simpsons*, rather than a must-see World's Fair exhibit.

So along with utopian and apocalyptic futures, we now also have ironic, jaded and wearisome futures. It's too soon to know if these mark an enduring cultural trend in depictions of the future, but they do suggest that the accumulating array of past depictions of the future weighs on the present more heavily than it has before.

PREDICTIONS FILE

Jamais Cascio

Environmental futurist and blogger at www.openthefuture.com

Highest hope: We will see a significant and lasting move away from preferring short-term advantage to long-term results, whether in economics, politics or technological deployment. This may be thought of as "responsibility", but it's bigger than that – it's a recognition that our choices matter.

Worst fear: We will continue to squabble over immediate crumbs while ignoring the devastating impact we're having on our own futures. This is most likely to manifest through the ongoing changes to our environment; we still have hope now of avoiding a civilization-threatening environmental collapse, but the point at which this will no longer be true is drawing near.

Best bet: We'll manage to avoid collapse, but doing so will require us to adopt risky and controversial technologies, and will heighten international tensions and conflicts. In short, we'll keep blundering our way from crisis to crisis, never quite able to see a way out of the cycle.

Organization, the World Bank, the International Labour Office, and so on. They all have their critics, but their data-gathering efforts are invaluable. We may not know what is coming next, but thanks to organizations like them and to the *State of the Future* report we certainly do know quite a lot about the state of the world right now.

FURTHER EXPLORATION

Gregory Benford Timescape (1980) Classic science-fiction novel about a struggling physicist in the 1960s who works out that the noise in his experiment is actually a signal from the future trying to tell the rampant consumers to stop wrecking the planet. The signal eventually creates an alternate universe, in which eco-collapse is avoided, but the original senders don't get the benefit. The fact that they were originally depicted as sending their signal from a futuristic 1998 adds piquancy.

Mark Brake and Neil Hook Different Engines: How Science Drives Fiction and Fiction Drives Science (2008) Fairly comprehensive overview of the history of science fiction, and how science has been woven into it. The "driving" part is less convincing.

Eric Dregni and Jonathan Dregni Follies of Science: 20th Century Visions of Our Fantastic Future (2006) A well-documented and profusely illustrated account of the last century's ideas about this one.

Herman Kahn The Coming Boom: Economic, Social Political (1982) The last of Kahn's characteristically optimistic forecasting efforts (published the year before his death) is out of print, but still obtainable. The Hudson Institute, which he founded, maintains a biography page and a selection of essays in the same vein at www.hudson.org.

Donella Meadows, et al Limits to Growth: The 30-Year Update (2004) A reprise of the argument of the original *Limits to Growth*, with new data and slightly more elaborate computer models. It concludes that some limits have already been breached – and we are in "overshoot".

Robert Rosenblum, et al Remembering the Future: The New York World's Fair from 1939 to 1964 (1989) A volume of scholarly essays, published for a US exhibition, which compares the two fairs' visions of the future.

Lawrence R. Samuel The Future: A Recent History (2009) Impressively wide-ranging and consistently sceptical scholarly history of twentieth-century visions of the future of America, with a few influential British ones included.

Oona Strathern A Brief History of the Future: How Visionary Thinkers Changed the World and Tomorrow's Trends are "Made" and Marketed (2007) A breezy guide to futurists past and present, and their methods.

Daniel Wilson Where's my Jetpack? A Guide to the Amazing Science Fiction Future that Never Arrived (2007) Whatever happened to underwater cities, food pills, X-ray specs, flying cars and, of course, jet packs? A jokey but technically well-informed guide to the toys boys still really want.

davidszondy.com/future/futurepast Fascinating compilation of images and commentary on frequently infeasible futures, divided into themes – cities, food, transport, war, and so forth.

www.millennium-project.org You can read all about the Millennium Project, and buy copies of the latest *State of the Future* report from their website.

www.paleofuture.com A diverting collection of press cuttings, magazine articles and illustrations showing what people in the past thought the future was going to be like.

www.wsu.edu/~brians/nukepop/index.html American historian Paul Brians has a riveting compilation of pop-culture responses to nuclear weapons from the 1940s onwards.

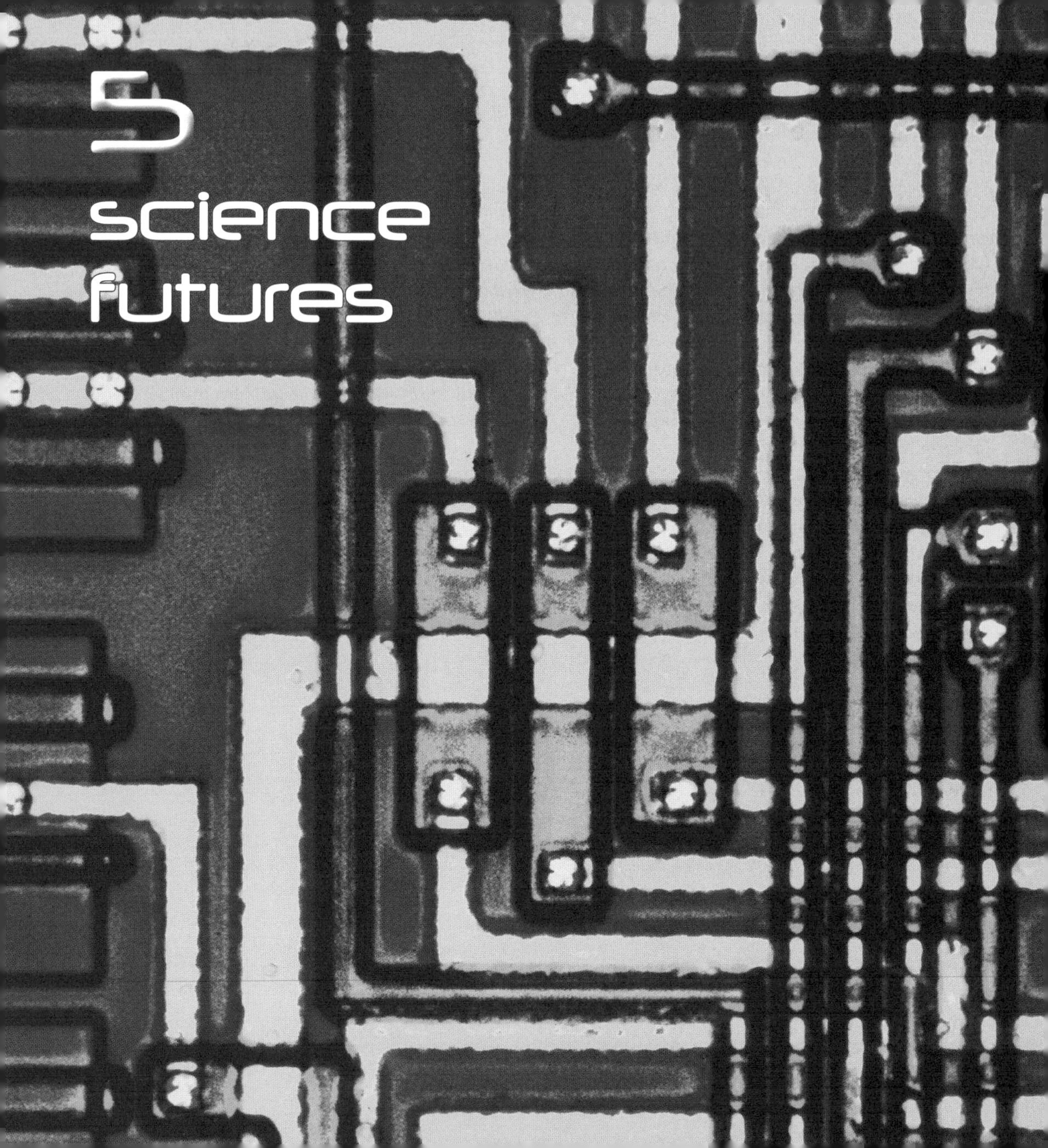

5 science futures

science futures

The next century will witness an even more far-reaching scientific revolution, as we make the transition from unravelling the secrets of Nature to becoming masters of Nature.

Micho Kaku, *Visions* (1998)

There will be a good deal of new science and technology in our future – everyone says so. The really new stuff will remain unpredictable (that's why people do research), but there is enough going on that is being developed from existing knowledge and techniques to get an idea of what is likely to be on offer. Some of it harbours remarkable possibilities for refashioning the world, and the materials, devices and even creatures we find in it. As the twenty-first century unfolds, we may reach a point where we have to make choices about augmenting or redesigning ourselves, and about how we relate to equally intelligent entities that we have brought into being. Like other aspects of our future, our attitudes to these prospects are coloured by experience of the costs and benefits of earlier technologies, and we face them with a mixture of hope and foreboding.

TECHNOLOGICAL HORIZONS

We most often hear about the great moments of discovery in science, or the triumphant unveiling of some new, revolutionary device. But most science, and most technology development, is a slow grind. That isn't a problem (unless you're the one doing it), but it does mean that many future lines of enquiry are already set. There will be some completely unexpected findings: some that make new things possible, others that ensure hopes are dashed. Timescales often get stretched or squeezed, and old goals are realized in new ways, or redefined to fit new situations or fresh understanding. But most research and development will continue to build on what is already being done, taking small, regular steps.

All of which means that we already know quite a lot about what is likely to come in the next few decades: most of the developments we shall see are already visible on the horizon. Beyond that, things get more speculative, and there is less agreement about what will be possible, but there is still plenty of evidence about what people would like to do. Current trends highlight four main areas as attaining potentially world-changing significance: genetics and biotechnology, computer science and IT, nanotechnology and brain science. And we'll look at each of them in turn.

BIOTECH FUTURES

Genetics and biotechnology have been the source of many promises since scientists first put together the tools for swapping genes around (in bacteria) in the late 1970s. But the real biotech boom is yet to come. The complete, detailed knowledge of genes that is now within our grasp could allow geneticists to reprogram biology, thereby eliminating many human diseases and greatly extending our lives.

Roughly speaking, scientists have spent the last half a century taking cells and their constituent molecules apart, and finding out how they are built. The result is that it is becoming possible to design modifications, and perhaps even completely new cells. The twenty-first century will see this move from analysis to synthesis bearing fruit – often useful, usually controversial. Past arguments about in vitro fertilization ("test-tube babies"), GM foods, embryonic stem cell research and the effects of biofuels on food prices give a taste of the kind of public furore biotechnology is sure to provoke over and over again. As geneticist J.B.S. Haldane said in *Daedalus* back in the 1920s, not without relish, a physical technology may be a blasphemy, but a biological technology is a perversion.

Our ability to pervert – or adapt – biology to our own ends is growing fast. One index is the speed at which scientists are reading DNA sequences, the "letters" that encode biological information in the genes. Working out what the genes say one bit at a time depends heavily not only

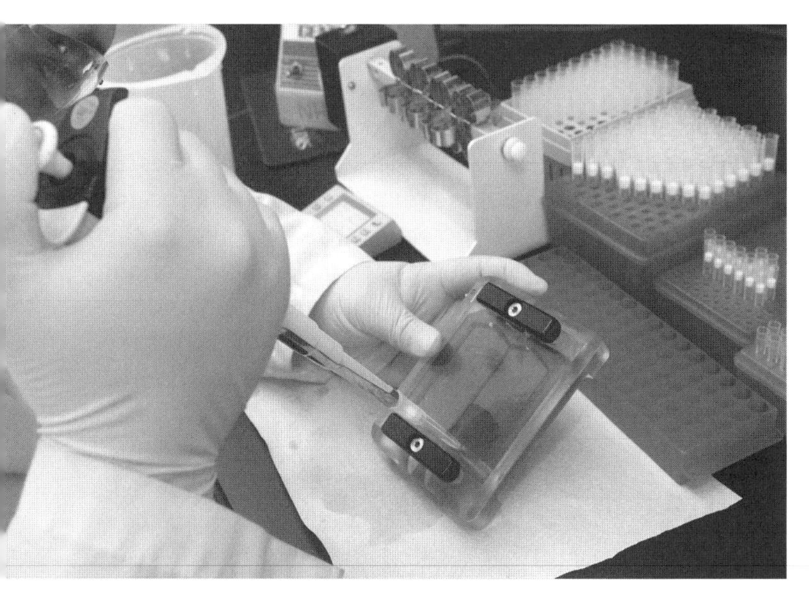

Beginning the process of DNA sequencing at the J. Craig Venter Institute in Rockville, Maryland.

Judith Light Feather

President of the NanoTechnology Group

Highest hope: Governments will recognize the need for introducing global nanoscience education to primary students. Most are opening Nano Centers and rushing to commercialize research, while ignoring the education platforms in their countries. It is unethical to change our world from the "bottom up" by building with atoms without teaching the science to the younger children who will have to face the consequences of our actions and live in a totally new environment.

Worst fear: Ignorance and complacency will continue to defeat any efforts in changing the education of our students.

Best bet: Nothing is on the horizon for at least a decade. All monies are being funnelled into research and commercialization without considering future consequences.

on computers, but also on many other physical and chemical techniques. The Human Genome Project, the international effort to map and sequence the human genome – the entire library of DNA held in every cell – was not just about compiling an accurate inventory of the three billion or so DNA "letters" each of us carries. It was also a technology-forcing project, making DNA sequencing faster and cheaper. The results were impressive. A report for the US government in 2007 found that the productivity of DNA-sequencing technologies had increased more than five hundred-fold in ten years – doubling every two years. The cost had gone down over the same period by a factor of a thousand.

Moreover, with faster reading came faster writing. DNA synthesis improved even more over the same period, increasing in productivity seven hundred-fold, and getting more accurate to boot. As with the better-known Moore's law for computers (see overleaf), these trends are expected to continue over the next decades. Compared with making computer circuits, we are only just getting started at reading and writing DNA. Today the genome of any species can be sequenced; soon it will be cheap to sequence all the DNA of an individual. In DNA synthesis, genome pioneer Craig Venter has already reprogramed a single bacterium with a completely artificial genome – copied from another, related microbe. In theory, that means that he, and other geneticists, can move on to writing new genomes more easily.

Science is moving beyond old-fashioned natural selection to shaping other species, and perhaps even human beings, to our own designs. True, there have been forecasts of the imminent arrival of genetic engineering since the early 1960s, but it still appears complex, unreliable and pretty risky. What's more, the much-publicized completion of the human genome sequence in 2003, while an extraordinary achievement, underlined how complex the interactions between genes are, and how difficult it is to muck about with cells without disturbing their finely poised internal mechanisms.

This, according to many, will lead to a move back to biology that looks at whole systems again, not just their smallest parts. Give that a few decades, plus a few tens of billions of dollars, and perhaps pounds or yen, and it is likely that it will become possible to steer those systems to develop in new ways. As discussed in Chapter 12, much of this work will initially be aimed at alleviating the effects of old age, and will move seamlessly from treatment to prevention, and then on to enhancements like longer life and improved memory and attention span.

The other side of current biotech developments that is being addressed with increasing frequency is synthetic biology. This is still biotech, but with an emphasis on building with artificial components, rather than modifying existing organisms. The boundary is not clear-cut, as most current projects are making biological bits and pieces (a suite of genes and the regulatory elements which make them work together, for instance) that need an existing organism to operate in. A bacterium, for example, can find itself making a plant-derived anti-malarial drug using a carefully constructed chemical pathway that it doesn't normally own.

Some scientists in the field want to build completely artificial organisms, partly in order to get a clear definition of the minimum set of components an organism needs to work properly. But the overall approach of this souped-up biotech has a great many practical applications. According to a report from the UK's Royal Academy of Engineering reviewing the prospects for the field:

> "The ultimate goal of synthetic biology is ... to design and build engineered biological systems that process information, manipulate chemicals, fabricate materials and structures, produce energy, provide food, and maintain and enhance human health and our environment."

Again, this sounds like a recipe for a potential bio-utopia, full of efficient, green technologies, but that overlooks reservations about the wisdom of manipulating organisms, and of opening the way to technologies that could be used to create new bioweapons.

COMPUTERS AND COMPUTER SYSTEMS

No one needs persuading that further advances in computer technology will be important, as we have been living through an explosion of computing power for the past five decades. The question is only how far it will go, and how fast.

The long-established trend here is in computer hardware, as described by "Moore's law". Gordon Moore, the founder of computer chip-maker Intel, said in 1965 that the number of transistors in a single microprocessor, hence the number of computing elements on a chip, was doubling roughly every two years, and that this trend was likely to continue. He was proved more or less right through the rest of the century.

In 2005, forty years after his original proposition, Moore joined many other industry observers in predicting that the doubling – which comes from making the circuits etched onto a wafer of silicon smaller and smaller – would come to an end. But not, he reckoned, until transistors were the size of atoms – in another ten or twenty years' time. So even without some as yet unknown breakthrough to take the trend even further, the world would have seen thirty doublings of computing power within fifty years or so. In more familiar numbers, that is more than a thousand million-fold improvement. Just one more doubling would make that two thousand million... that's how exponential growth works.

The amount of computing power that can be packed into a small space increases with the discovery of more exotic materials and ever more precise ways of etching electronic circuits. The limitations on speed are partly to do with the physical size of the circuits (shorter connections are speedier), and partly to do with factors such as how much heat is generated and how quickly it can be dissipated.

The physical limits of current designs may be near, but there are already ideas about how to go beyond them, perhaps with quantum computers. These exotic beasts would exploit the properties of quantum mechanics, which allows physical entities to exist in more than one state at a time – the states are said to be in superposition. Apply this idea to computer processors which use atomic-scale memory and processor elements, and your computer can be set up to do many calculations at once. So far, this is largely theoretical, because the technology for keeping the superposition going is too hard to devise, but there are dozens of groups trying out ways of doing it. One of them might even succeed.

In the meantime, computers will carry on getting more powerful. That will go along with advances in software (which tend to be much slower than in hardware) to improve robotics and, perhaps, create computer systems with more of the capacities we call intelligent. Predictions of the advent of genuine artificial intelligence have so far all proven optimistic, but that doesn't mean it's never going to happen. More immediately, computers will continue getting smaller, cheaper, more universal and more connected, which will probably produce more obvious changes over the next few decades than increases in ultimate computing power or intelligence.

No headers allowed: the annual Robocup aims to develop a team of highly intelligent autonomous robots that can take on the best of the world's footballers by 2050.

When objects dissolve into information

Another forthcoming development in IT will be the presence of a tiny amount of intelligence and memory in all manner of everyday objects and consumer goods. Ultra-cheap, ultra-miniaturized computers, sensors and communication devices will soon become ubiquitous to the point of near-invisibility. Anything we make, or choose to enable, will be able to record its own history and relay it to other objects. This leads to the so-called "Internet of Things" – sometimes also called ubiquitous computing or ambient intelligence – in which the descendant of the World Wide Web incorporates most of the devices around us. Our belongings will talk to each other as well as to us, so your fridge won't just contain food marked with use-by dates, but might also tell you when it's starting to go bad, based on readouts from the packaging, or your recycling might send a message to the collection service when it needs emptying. (See Chapter 16 for more examples.)

THE HISTORY OF STUFF

Science-fiction writer and futurist-turned-design-guru Bruce Sterling gave us his personal rundown of the stages in the evolution of "stuff" in his booklet *Shaping Things* (2005). Created objects, he says, began with **artefacts**. These were the first and generally simplest artificial objects, made and used by hand (think flint hand axes). People made them one at a time, and they learned how to do so as they went along. People who only had artefacts were hunters and farmers.

Then came **machines**, which had moving parts and motive power. They required technical support, and people who wanted them became customers. Machines eventually led on to **products**, which are mass-manufactured things pouring off an assembly line. They are all alike, and are distributed to a new kind of customer, who we call a consumer. This is where stuff had got to by the early twentieth century.

Then, in Sterling's lexicon, came **gizmos**; objects that appeared some time in the late 1980s. You can spot a gizmo because it has lots of features that you can alter to your own preferences and functions you have to learn how to use, and along with it come regular messages about bugs, fixes and upgrades which keep demanding decisions from the lucky owner. Said owner has moved beyond the comforting simplicities of being a consumer to become an end-user. This, of course, describes where we are now, and how we relate to items like the computer this book was written on, or the eReader you might be reading it on. But what comes next? Ah, says Sterling, that would be **spimes**.

Spimes? What are they when they're at home? They are the next stage of gizmification, in a way. But their primary existence, for Sterling, is as information. There is a manufactured object in there somewhere, but the keys to its use, even its very existence, are in the information which preceded it, surrounds it and which will probably survive its physical demise (or, preferably, recycling).

Sterling introduced spimes to the world. They don't quite exist yet, but they should be here in 25 years or so, he reckons. Their origins can be seen now in developments like Radio Frequency Identification Tags (RFIDs, pronounced "arphids"). At the moment, these keep track of goods, or stray dogs. But imagine them being packed with more and more technology – sensors, timepieces, GPS, records of use – and communicating with each other over the "Internet of Things". Then the world of created objects has moved beyond gizmos (which, like all the other earlier stages, continue to co-exist with these wondrous new spimes) into a new and information-saturated regime whose properties we will be exploring around the time the planet really starts warming up.

Sterling enthusiastically logs developments like this, and many others, on his blog: **www.wired.com/beyond_the_beyond.**

All this implies a fundamental shift in our human-made environment. Up until now, we have had to learn about the objects we have made – how to use them, what skills might be needed to get the best out of them, how to operate, maintain and repair them. All that will probably still go on, but with a new ingredient: the objects will also be learning about us.

Technology-oriented commentators tend to agree that this is the future of stuff. They describe an important new stage in the evolution of our relationships with the objects we create. According to the co-founder of *Wired* magazine, Kevin Kelly, this is the latest major transition in the organization of knowledge:

> "We are in the midst of a movement where we embed information into all matter around us. We inject order into everything we manufacture by designing it, but now we are also adding small microscopic chips that can perform small amounts of computation and communication. Even the smallest disposable item will share a small thin sliver of our collective mind."

Mike Treder

Managing director of the Institute for Ethics and Emerging Technologies, Hartford, Connecticut

Highest hope: Powerful emerging and converging technologies, such as genetic engineering, artificial intelligence and nanotechnology, will be managed for optimum safety and maximum benefit to all humanity, not just a privileged few.

Worst fear: Continued escalation of wealth disparity combined with accelerated devastation of the biosphere from climate change will result in an overwhelming gap between the haves and have-nots.

Best bet: Muddling through, with many terrible disasters – both natural and manmade – and strong opposition to change from entrenched special interests. Wars, famines, pandemics and also heroic efforts to make the world better for all. Eventually, though, a place that will look a lot like today, just shinier in some places and dirtier in others.

PREDICTIONS FILE

NANOTECHNOLOGY

Of the three areas of technological development most often touted as likely to spawn twenty-first-century wonders, nanotechnology is the hardest to get a clear fix on. Unlike bio and info, the "nano" prefix does not refer to a particular area of technology, it just means teeny-tiny things – specifically, things measured in nanometres, or a thousand millionths of a metre. This is the size range of atoms and molecules, which range from fractions of a nanometre for single atoms up to a couple of nanometres across for a strand of DNA. A strand of human hair, on the other hand, is 80,000 nanometres across.

Real nanotechnology already exists, in the form of techniques used to produce materials of nanodimensions. Some of these minuscule materials have interesting or useful properties beyond those of the same stuff in larger lumps. For instance, you've probably seen adverts for socks impregnated with nanoparticles of silver which kill bacteria and help suppress foot odour and prevent infections.

But there is a big difference between making nanomaterials and nanoengineering, that is designing and building things from the bottom up. There are two famous visions for where nanotechnology might go in the future. One is of ultra-small robots that can repair damaged cells from the inside. Make them smart enough, and a horde of nanobots in your blood could keep watch over your entire body and fix any problems as they arose – making you live forever. The other is molecular manufacturing, in which something is built up literally atom by atom, the position of each atom specified precisely.

Once there are reliable ways of picking and placing atoms, you could build literally anything. Tales of nano-futures tend to feature a general purpose "molecular assembler" which takes in basic raw materials and delivers, say, a steak, medium rare, with fries, far more conveniently than starting out with a cow and a field of potatoes. A steak – for some reason it's nearly always a steak – is a good example, because the steaks we eat now *are* built up atom by atom, using what nanoenthusiasts like to call "nature's own nanotechnology". If natural selection can evolve an apparatus for turning grass into beef, we already have proof of principle for molecular manufacturing.

This is what a nanobot might look like, removing a deposit from a blood vessel wall – if the other molecules don't disable it first.

Does that mean that we can look forward to a molecular assembler of our own design that will one day be able to make anything that can be specified at an atomic level (including, of course, copies of itself)? Opinions differ, though *Star Trek* fans may already be dreaming of owning their own replicator. There is certainly plenty of activity in nanotechnology, especially in medical devices, drug delivery and electronics. But there are quite a few arguments as to why the more far-reaching visions are still a long way off.

Let's take the robots first. There they are, these miniature mechanical marvels, with their on-board power and control systems that somehow coordinate the whole swarm. But if they are inside a human body, or entering a cell, they would have to push aside water molecules, which is a hard job on the nanoscale – like swimming in tar – while also being battered about by other fast-moving molecules. They will also tend to stick to passing proteins. Our poor nanobots are struggling to operate in a sticky, obstacle-filled world that will probably disable them in short order.

Outside the body, atomic-scale machines would still have to reckon with physical and chemical problems that do not arise at current levels of miniaturization. Designs for molecular gears and axles take account of known atomic sizes and bond lengths and angles. But if these are even slightly out of

NANO-VISIONS

Some of the early technical treatises on nanotechnology read like science fiction, Eric Drexler's *Engines of Creation* (1986) being the best-known. Since writing that book, nanoprophet Dexler has been busy working out the details of precisely how engineering might slowly move in the direction he outlined. Science-fiction writers, on the other hand, have gleefully skipped over the hard graft of making anything work at a molecular level, and jumped straight into worlds in which nanotech magic delivers all the promises they can imagine, good and bad.

Neal Stephenson The Diamond Age (1995) A complex tale about the education of a young girl by a super-intelligent interactive book, set in a future where all the nanovisionaries' hopes have been realized. For instance, diamonds are commonplace because matter compilers can make them as easily as anything else, and urban "immune systems" defend the cities with swarms of airborne nanomachines. The narrative centres on the conflict between a centralized, quasi-Victorian tribe and two other cultures that make different use of the material abundance offered by nanotechnology.

Ben Bova Moonrise (1996) and Moonwar (1998) Two novels by a master of "hard" (i.e. kind of realistic) science fiction, set in a world in which nanotechnology supports life on a moon colony, but has been banned on Earth because of fears of new technology standing in the way of progress (tut, tut). Nasty political conflict ensues.

Michael Crichton Prey (2002) Does for nanotech what *Jurassic Park* did for cloning ancient DNA. A didactic techno-thriller in which Crichton (who got preachier as the years went by) warns us about the hazards of killer swarms of escaped nanobots. Big chunks of explanation, thinly drawn characters and a moderately tense narrative if you can take the premise seriously.

Kathleen Goonan The "Nanotech Quartet": Queen City Jazz (1994), Mississipi Blues (1997), Crescent City Rhapsody (2000) and Light Music (2002) A four-novel sequence set in a near-future US, in which half the population have disappeared into mysterious high-tech cities that have been "enlivened", and the others live outside, suspicious but fascinated by what they imagine is going on within, while also trying to evade the nanoplagues that periodically sweep the landscape. These are richly imagined tales, but life gets pretty confusing once you get into the cities, a common feature of even the best attempts to imagine where accelerating technology might lead. A narrative in which no one ever dies, but their revived form can look quite different, and is as likely to be virtual as real (in our old-fashioned sense of the word), becomes increasingly hard to follow if you are wedded to conventions like people living, growing older or dying.

Nancy Kress Nano Comes to Clifford Falls (2008) A short story as satisfying as most novels, this portrays the mixed blessings of rolling out a technology that supplies everyone's wants and needs without effort, employment or, it turns out, satisfaction. Find it alongside Kress's other tales of the near-future in the story collection of the same name, or as a podcast at escapepod.org/2006/10/12/ep075-nano-comes-to-clifford-falls sync with their optimum energy configurations, the atoms in your beautifully crafted nanodevice may decide to spontaneously rearrange. A counter-argument is that there are already nanomachines which work reliably inside cells. The wavy cilia that some cells use to move things past their surfaces are powered by a wonderfully neat rotary motor, for example, made out of protein components. The bits and pieces self-assemble quite happily in their watery environment, along with the cell membrane in which they sit. And cells have evolved many other, even more complex pieces of machinery, for building large molecules, repairing DNA and so forth.

So perhaps, one day, these two realms can be brought together, and biological principles can be used to build nanodevices to our own design. But there are so many stages to go through between what we can achieve now and our ultimate visions of molecular manufacturing or nanobot cell repair that it is hard to sketch plausible pathways from one to the other. The road may not be impossible, but it will certainly be long.

The principal nanoprophet, Eric Drexler, began making detailed arguments about the possibilities of nanoscale engineering in the 1980s. His work drew extreme reactions, and most considerations of nanotechnology have discussed his ideas. One pole of the spectrum is marked by science writer Ed Regis's book *Nano!* (1995), which basically endorsed everything Drexler said and suggested that the world will inevitably be transformed by a new collection of super-technologies. At the other, sceptical, end is David Berube's *Nanohype* (2006), which debunked a catalogue of nanotechnologists' wilder claims.

There is, needless to say, a middle ground, and interestingly Drexler made a bid to occupy it himself. In an updated edition of his 1986 book *Engines of Creation* (2006), Drexler claimed that his original ideas had been misunderstood, and that his notion of molecular manufacturing was not about nanobots as miraculous self-replicating entities. These, he emphasized, were creations of popular culture, not technological forecasting, though he continued to believe that nanoscale manufacturing was a worthwhile goal. In the twenty

PREDICTIONS FILE

Ray Kurzweil

Inventor and futurist, author of *The Singularity is Near* (2005)

Highest hope: Once we achieve the full range of human intelligence in a machine (which I believe will happen by 2029), the machine will necessarily soar past the human because of the advantages of non-biological intelligence. But this will not be an alien invasion of intelligent machines; we will merge with the intelligent technology we are creating. My highest hope is that we will overcome human suffering and expand our creativity a billion-fold.

Worst fear: The existential risks inherent in the exponentially growing information technologies, such as a "G" (genetics of biotechnology), "N" (nanotechnology) and "R" (robotics or artificial intelligence), will manifest themselves in the destruction of our civilization. We already have existential risks from twentieth-century technologies, such as the many thousands of thermonuclear weapons that still exist and are on a hair trigger.

Best bet: That the worst-case scenarios will not happen, but I am less sanguine that we will be spared painful episodes. After all, we've had them in the past: fifty million people died in World War II for example.

years since his original manifesto appeared, he suggested, progress in nanotechnologies – especially those based on biological molecules such as DNA and proteins – meant that the time was ripe to "move to molecular engineering at a new level".

Dexler's vision clearly still extends to a pretty magical sounding device. In this latest version:

> "Technical studies indicate that nanofactories, ranging to desktop scale or larger, will be able to convert simple chemical feedstocks into large, atomically precise products cleanly, inexpensively and with modest energy consumption. They indicate that a 10 kilogram factory will be able to produce 10 kilograms of products in hours or less – a stack of billion-processor laptops, a package containing a trillion cell-sized medical devices, or a roll containing hundreds of square meters of tough, flexible stuff that converts sunlight into electric power."
>
> *Engines of Creation* (2006)

In short, the visionary-in-chief still believes that nanotechnology will completely transform industry, energy and the economy. If it works, it will. Meanwhile, its appeal is explored by as many science-fiction writers as policy-makers (see box on p.87).

BRAIN SCIENCE

The fourth and final item on every list of potential world-changing research is neuroscience, and associated brain-related technologies. Future developments might be better drugs based on greater knowledge of how brain-messenger chemicals (neurotransmitters) work, or perhaps neural implants to fix or improve brain function. The latter would depend on finding ways to link artificial components – probably microelectronic ones – to neurons. However, all this depends on improving our understanding of an organ which remains, in many ways, mysterious.

Ambitious plans are underway to accelerate this process by what is fashionably known as "reverse engineering the brain". This is a way of blending artificial intelligence research with neuroscience, so that machines designed to have some semblance of intelligence are based on biological models. A key approach is to analyse the neural connections in a small piece of the brain, say one of the millions of cortical columns in a rat brain, each of which contains six layers of more than four hundred different types of neurons. Recording the properties of each of those neurons, and their connections (the synapses), opens the way to simulating this small piece of neural tissue on a computer. Run the whole thing on a supercomputer, and its behaviour, or at least its electrical inputs and outputs, can be compared with the real thing.

This is just one of a range of approaches that may gradually build up a better picture of the inner workings of the brain. The ultimate goal of being able to understand every nuance of how the brain works is supremely ambitious. According to a report from the US National Academy of Sciences on grand challenges for engineers:

> "Discovering how the brain works – *exactly* how it works, the way we know how a motor works – would rewrite almost every text in the library. Just for starters, it would revolutionize criminal justice, education, marketing, parenting and the treatment of mental dysfunctions of every kind."

This is a sales pitch, designed to recruit young scientists and appeal for research funds, and the brain doesn't exactly work like a motor. Still, the amount of effort going into projects like this will yield some results, if not necessarily a complete understanding of the most complex object in the known universe.

Commercial interest in brain science is building too. According to Zack Lynch, Executive Director of the Neurotechnology Industry Organization, neurotech is part of a "sixth wave" of technological innovation that will succeed information technology and come to the fore between 2010 and 2060. Building on advanced biochips and brain imaging, it will develop tools that can influence the nervous system and the brain. The applications will be wide, he predicts, enabling people "to consciously improve emotional stability, enhance cognitive clarity and extend sensory experiences".

THE GREAT CONVERGENCE?

One interesting feature tying all four of these upcoming technologies together is that they all involve working with very small stuff. So perhaps it's not surprising that developments in one area may aid developments in another. For instance, genetic analysis can help us understand the brain, and in the future nanoprobes may help us record information from neurons, or interface brains with computers. Potentially, there are lots of ways they can converge, and some official reports just refer to a general "nano-bio-info-cogno" (NBIC) technology as the thing to watch, although that covers such a huge range of work it's hard to make much sense of it.

This radical kind of convergence adds to other kinds of convergence that mark technological developments. The first, in some ways, is an old story in the history of technology: sometimes clever things work even better together. Different technologies (and different sciences for that matter) often come together to make things possible that wouldn't happen at all if they weren't all working at once.

The idea is explored at length in writer and physicist Stanley Schmidt's *The Coming Convergence* (2008), a future-oriented work that mostly reviews aspects of the past. Some of his examples are classic case studies in innovation. Steel frame construction made it possible to build skyscrapers; electric-powered safety elevators made it possible for people to actually use them. X-rays provide images of the inside of the body; computer processing allows a series of X-rays to be built up into a 3-D scan. Solid-state electronics, miniaturized with the aid of the many other technologies used to manufacture microchips, gave the world personal computers; fibre optic cables – often connected to old-fashioned copper wires at local level – connected them all together. New kinds of software allowed anyone to search and retrieve files stored on computers all over the world. And good old electromagnetic waves allowed the whole system to be accessed from laptops that are free to roam without wires.

Then there is the other kind of convergence – between humans and their technological creations. Again, this isn't completely new. Since early man first wielded a hand axe, we have used technology to extend our abilities and shape the environments in which we live. But nowadays man-made devices pervade almost every area of our lives. The amount of technology we directly depend on to maintain health and normal body functions tends to increase as we age, from childhood vaccinations through spectacles and antibiotics to false teeth, hip replacements and heart pacemakers. The next stage, enthusiastically trailed in science fiction, is a more complete merger between humanity and technology, in which our inventions work directly with our physiology. In other words, we become cyborgs – cybernetic organisms.

This cyborg vision leads back to the convergence of technologies investigating the small and the complex. If anything is going to accelerate the advent of true cyborgs, it will be the emergence of technologies that bring together the NBIC developments in new devices and interconnected systems. Advocates of this perspective put forward their thoughts in a 2002 report for the US National Science Foundation on "Converging Technologies for Improving Human Performance". They foresaw potential in a host of areas including: "improving work efficiency and learning, enhancing individual sensory and cognitive capabilities, revolutionary changes in healthcare, improving both individual and group creativity, enhancing human capabilities for defence purposes, reaching sustainable development, and using NBIC to slow cognitive decline that is common to the ageing mind". Quite a list. (For more on how technological advancement and convergence might affect future health and even lead to super soldiers, see Chapters 12 and 13 respectively.)

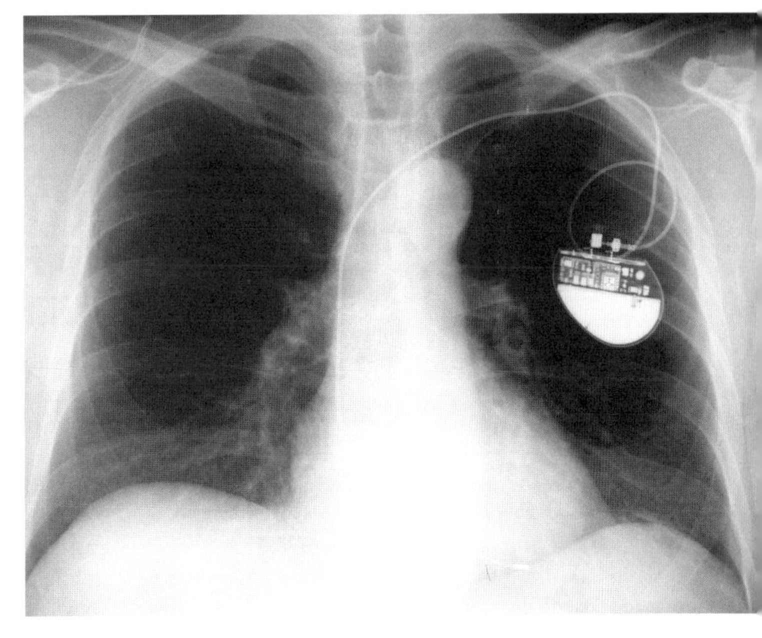

Precursor to the cyborg: an X-ray of a heart pacemaker shows physiology and technology merging in order to maintain human health.

R&D: WE AIN'T SEEN NOTHING YET

The four technological horizons just outlined are some of the more predictable (though not certain) sources of innovations that will transform our lives – or even the form our lives can take – in the future. But there will be others.

It has been said that the greatest invention of the nineteenth century was the method of invention. And science and technology have grown exponentially over the last couple of centuries as that method has been enthusiastically applied. Since the end of World War II, in particular, there has been a huge expansion in research and development, with both governments and industry getting in on the act. The result has been an explosion of new knowledge, and it's set to continue.

Like other twentieth-century developments, the growth of science and technology has been uneven, to say the least. According to UNESCO, the world spent $1130 billion on research and development (R&D) in 2007. As you can see from the diagram above, just over a third of this amount, 34.7 percent, was spent in North America, but Asia was also a big spender, accounting for 32.7 percent, up from 28 percent in 1997, mainly due to rapid growth in

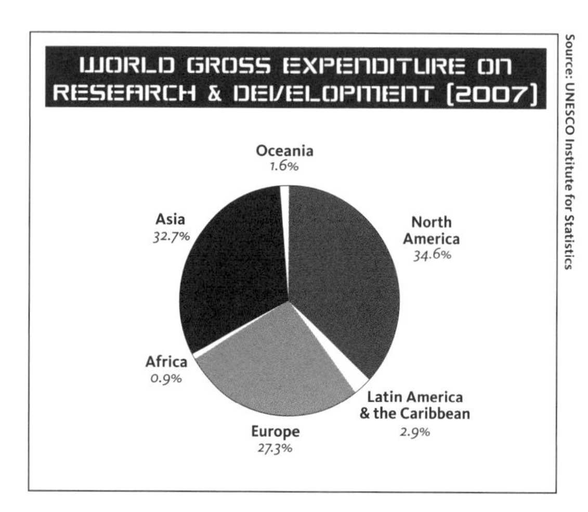

The balance of global research and development expenditure is shifting. Asia is now placed second, just behind North America.

China. At the other end of the scale, Africa registered just 0.9 percent. China has been expanding fast in the years since, but the overall balance is otherwise little altered. A similar split is apparent in other measures of the scientific system, such as numbers of researchers, papers published and patents granted.

Equally significant for the future is the size of the system. Not only is there vast spending on new science and technology, there are also large numbers of people doing it: 7 million around the world in 2007 (up from 5.7 million in 2002). There is an enormous, growing effort in applied ingenuity, devoted to trying to solve interesting problems. The result is that humans' collective capacity for useful invention is greater than it has ever been. It's not just that more is known (what policy obsessives call the "knowledge base"), although that is true. It's that there are more, better educated, people to use it than ever before, and they can get their hands on more resources to develop new ideas. That doesn't mean that all the world's problems can be solved, but it does mean there could be more options for trying to tackle the difficulties that inevitably lie ahead.

AN AVALANCHE OF TECHNOLOGY

More options is the best way to think about the forthcoming "avalanche of technology" (to borrow a phrase from futurologist and IT guru James Martin). Future-gazing often gets fixated on technology because it sounds more concrete, and can be easier to imagine, than more elusive social or cultural changes. Furthermore, it is often claimed that science and technology are moving faster and faster, because new knowledge is arriving rapidly and the gap between discovery and application is getting shorter. Then there is convergence, which makes it seem as though all technology is going to blend together, and reinforce a trend towards the high-tech transformation of our lives. Finally, there is a temptation to assume that the forms technology will take are preordained, and all people can do is adapt as best they can once they get here.

History, however, that indispensable guide to the future, suggests there are other ways of looking at technology. We need to remember that technological transformations in the past are at least as important as anything going on now. The peak of really important innovations came between around 1860 and 1914, when we saw the arrival of long-distance communication (by phone and telegraph), generators for electric power, the motor car, radio, powered flight. Some things are fast-moving now, as Moore's law describes, but the speed of change is easy to exaggerate in other areas. As historian David Edgerton emphasizes in *The Shock of the Old* (2006), the new technologies that attract all the attention typically coexist with much older ones which carry on unchanged and often unnoticed. World War II, for example, delivered the awesome world-changing power of the atomic bomb; it also saw the deployment of several million horses by the German forces alone.

This is not to say that things do not change, just that we should look closely at what *actually* changes rather than assume that innovation sweeps all before it. Invention and innovation are different, and many lovingly nurtured inventions fall by the wayside. Moreover, confident assertions about what new technology will do for us in the future have to face several challenges.

First, wishing for something does not make it possible. If it did, we would have had cheap, long-life, high-power, large-capacity batteries decades ago. These do not exist because the laws of physics have (so far) always imposed compromises that mean we cannot have all these elements at once. Second, developing particular kinds of technology does not dictate how they will be used. It often looks, with hindsight, as though certain consequences of an innovation were inevitable (what historians call technological determinism), but a closer look at the history of technology suggests this is not the case. Technology creates possibilities for new ways of doing things, but how it gets used is not preordained. The possibilities that are eventually realized are chosen, even fought over, by the

ARE SOME TECHNOLOGIES UNCONTAINABLE?

Claims that we are seeing the dawn of a new technological era are often accompanied by enthusiasm for the new possibilities opening up, but monumental technologies can also have monumental consequences. One of the strongest expressions of doubt about the path future technology should take came from computer scientist Bill Joy, in his April 2000 *Wired* magazine article, "Why the Future Doesn't Need Us".

Joy argued that new technologies often have unintended consequences, and the most compelling technologies of the twenty-first century have a common feature that makes those consequences even more worrying: they can self-replicate. Robots, engineered organisms and nanobots can potentially all create copies of themselves. In Joy's own words, "A bomb is blown up only once – but one bot can become many, and quickly get out of control." He also argued that, unlike the vast facilities and rare raw materials required to build nuclear weapons, these new technologies could fall within the reach of small groups or nefarious individuals.

The fact that this warning came from the Chief Scientist of Sun Microsystems, whose machines keep the Internet going, gave it added force. Joy proposed that developments in these fields be monitored, and that scientists should be prepared to stop work if they believed the worst might happen. A decade later, it's fair to say that none of these technologies looks much nearer realization, but the work goes ahead as strongly as ever. The debate he tried to highlight is sure to be reiterated in years to come.
www.wired.com/wired/archive/8.04/joy.html

people who want to get the benefit. A near-certainty about our future is that, along with new technology, we will have more choices and, most probably, more fights about those choices.

FURTHER EXPLORATION

David Edgerton **The Shock of the Old: Technology and Global History since 1900 (2006)** A wide-ranging look at how new technology coexists with older ways of getting things done.

Richard Jones **Soft Machines: Nanotechnology and Life (2007)** Well-balanced account of what may, and may not, be possible for technologies that work at the molecular level. Although Jones is a physicist, he takes inspiration here mainly from biology.

Mihail C. Roco and William Sims Bainbridge (eds) **Converging Technologies for Improving Human Performance: Nanotechnology, Biotechnology, Information Technology and Cognitive Science (2002)** A visionary (or overblown, depending on your point of view) report from the US National Academy of Science on the way new technologies of the very small will gain power by working together.

Royal Academy of Engineering **Synthetic Biology: Scope, Applications and Implications (2009)** A carefully compiled and argued report on the possibilities for the next 25 years (downloadable from www.raeng.org.uk).

Bob Seidensticker **Future Hype: The Myths of Technology Change (2006)** A software engineer's take on the history of technology, and why it's not as revolutionary as enthusiasts habitually claim.

www.engineeringchallenges.org This site outlines the "Grand Challenges for Engineering", as defined by the US National Academy, in an attempt to mesh research and development with some of the larger problems facing the world in the coming decades.

www.techcast.org One of the more elaborate rolling efforts to poll experts and synthesize their predictions about which gizmos will appear when, though you need to pay a subscription to access most areas of the site. (The 2008 book by William Halal, *Technology's Promise*, was distilled from these forecasts.)

6 population

population

Within 10 years, for the first time in human history there will be more people aged 65 and older than children under 5 in the world.

Richard Suzman, US National Institute on Aging, 2008

The twentieth century saw unprecedented increases in world population. Along with these came fears of a Malthusian crisis (the theory that population increases exponentially, but food supply cannot grow at the same rate, making famine the inevitable outcome; see p.22). That concern faded to some extent as global food production kept pace with the rising numbers, and birth rates began to level off. But the planet's population is still rising, albeit at uneven rates in different regions. The resulting changes will be a major influence on the economic, environmental and political shape of the globe as this century unfolds, as will another inescapable outcome of current population trends: with people on average having fewer children and living longer, humanity as a whole will grow older.

This chapter, and the following nine, are all concerned with the medium-term future: the next fifty to a hundred years. They review the large range of efforts to fathom – by imagination, speculation, prediction, forecasting or just plain guessing - what will matter most in the world for the next few generations, and what it will be like to live in that world. These, like all such attempts to divine the future, are doomed to be unreliable. But population projections are a good place to start because they're more reliable than most, and the patterns of human life unfold at a steady pace. (In 2010, all the people who will be teenagers in 2020 have already been born.)

On the other hand, quite small changes in key variables, most obviously the birth rate, can make a big difference to estimates of overall future population, and such changes happen quite often. Even a data-rich field like demography is also a reminder that forecasts are subject to constant revision. That does not mean, however, that they have no value. Having some notion of future population is essential for any sensible look at the future.

THE NEXT POPULATION EXPLOSION

The "hockey stick" graph – so named for plotting a long, fairly flat line that suddenly zooms toward the top of the chart – showing global average temperatures soaring upwards in the twentieth century is a frequent target of contrarian bloggers and global warming denialists. But no one disputes the earlier hockey stick graph tracking total human population.

The current world population of a little more than six and a half billion is the result of a remarkable population explosion spanning the last hundred and fifty years. The number of humans alive doubled

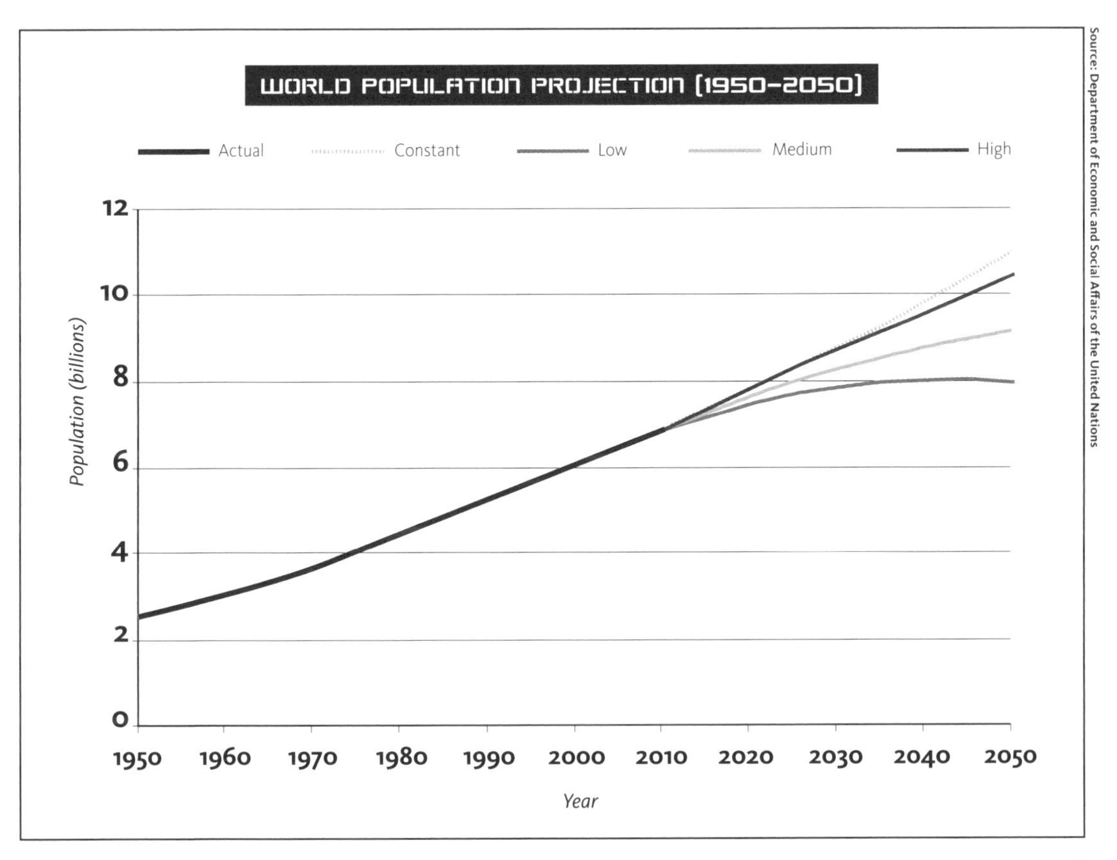

World population levels over the next four decades, on four different assumptions about average birth rates – which are currently declining.

THE POPULATION BOMB

The population explosion which got so much attention in the 1960s and 1970s was real, and extrapolating from the trend in numbers then was certainly scary. You did not have to be an uncritical disciple of Thomas Malthus (see p.22) to believe that growth rates had to taper off. Yet the lower overall rates of increase the world has shifted to more recently are not sustainable for any appreciable length of time either. Even with relatively modest annual percentage rises, when these rates are compounded the power of exponentially accelerating trends still kicks in, and the regular increases multiply to fantastic levels.

The UN population division, largely for the exercise, produced population curves for the next three hundred years in 2004. That is a huge span and the results were debatable, to say the least. But they did calculate one set of global population figures which assumed that the average fertility rate remained constant after 1995–2000. Here is the result:

- 2050 12.8 billion
- 2100 43.6 billion
- 2150 244.4 billion
- 2200 1775 billion
- 2250 14,783 billion
- 2300 133,592 billion

There is much debate about the "carrying capacity" of the planet, but on any assumption it seems likely to be quite a lot less than one hundred trillion. Even if our descendants discover a method of interstellar travel, unless the energy and materials costs are next to nothing, the annual number of emigrants is highly unlikely to reduce such massive Earth-bound populations much. This is one bomb which was always going to fizzle out.

between 1850 and 1950, and doubled again between 1950 and 1990. In the 1960s, the world population graph was steepening so fast it looked as though it would end up climbing vertically – clearly an impossibility. Since then, the rate of increase has slowed down. Going by current trends (that is, without a massive unexpected die-off), the total will probably stabilize some time in the middle of this century at nine billion, give or take a billion.

The range of projections is still wide – the margin of error is equivalent to rather more than the population of the globe less than two hundred years ago. Yet the global total, and how and when it settles down, is crucial. No issues are more important than how many people will need to be fed, provided with clean water, housed, educated and employed in the future, and how much other goods and natural resources they will consume. Plenty of such problems shrink or expand dramatically depending on the number of people on the planet. So it is probably a good thing that as global growth rates have slowed in recent decades, old projections have been torn up and recalculated regularly, with the future total revised downward by billions (see box). This easing of total population pressures, compared with what was once expected, now looks set to continue, though no one can tell how far it will go. Even so, barring calamity, the immediate future is still set for large overall population increases. The UN's mid-range estimate of nine billion by 2050 means that world population will have shot up more than fourfold from two billion in just one hundred years.

CRUNCHING THE NUMBERS

The best estimate of current population is around 6.7 billion, so the mid-range projection indicates an extra 2.5 billion people by mid-century. Population by then will still be increasing, though by a relatively modest thirty million a year, less than half the annual gain at the time of writing. The increase will almost all be in the less-developed parts of the world, which will have almost 8 billion people, compared with 5.4 billion now. The population of the rest of the globe will stay roughly level, at 1.2 billion. In fact, it is likely to go down in many places, and will only be maintained by immigration.

These mid-range estimates assume that the overall fertility rate (the average number of children per woman over a lifetime) continues to decline – from 2.5 children per woman to just over 2. If it stays where it is, the 2050 projection rises to 10.8 billion. If it goes down to, say, 1.5, the result is a mid-century total of fewer than 8 billion. The overall fertility decline is still a big assumption. To reach these averages, fertility in the 50 least developed countries has to decline from a current level of 4.6 to the present world average of 2.5. If that decline, matching that already recorded in the rest of the world, does not appear on schedule, those countries alone would be home to 10.6 billion people.

Population policy

Governments have some influence on population size, if they wish to use it. The "one child" policy in China has been effective in reducing population (see p.105 for possible drawbacks), as have similar measures taken in India including, at times, compulsory sterilization. Other countries, most notably Indonesia, have achieved similar results with less draconian measures.

More impressively still, Iran made a superfast U-turn from pro-natal policies instituted after the fall of the Shah in 1979. In 1987, the government decided that the population growth rate was too high, and managed to bring the total fertility rate down from a high of 5.2 to 2.6, halving it in a mere eight years. This was done through a combination of family planning propaganda, promoting a voluntary two-child policy, contraceptive clinics, and efforts to increase female literacy. On this evidence, if a country really wants to slow population growth – and most rapidly expanding nations do in the end – it can.

On the other hand, nations seeing shrinking numbers of citizens may try to persuade people to have larger families through exhortation and cash handouts. But the evidence so far is that once women experience the benefits of having fewer children (or even none), most are reluctant to devote large chunks of their lives to bearing and rearing kids. Female emancipation, in other words, rests crucially on choice in the number of children, and once achieved is hard to reverse.

Population trends and uncertainties

All these uncertainties mean that any forecast for more than a decade or so ahead could be wide of the mark. However, there are some trends which are unlikely to shift. Population *will* grow for decades to come, at least. Even if the decline in fertility accelerates through education, contraception, and emulation of the habits and choices of the affluent, world population growth will continue until 2050 and possibly beyond. The number of women of childbearing age will rise, and average life expectancy is increasing and (we hope) will go on doing so.

Fertility rates, life expectancies and the number of people in different age groups vary widely between countries, but all of this detail about current populations can be used to significantly enrich a picture of the near future. Knowing how many people live now,

FUTURE FAMILIES

Demographic trends, as well as some other notable recent developments, allow a few informed predictions about the likely shape of future families. They will still vary in lots of ways, and be shaped by conviction, circumstance, culture and personal preference. But on average, some probabilities look certain to shift.

The two biggest trends are fewer children and longer lives. That means more generations alive at the same time, and more children with great-grandparents. More parents will have parents of their own who can help with childcare, or who need help looking after themselves. As more people in developed countries, especially in the middle classes, defer child-rearing, children may come to be minority members of families filled with much older adults, who dote on them or dominate them. Perhaps more children will be like the "little emperors" said to have been fostered by China's one-child policy – indulged small boys whose parents and grandparents pander to their whims.

As families reach up and down the generations to grow, in a sense, narrower, or maybe taller, there will also be forces at work to make them broader. These are more linked to shifts in values than demographics, but are part of the mix which helps shape childbearing and child-rearing.

This trend varies more from country to country than the other tendencies detailed here. But there are certainly countries where the notion of a family with two parents who stay together to raise one, two or a few children, and are still around to enjoy the grandchildren, begins to seem like a quaint historical model. For one thing, staying married is less and less popular. In the UK, indications are that 45 percent of marriages will end in divorce. The figures for the US are similar. And while some divorced people remarry, divorce rates are considerably higher for second and third marriages.

At the same time, actually getting married is a less popular option. Living together, declining to marry after the arrival of children, and single motherhood are all more acceptable choices than they were a few decades ago. The marriage rate in the UK and US has fallen by half since the 1970s. The annual number of UK marriages is now at its lowest level since records began – which was in the 1890s, when the country had a population half its present size.

That means lots more stepchildren, stepparents, and half-siblings. Add the increase in life expectancy, and there are more stepgrandparents, too. So more people will be part of increasingly complex networks, linked by kinship but also by other, less enduring associations.

Recent technological breakthroughs loosen things up still further. In vitro fertilization and artificial insemination make genetic parenthood an alternative to gestational parenthood. Surrogate motherhood and prenatal adoption can to an even greater degree mix up roles which used to go hand in hand. There are an estimated five hundred thousand frozen embryos stored in the US as by-products of advanced fertility treatments. Who gets to use whose gametes depends on what is sanctioned in any particular society, as does the growth of same-sex marriage or same-sex parenting. All these factors are reshaping families into units whose structure is less of a given, but where all, or almost all, is negotiable, and may be revised or reconfigured later on. How that might work out in more detail is discussed in Chapter 15.

where they live, how old they are and where fertility trends are heading leads to what are likely to be reasonably good broad predictions for the next couple of decades. After all, most of the women who will have children during that time frame have already been born.

Population forecasts do not tell us how people will live, but they do illuminate some of the conditions that will shape their lives and alter the societies in which they live. When population shifts occur alongside changes in economic power, the political map changes too. Europe, for example, looks set to become less and less important in the future, as it is one of the regions at the leading edge of another global trend – ageing.

THE GREYING OF THE GLOBE

The distribution in age of the world's population is changing fast. Half the increase in world population expected between 2005 and 2050 will be among the over-sixties, while the number of children aged one to fourteen will actually go down slightly. In the more developed countries, the over-sixties segment will increase from 245 million to 406 million, while under-sixties reduce from 970 million to 840 million. In the world as a whole, the number of over-sixties will likely almost triple, from 670 million in 2005 to 2 billion by 2050. So although the populations of the wealthy world began ageing earlier, as it were, and will live longer, the vast majority of older folk will still live in developing countries. In 2005, they had 64 percent of the over-sixties, but this will increase to almost 80 percent by 2050.

The same is broadly true for the "old old" – those over eighty – except that their numbers will increase even faster. There were 88 million people over 80 in the world in 2005. By 2050, there are likely to be more than four hundred million of them, of whom seventy percent will be living out their time in the still-developing countries.

There have always been a few, perhaps wise, elders. But this boom in the elderly population, combined in many countries with a static or reducing number of young people, is new territory for humankind. As the Washington-based Center for Strategic and International Studies said in 2008: "Global ageing is not a transitory wave like the baby boom that many affluent countries experienced in the 1950s or the baby bust that they experienced in the 1930s. It is, instead, a fundamental demographic shift with no parallel in the history of humanity."

A useful indicator of the structure of a population is the median age – the number which leaves half the people above it, and the other half below. Median ages have reached 30 or so in some fortunate times and places, but the median age of the world population is projected to increase from 28 to 38 between 2005 and 2050. By mid-century, many countries in Europe and Asia are expected to have median ages above fifty. Apart from Japan (with a median age of 56), they include South Korea, Italy, Spain, Germany, Austria, Hungary and Greece. The forecasts are pretty solid, as the people who will be old in the middle of the century are already around. Even if people suddenly started wanting lots more children, the effect on the figures would be slow because of the limited number of women of childbearing age.

A different way of marking the same trend is to look at life expectancies. The global average in 2005 was 67, up from a non-pensionable 58 in 1970. It could reach as high as 76 by 2050, with a top figure of 83 in the more developed countries. (The average ignores differences between the sexes – women live around five years longer than men.)

Although the trend towards adults, and especially women, living longer is general, it's more pronounced in some countries than others. In fact, the more you investigate individual countries, the more it becomes clear that history and circumstance create important differences between them. These will affect the geopolitics of the first half of the twenty-first century in ways which will work themselves out gradually as the world moves along the path to what ought to be a level population, but at different rates in different places. The latest projections from the UN are easily available, and you can check the details for any country or region of interest online at esa.un.org/unpp. Here are some of the most important cases.

EUROPE: AN AGEING CONTINENT

The global picture shows a population growing, but also ageing. Yet the distribution of these changes is going to be very uneven for the next several decades. Some countries will find they have "youth bulges" as their populations rapidly increase. Others will be getting smaller, and their people older. That is the prospect for the countries who once ruled the world, in Europe.

The continent's politicians won't tell you this, but however brilliantly led, innovative and hard working the folk who live there, Europe looks set for a continuing decline in power and influence over the next few decades. In 1900, Europe accounted for about forty percent of the world economy, with nearly twenty percent of the population. By 2000 its economic share was below 25 percent, and its population less than 10 percent. But there is a larger shift to come. Quite a few European countries are now registering birth rates below replacement levels. If current trends continue, the European countries will have perhaps five percent of global population by 2050, and around ten percent of global economic production.

Demographer Paul Demeny points out that this will be accompanied by continuing increases in population in countries on Europe's southern and southeastern borders. In 2005 the 25 nations of the European Union had almost as many people as their (mainly Muslim) hinterland, which stretches from Morocco, Libya and Egypt in North Africa to Turkey and the former Soviet territories of Belarus and the Ukraine. But the ratios will move from 1.39 times as many people in these nearby countries to 2.75 times as many by 2050, if you add up UN population projections for all the countries in question.

That shifting balance needs to be viewed in the light of the other inescapable feature of those populations. Europe's decline in fertility, along with increasing life expectancy, implies more old people in the population, which will strongly affect all levels of society.

The change in European fertility rates has been unusually large. One anthropologist calls it a "population implosion". Western Europe's fertility rate fell below replacement level in the 1970s, and it just kept falling. By 2005, the average lifetime number of births in the 25 members of the European Union was just 1.5 per woman. There have been some recent increases in birth rates in a few countries – notably the UK and France – but these do not alter the overall European picture much so far.

So the population shrinks, and gets older, on average. Half the population of Western Europe will be over fifty by 2030. That means fewer people in the workforce and paying taxes, and each worker supporting probably twice as many elderly people as they do now. That's well beyond the level where pension systems fall apart. Robert Shapiro's book *Futurecast 2020* (2008) explains the detailed differences between the pension systems of Japan, the US, China and Europe, and their prospects of surviving. In Europe, Britain is slightly better placed than Italy, France and Germany, for example – but largely because its state pensions aren't much above poverty levels.

Societies with fewer children and more, longer-lived elders will be dealing with many other changes. There will be more generations coexisting in most families, with grandparents and great-grandparents still around. Many families will be "extended" as divorce and remarriage become more prevalent and "blended" families more common (see box on p.100), and longer lifespans may mean that staying married "until death do us part" may become just too damn long for most people. In such families, which have been repeatedly reconstituted and include a wide spread of ages, incomes and degrees of dependency, who cares for whom and who pays for whom will be constantly renegotiated.

As these changes unfold, ideas about when and how long to work, when it is best to add to your education, and when (if at all) to have children will also be shifting. Skill shortages are likely to lead to more efforts to retrain older workers. And planning for health, housing, transport and other services will be oriented more and more to the needs of older customers. The expense involved for both governments and citizens will probably be substantial, and possibly massive.

In Europe, particularly, one likely response to these demands will also take some dealing with, politically and culturally. Immigration from the still rapidly growing, and on average much younger, populations of the surrounding nations could remedy labour shortages, as it has in the past, and help care for the frail elderly citizens of an ageing continent. It would also mean more rapid increases of the already growing Muslim minorities in those countries. At the outer limit, recruiting enough immigrant labour

PREDICTIONS FILE

Joe Pelton

Former dean, International Space University and author of *MegaCrunch: Ten Survival Strategies for the Twenty-first Century* (2010)

Highest hope: We will have managed to stabilize global population and even to nudge it back toward the five billion level. This would be crucial to our reducing carbon dioxide levels to 350 parts per million, making a key response to climate change and helping us with the huge problem of technological unemployment that will come with the continuing onset of "super automation".

Worst fear: Due to overpopulation, we will start on a runaway path to global heating of the Earth's environment, plus massive levels of unemployment and conflict due to climate change as well as super automation via self-aware and increasingly smart machines.

Best bet: We will just manage to squeak through. But the survival of the human species may be due to technical or economic solutions we have not yet defined. This might be by "painting the clouds white", developing a new species of oxygen-breathing and carbon dioxide-absorbing algae, or new forms of birth control. It may also be due to restructuring free market capitalism to optimize survival and sustainability rather than the amassing of material wealth.

to compensate fully for the projected decline in Europe's working age population might mean that new immigrants and their descendants would constitute more than a third of the European population by the middle of the century.

JAPAN: A GERIATRIC SOCIETY?

Europe has many countries with decreasing fertility rates and ageing populations, but Japan is some way in the lead. The total population could drop from 128 million in 2005 to just 102 million by 2050, with most of the fall in the two decades before mid-century. At the same time, the country has a notably high life expectancy; forty percent of the population could be over sixty by 2050, and fourteen percent over eighty. These future Japanese will inhabit what it is probably fair to call a geriatric society, with more people over eighty than children under fifteen years old.

A senior citizens' home in Tokyo: Japan's high life expectancy and low birth rate are leading to a rapidly ageing population.

Other countries may solve some of the problems this poses by importing labour, but immigration to Japan has been very low in the past. Among the thirty countries in the Organization for Economic Cooperation and Development (OECD), almost all of them among the world's most developed and affluent, the proportion of foreign-born labour ranges as high as sixty percent – in Luxembourg – though ten to fifteen percent is more common. But Japan has by far the lowest, at just 0.2 percent. The Japanese enthusiasm for robots begins to look easy to explain.

Russia: a superpower implodes

Europe and Japan may be shrinking but Russia, or the Russian Federation as the post-USSR nation is known, is on a course of even more rapid decline. Along with the Ukraine and other predominantly Christian nations of the former USSR, it has a unique combination of low fertility rates and *falling* life expectancy. Astonishingly, the population is projected to fall from 147 million in 2005 to only 107 million in 2050.

The decline is due to a combination of factors, all hard to reverse, including poor infant and maternal health and high levels of alcoholism, drug abuse and sexually transmitted diseases. The growing incidence of STDs is especially ominous as it will inevitably lead to a rise in AIDS cases, which could make things even worse. The government has put the demographic crisis as the number one problem facing the nation for years, but the complexities of the causes mean effective action to boost birth rates or life expectancies looks unlikely.

Surprisingly perhaps, even with its relatively poor life expectancy, Russia will still see its older citizens comprising a larger slice of the population, and its younger ones a smaller portion. The nation's low birth rates are simply outweighing the high death rate. In 2005, the country had just fewer than 20 million people aged over 65 and 24.5 million aged 15–24. By 2050, the over-65s are expected to number around 25 million, while there will be just 10.8 million 15–24-year-olds. Despite its enormous natural resources, especially oil and gas, the country's days as a globally significant power appear to be numbered.

China: an ageing giant

The world's most populous nation, China, also has an unusual age profile thanks to the coercive one-child policy imposed in the 1970s. This was an effective means of population control, and the country's total is expected to rise relatively slowly in the next half century, with 1.3 billion people recorded in 2005 and a mid-range projection of 1.4 billion in 2050. But the government's mission to continue raising living standards and improving life chances for those 1.4 billion will be made much harder by the shift to an older population, which is happening at a relatively slower pace than it is in the other countries facing this transition.

In 2005, 11 percent of Chinese people were 60 or older, but the proportion is expected to increase to 31 percent by 2050 – 5 percent more than in the US by that date. Over the same period, the proportion of the population of working age will fall from around 68 percent to only 53 percent. Add the fact that the vast majority of families have no pension plans, and a generation who were raised in one-child families in the late twentieth century can look forward to finding themselves responsible for two parents and, very often, four grandparents.

The country will also be dealing with a different product of the one-child policy – an increasing surplus of men over women. This results from the widespread use of abortion to ensure the one child was a boy. In 2005, 118 boys were born for every 100 girls, a rise from 110 per 100 in 2000. In some regions, Chinese statistics show 130 boys for every 100 girls. By 2020, the country expects to have thirty

A GREAT POWER SHIFT

Demographics + economics = geopolitics is too simple an equation, but forecasting both at the same time is part of trying to figure out who will be top nation, or whether there will be one, later this century.

America's defence technology will give it a military edge for a long time. But in terms of global influence, China looks set to move ahead. Assuming that China's superfast economic growth – as much as ten percent a year – will continue for a few more decades, the size of its economy will dwarf any other country. Low estimates put it at twice the size of the US economy by mid-century. High estimates suggest it could be nearly three times the size of the US as soon as 2040, and account for forty percent of global GDP (gross domestic product).

There are several reasons to believe the extrapolation is realistic. First, the expansion is fuelled by China's own home market, rapidly increasing education standards and the government's commitment to massive investments in infrastructure. Second, there's the similarity of the country's trajectory to the paths followed by other, smaller East Asian states which developed at speed, such as South Korea and Taiwan, and the general trend for more recently modernizing nations to grow faster. Reasons to think the rate of growth, at least, will tail off are the ageing population, and increasing problems caused by pollution and water shortages. But it seems highly likely China will be the biggest economic player, if not overwhelmingly dominant, and will play a larger role in international organizations and global politics. According to Georgetown University economist Albert Keidel, "China's success will end America's global economic pre-eminence".

China probably won't assume America's place as a superpower, though. The US can (and does) wield its military power anywhere in the world. China is developing its armed forces, but seems more likely to use them to increase its regional influence. It has no overseas bases, for example. And its efforts to wield more influence in world affairs will have to contend with its growing neighbour India, the second candidate for a new future superpower.

India has almost matched China's growth rates in recent years, and its higher birth rate means it is likely to overtake China in population. Most projections suggest it too will have a larger economy than the US by mid-century, and will be second only to China in GDP. Unlike China and the US, India and China share a border, and have territorial disputes which have soured relations between the two at times. But both will also be competing for resources more widely with the US and the even older, ageing superpower represented by the nations of the European Union.

The relative industrial strengths of the two new, contending superpowers in different sectors, their military capabilities, and their government policies (democratic or otherwise) are all hotly debated by policy analysts. But most share the overall conclusion that, unless there is general economic calamity, there will not be one superpower in the middle of the twenty-first century, but several. So perhaps there won't really be a superpower at all, but a world where power and influence are more dispersed than they were in the second half of the twentieth century.

million more men than women, even though in China, as elsewhere, women typically live longer.

A study by Therese Hesketh of London's Institute of Child Health and a Chinese colleague in 2006 pointed out that the extra twelve to fifteen percent of young men in parts of China and India mostly appear in societies where marriage is nearly universal and strongly linked to social status. They add that many of the unmarried men are poorly educated rural peasants, probably having to watch eligible females "marrying up". As an additional negative by-product, in India there are already surveys indicating that murder rates in individual states correlate with the sex ratio – the more males, the more violence.

There is also evidence, according to Hesketh, that when young single men are thrown together, organized aggression increases, and they may seek "military, or military-type organizations". The future for organized crime, and perhaps even war in or between countries with an abundance of frustrated young men, sounds assured. The hope is that the male surplus will gradually disappear as stable population numbers allow more families to have two children again. The Chinese government is also promoting the shift, under the banner "care for girls". Even so, the sex ratio of newborns is unlikely to return to normal much before 2050.

India: the population superpower

India, which became the second nation with more than a billion people around the turn of the century, is set to overtake China as the most populous country. It is forecast to add half a billion by 2050, reaching a total of 1.6 billion. The increase is larger than the total Indian population in 1965. It will provide plenty of eager young workers, if they can be fed and educated – a demographic pattern which can lead to economic dividends. As in China, the sex ratio is skewed toward boys, especially among wealthier people with access to ultrasound scans.

Nigeria: doubling and then some

Nigeria is the largest of a number of countries whose fertility rates remain high enough to ensure that the population will double by mid-century, provided death rates remain at current levels. The 141 million Nigerians in 2005 are expected to increase to 299 million by 2050, lifting the country further up the list of the world's top ten nations by population.

Even here, the shift toward a larger proportion of elderly people will be noticeable. While the number of people aged 15–24 is projected to rise from 29 million in 2005 to 50 million by 2050, the number of those over 65 will climb twice as fast, from a little more than 4 million to almost 17 million.

But the rapid increase in young people is just as important. Nigeria is the largest of a group of West African countries which could fail to reap the "demographic dividend" which results from a large number of young people entering the labour market. That only works when there are good educational opportunities, with jobs to follow, as has happened in a number of Asian countries. If they are not so easy to come by, the result is quite different. The less positive outcome seems likely in Nigeria, which is expected to see an increase in the number of people of working age (15–64) from 75 million in 2005 to 125 million by 2025 and nearly 200 million in 2050. That increase will come in a country where four people out of every ten are already out of work – not a recipe for political stability.

Other countries expected to see large population increases include Pakistan, Bangladesh, Iran, Iraq, Egypt and Afghanistan. In many of them, the current age structure is likely to produce a "youth bulge" in the next few decades. In Afghanistan, for example, the number of people aged 15–24 is projected to increase from 4.9 million in 2005, to 10.7 million by 2030, and 16 million by 2050.

THE US: STILL GROWING

Among the developed nations, the dynamics of the US population appear exceptional. There will be ageing, certainly, with a threefold increase in the number of over-eighties by 2050, but there is no sign yet of population decline. And all of these senior citizens will be expecting high standards of health care if they become infirm, which is bound to place an enormous financial burden on both individuals and the government. Whatever health-care reforms may be passed in the near future, they could soon become inadequate considering how quickly the composition of the population is changing. (For more on future health, see Chapter 12.)

The US population is projected to increase from three hundred million in 2005 to four hundred million by 2050. Those UN figures include an allowance for net migration of a little more than one million a year. As the US authorities catch well over a million people a year crossing the Mexican border, almost any other guess is as good, but it's probably better to make it a higher one. Whatever the composition of the population turns out to be, the result is that, uniquely among the developed countries, the United

10 LARGEST COUNTRIES BY POPULATION

	1950	2009	2050
1	China	China	India
2	India	India	China
3	USA	USA	USA
4	Russian Federation	Indonesia	Pakistan
5	Japan	Brazil	Nigeria
6	Indonesia	Pakistan	Indonesia
7	Germany	Bangladesh	Bangladesh
8	Brazil	Nigeria	Brazil
9	United Kingdom	Russian Federation	Ethiopia
10	Italy	Japan	Dem. Republic of the Congo

Source: Department of Economic and Social Affairs of the United Nations

Current projections indicate that India will overtake China as the most populous country before 2050, while the Russian Federation and Japan's population decline sees them drop out of the top ten.

AFRICAN FUTURES – NEW HOPE FROM CHINA?

Africa, or at least the nations of sub-Saharan Africa, are often depicted as mired in food shortages, disease, weak government and civil wars. All of these do exist, but they obscure the fact that there have been positive developments in many African countries. There is great variation across the continent, but according to World Bank figures, the proportion of Africans living on less than $1.25 a day fell from 58 percent in 1996 to 50 percent in 2009 – not a great reduction, but not an increase either. The prevalence of HIV and AIDS stabilized, and primary school enrolment increased. Some countries, such as Ghana, were on target to halve poverty by 2015. Economic growth was strong, and government stability was spreading.

More than twenty African countries now hold democratic elections, compared to just three in the 1980s. The new, annual Ibrahim Index, compiled by the Ibrahim Foundation (named after telecoms billionaire Mo Ibrahim) shows governance improving in two thirds of African countries. On the other hand, endemic malaria, challenges from climate change and expected population increases (from 800 million in 2009 to perhaps 1.5 billion by 2050) are continuing problems. The world recession stalled growth rates across the continent, but less than in other regions, and with a faster recovery. In common with the burgeoning economies of China and India, much of Africa's economic growth over the past decade has been due to increases in domestic demand. New technology can bypass some infrastructure needs, with mobile phones being the best example. Around a third of the population already own one, and they are used extensively for transferring money and managing business. New energy technologies, such as micro-solar power, also have obvious potential in a continent with widespread power shortages. There's also oil and mineral wealth in many countries to fuel development, if the cash it attracts can be turned to the right ends. The two nations that have recently shown the most interest in firmer ties with African countries are the future superpowers China and India. Some analysts see this as a switch from a predominantly north–south economic axis of trade and investment, globally speaking, to south–south trade. At the moment, the scale is still small, but growth rates are high. African exports to China increased by 48 percent annually from 2000 to 2005. Oil and mining are the main industries involved, and the whole trend could be seen as a resource grab by new-style colonialists interested in taking a share of Africa's riches. If that's true, however, cheap loans – running into billions of US dollars – are more helpful instruments of colonialism than armed force. Efforts to tie these deals to projects to improve infrastructure, education and the establishment of other industries could be a big boost to African development, especially as China and India increasingly find themselves competing for favour with the same governments. Growing Chinese and Indian demand for materials ought to keep prices high, and benefit African producers.

More ambiguous is the other new trend in investment in Africa – countries from outside the continent buying up land for food production. British newspaper *The Observer* estimates that as much as fifty million hectares of land, equivalent to roughly twice the area of the UK, has been leased by overseas governments or private investors in recent years. The land typically comes with water rights, and is used to grow crops for export, even from countries where there are food shortages. Such deals could work positively, but only if they are negotiated so that the investment brings employment and know-how to Africa, rather than just using it as a source of perishable goods.

States' population ranking will stay the same – third – according to the UN's projections for the next few decades (see table on p.108). Whatever other challenges there may be to the United States' position in the world – economic, military or cultural – they will not be due to any decline in numbers.

MIGRATION

At any given time, the numbers and ages of populations provide the main basis for working out how many people will be trying to make a living in any particular country in the decades to come. But will they stay where they started out in life? There are many predictions of large-scale geographical shifts of population in this century, with economic migrants and refugees from climate change or other calamities adding to the regular outflows caused by war, civil unrest and political oppression. How does this affect where people end up?

At the moment, the International Organization for Migration (IOM) estimates that there may be two hundred million environmentally induced migrants by 2050, but also believes that most of them will move within countries rather than between them. If the worst kind of climate calamity unfolds in the latter half of this century, then further population movements could ensue on an unprecedented scale. Drought and expanding desert could lead to a mass migration from the inland parts of Africa, for example, leading to "Australianization" of the population – a massive concentration on the coast and a largely uninhabited interior – but involving much larger numbers of people.

Until then, however, the overall impact of migration will be small, even though numbers are increasing. The number of migrants in the world went up from an estimated 76 million in 1960 to 175 million in 2000, and rose again to 190 million in 2006. But even the latest total is only three percent of the global population.

The UN's estimates for future migration look pretty conservative. They point out that immigration's significance in more developed countries has increased as fertility has declined, and is even becoming necessary in some cases to maintain population. And the UN's best guesses suggest the major net gainers of immigrants will be the US, at 1.1 million a year, and Canada, at 200,000, followed by Germany, Italy, the UK, Spain and Australia. They would come mainly from China, losing 329,000 a year on average; Mexico, saying goodbye to 300,000; India, at 241,000; and Indonesia, 164,000.

All of this would only add up to a population gain of a little more than one hundred million between 2005 and 2050 in the more developed countries. Is that a lot or a little? A lot, of course, if you think of the shift of one hundred million people – that is a fair-sized country on the move. But in relation to the total global population in 2050 of nine billion, it is only a modest percentage of the possible error in the projections. Migration is sure to be a hot topic, partly because more countries will seek immigrants to remedy labour shortages. The future Muslim population of Europe and the Hispanicization of the US are already live issues in political debate. On the best current estimates, however, the overall pattern of migration is not going to alter the global population picture very much.

EXPLODING CITIES

Most of the rising population will be in the developed world, and most of the increase in numbers will be soaked up by the cities. More than a million people worldwide move from rural to urban areas every day, and the global city-dwelling population topped fifty percent of the total for the first time in 2008. The UN projects that seventy percent of people will live in cities by 2050, a bit less than the percentage living in cities today in Australasia and North America. The total population in cities would be 6.4 billion. Asia (1.8 billion more in cities) and Africa (0.9 billion more) would see the largest urban expansions. As a result, the world's rural population will actually begin to go down in about 10 years' time, with 0.6 billion fewer rural inhabitants expected in 2050 than today.

The migration into cities leads to two other trends. One is an increase in the number of "megacities" – there are likely to be nineteen cities with

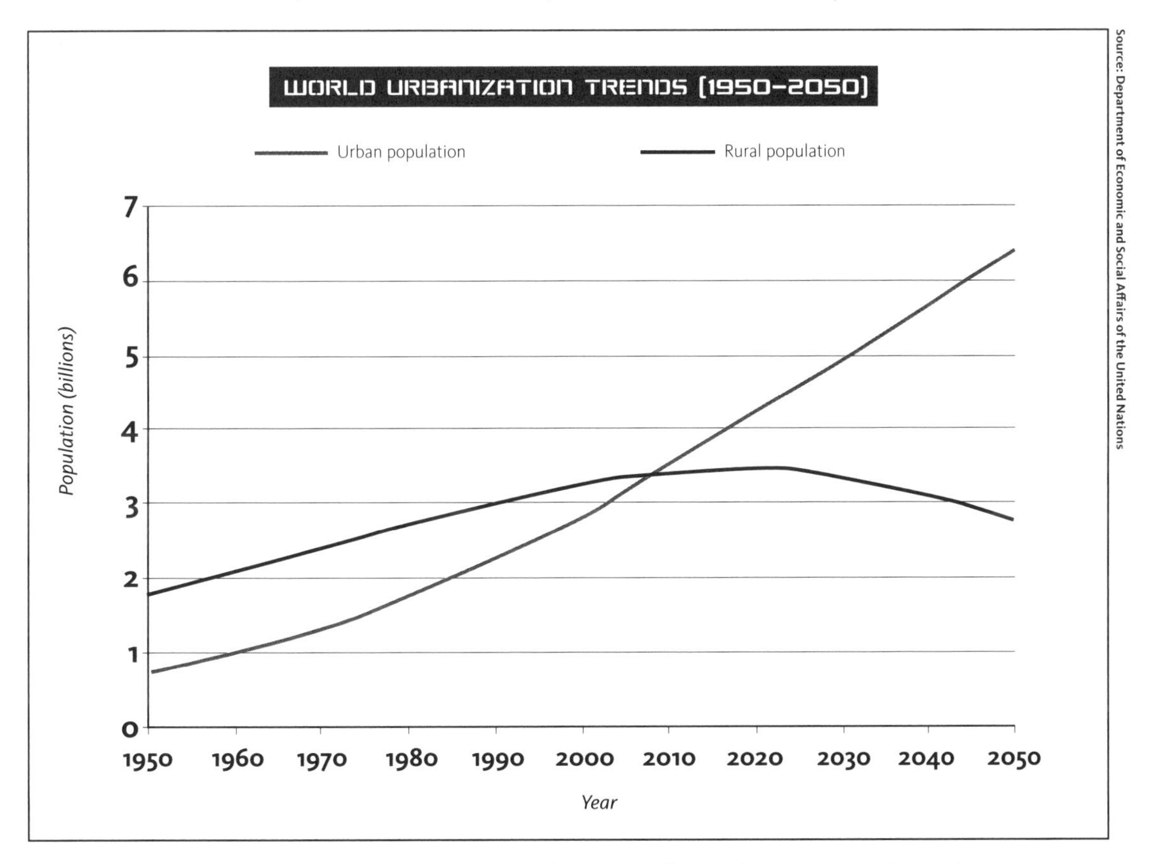

As of 2009, more than half the world's people live in cities, but the migration from rural areas is projected to continue apace.

CAN FUTURE CITIES BE GREEN?

We tend to think of cities as sites of intense consumption, suffering under the sheer weight of people, buildings, traffic and pollution, but contrary to popular opinion, mass migration to cities could be a good thing.

They'd have to be the right cities, shorn of sprawling suburbs where every trip means getting in the car. The World Future Council's 2009 report "A Renewable World" quotes a 1932 vision from a US urban planner as an example of what we need to leave behind: "The future city will be spread out, it will be regional, it will be the natural product of the automobile, the good road, electricity, the telephone, and the radio". Clearly this is a recipe for cities like the vast, soulless, energy-profligate conurbations found in much of North America.

Environmentalist Stewart Brand argues, in *Whole Earth Discipline* (2010), that future cities must be well supplied with high-density, mixed-use areas where people can walk much of the time and use mass transit for other journeys. If the density is high enough – with people living in apartment blocks, not houses – it is cheaper and more efficient to provide services than if the same people were scattered throughout the countryside.

Most of the twenty-first century's new urban dwellers will not live in cities like these, at least at first, but in the slums of the Southern Hemisphere. Viewed from the comfort of the developed world, they are pretty terrible places. But they are also places full of people working their way out of poverty as fast as they can. They offer freedoms denied to villagers, especially women, and foster millions of micro-enterprises. And they can exact a lower environmental cost than if people stay in villages, trying to get by as subsistence farmers. Mumbai has as many as a million people per square mile. But even New York is, by some measures, the greenest city in the US because it is so tightly packed. New Yorkers use around half as much electricity per head as people living in more sprawling American cities, despite all those elevators, and their carbon footprints are less than a third of the US average.

As cities grow worldwide, the space people take up will shrink. In developed countries, urbanization has mostly levelled out at eight percent of their populations. If other countries follow suit, four fifths of the population might live on just three percent of the land by mid-century. There will still be huge land use issues to tackle, because of those cities' demands for food, materials, energy and waste disposal. An old city like London, for instance, needs two supertankers' worth of oil *every week* to serve 7.5 million people.

Optimists like Brand see the problems being solved partly by cities' other qualities as centres of innovation and places that can easily learn from one another. Urban farms, smart grids, rooftop solar power and rainwater collection can all help. Some wealthy countries may be able to build new cities from scratch which show what can be done – Masdar City in Abu Dhabi, planned to provide carbon-neutral living for fifty thousand people, is one smallish example.

Elsewhere, much larger cities with little existing infrastructure may be able to move directly to new, greener ways of doing things, as they turn ramshackle, unlicensed settlements into urban areas fully integrated into city life. Each of those hundreds of cities of half a million people can be a site for new experiments in urban living, with planners and administrators staying alert for the best solutions. If so, future life being city life is a hopeful prospect.

more than twenty million people before the middle of this century. In contrast, in 1950 there were just two cities – New York and Tokyo – of more than ten million. Such vast agglomerations of people in a small space tend to attract a lot of attention, and to dominate their surrounding region. But growing urbanization also increases the number of cities of all sizes – and the megacities, while impressive, will

Mumbai, India, has as many as a million people per square mile – and more than a million of them live in central Dharavi, Asia's second largest slum.

PREDICTIONS FILE

Peter Barrett

Professor of Management in Property and Construction, University of Salford

Highest hope: That we will create built environments that truly support the health, wellbeing and full potential of individuals, including schools that really do facilitate learning, hospitals that provide healing environments that complement medical treatments, and so on.

Worst fear: That we will continue to build with a short-term perspective, ravage the environment and progressively lose the capacity even to meet people's basic housing needs.

Best bet: We will get to the point where the environmental problems are so obvious that legislation will drive innovation that will improve (but probably not solve) these issues. At the same time, education and awareness will increase the positive impacts that can be produced through the built environment.

still only be home to a minority of urban dwellers. There will be a larger set of smaller cities of one to a few million residents, and a still larger complement of yet smaller cities more similar in size to most of those built before the nineteenth and twentieth centuries. The UN population experts say that more than half of the urban population will still be in small urban centres with fewer than half a million inhabitants. A bit of maths suggests there will be six or seven thousand of them around the world.

HOW MANY IS TOO MANY?

The history of debates about population and birth rates is coloured by alarmism, racism and eugenic ideas about the threat of excessive breeding among the lower orders. When the Malthusian argument was restated vigorously in the 1960s and 1970s, many were uneasy that analysts in the rich countries were warning of population increases in the poor world.

Nevertheless, more recent environmental concerns, especially over global warming, have revived arguments about whether there are too many people on the planet, and what to do about it. The discussion now tends to be couched in terms of the Earth's carrying capacity – the number of a species a particular area can support without damaging the habitat and reducing the population. Like the older Malthusian predictions, the current debates acknowledge that while the Earth presumably does have a carrying capacity for human beings, no one knows what it is. Supposedly scientific estimates range from two billion up to forty billion, partly because what individual human beings consume varies enormously.

Some campaigners urge reductions in population well below the current UN projections. The UK-based Optimum Population Trust has some of the most detailed estimates and targets. Their methods are modest – they recommend making contraception available to everyone who wants it, promoting sex education and encouraging people to limit their families to two children. Their targets, though, are stringent. They estimate the sustainable population of the UK, at lower consumption levels than we currently reach, at 20 million – when it is forecast to increase from 61 million to 77 million by 2050. And their target for the world is, on differing assumptions, as high as 5 billion or as low as 2.7 billion.

This kind of estimate is far from the likely reality, which prompts the question: can a world with nine billion people avoid population-induced calamity? The answer is yes. But major efforts are needed to avoid several kinds of potential disasters, as the following chapters describe.

FURTHER EXPLORATION

Stewart Brand Whole Earth Discipline (2010) This eco-pragmatist manifesto devotes two chapters to evidence showing that while cities may be home to vast numbers of poverty-stricken people, they're also their best chance for escaping poverty.

Harry G. Broadman Africa's Silk Road: China and India's New Economic Frontier (2007) Commissioned by the World Bank, this book looks at Africa's economic development in the context of increasing trade with the rising powers in Asia.

David Coleman Europe's Population in the Twenty-first Century (2010) A demographer's view of the shifting population structure of Europe, and its effect on European countries' global position and influence.

George Magnus The Age of Ageing: How Demographics are Changing the Global Economy and the World (2009) Comprehensive review of population ageing in different countries, focusing on economic effects.

Fred Pearce Peoplequake: Mass Migration, Ageing Nations and the Coming Population Crash (2010) A detailed look at population trends in the coming decades, emphasizing the effects of declining birth rates in (some) developed countries.

www.iiasa.ac.at/Research/POP/proj07 The International Institute for Applied Systems Analysis (IIASA) in Vienna has a detailed population model (including diagrams) giving what it calls "probabilistic projections" for countries and regions to the end of the century.

www.prb.org The Population Reference Bureau is a US nonprofit organization whose site is a good jumping-off point for population info on the Internet; leans towards environmental and development issues.

7

energy and the climate crunch

energy and the climate crunch

Energy is the hardest part of the environment problem; environment is the hardest part of the energy problem.

John Holdren, Harvard University

The debate launched in 1972 by the book *The Limits to Growth* – which examined how much our population can grow, how many resources we can consume and how much we can continue to industrialize and pollute the environment before our very existence is endangered – goes on. But it's become increasingly clear that at least one dimension of growth, and the way we have achieved it, has limits which threaten dire consequences. The least dispensable fossil fuel, oil, is in limited supply. More crucially, dumping the waste from burning fossil fuels into the atmosphere is changing the climate. There is little doubt that there needs to be a massive shift in global energy systems this century. But how far, and how fast?

CARBON DIOXIDE EMISSIONS: THE REAL LIMITS TO GROWTH

We depend heavily on fossil fuels – the unholy trinity of oil, gas and coal. Generally, the more developed our economies, the heavier the dependency. We are burning "old" carbon (carbon laid down long ago) on a scale large enough to alter the composition of Earth's atmosphere, subtly but undesirably. This accelerates the "greenhouse effect", which traps the sun's heat and leads to global warming. It's not all bad – without some greenhouse warming, our planet would be a chilly and inhospitable place. But you can have too much of a good thing. What that means is uncertain, but the potential risks are too high to ignore. In this area at least, the "limits to growth" will probably be met sooner rather than later. If the book had been titled *The Limits to Carbon Dioxide*

Emissions, there would be hardly any room for debate about its conclusions.

Potential resource shortages and pollution problems always have to be considered in planning for the future, but carbon dioxide stands out from the others. Carbon dioxide isn't the only problem – methane, especially from agriculture, is a yet more potent greenhouse gas, and also increasing, though it doesn't stay around as long. But carbon dioxide is the pollutant increasing at the greatest rate. The carbon dioxide already in the air, if left to circulate, will go on warming the globe for a couple of hundred years.

The relationship of carbon dioxide emission limits to energy use makes the problem fundamental, and global. Energy is the master resource in any society. If a country has enough, then other problems – like growing enough food for an increasing population, producing other goods and cleansing pollution – are all easier to deal with. But energy technologies are woven into the fabric of life in ways which make them hard to change.

Cutting carbon emissions is yet more difficult than reducing the concentration of some other pollutants. As a point of comparison, the discovery in the 1970s that supposedly harmless, inert chemicals – chlorofluorocarbons (CFCs) – were significantly reducing the concentration of ozone in the stratosphere, and with it our protection from hard ultraviolet radiation, revealed a global problem. But cutting out the chemical culprits meant finding alternatives for things like aerosol propellants, refrigerants and fire retardants. This was no big deal, except for the manufacturers.

Reducing greenhouse gases will not be like that. The twentieth century's achievement of what most of those who got some of it considered the good life – or at least a better one – depended fundamentally on fossil fuels. The big question is: can we possibly have one without the other?

A BRIEF HISTORY OF ENERGY

There has always been plenty of energy around on Earth, largely courtesy of the enormous, luminous fusion reactor which is our own star, the sun. The problem is getting hold of it in ways which can do useful work.

ENERGY PRE-CIVILIZATION

The first ways of doing this evolved naturally. All organisms need energy from somewhere to maintain their complex structures in the face of nature's tendency toward disorder (see entropy, p.29). The earliest living things probably relied on chemical energy derived from breaking down large molecules into smaller ones. By 3.5 billion years ago, some of them had found ways of using sunlight for a form of photosynthesis. These first eaters of the sun could capture packets of light energy – photons – and use them to split simple molecules. They needed molecules which are chemically primed to give up an electron, which is needed to drive the manufacture of sugars from carbon dioxide (the synthesis in photosynthesis). Initially these electron donors are thought to have been hydrogen, hydrogen sulphide or iron. Modern proteobacteria – which probably resemble earlier life forms – subsist on hydrogen sulphide.

None of these electron-rich compounds were easy to come by in the ancient seas where these organisms lived. So the overall productivity of the biosphere was still pretty limited, comprising perhaps a tenth of that of the modern oceans. But there was another useful compound present in much larger amounts: water. A little more than 2.5 billion years ago the cyanobacteria appeared – the first organisms which could grab enough energy from sunshine to split water into hydrogen and oxygen.

This was the biggest news since the origin of life. There was a slight problem, however, as oxygen was poisonous to other existing life forms. When all the iron in the oceans had turned to rust, the gas accumulated in the atmosphere, which then became hostile to more old-fashioned bacteria. Over time, though, other organisms evolved which could burn (technically, respire) already existing organic matter – plants and other animals. Breaking them down with the aid of oxygen yielded lots more energy than any other way of life. Eating plants, or eating the plant-eaters, enabled some creatures to grow more complex. Some of them – already relatively large and mobile – also developed big brains, which may have been dependent on them eating meat as it is an especially energy-rich foodstuff. All, though, ultimately depended on photosynthesis for energy.

Energy in modern times

When we big-brained creatures invented culture, all our efforts were still underpinned by energy from the sun. Barring some heroic genetic engineering to create plant-humans, our use of photosynthetic energy capture is necessarily indirect. The amount that individual humans could divert grew slowly, in stages. Hunters and gatherers just used what they ate to fuel the next round of hunting or gathering. Then they burned wood for heat and cooking. Early agriculture increased control over photosynthetic energy through planting, harvesting and perhaps storing grain, and herding herbivores. Using draft animals' or human slaves' muscles could boost production and allow some folks to get civilized, but both animals and slaves were ultimately grain-powered. Later on, a few water wheels and windmills provided concentrated sources of power, but did little to change the overall dependence on what we now call biomass. Heat still came from burning wood, straw, charcoal, peat or dung.

PREDICTIONS FILE

Ron Oxburgh

Geologist, former chairman of Shell UK

Best hope: That the major countries of the world will have realized that cooperation on managing global emissions is the only way to retain a terrestrial climate that is recognizably close to what we have today and to which current life on Earth is reasonably well adapted.

Worst fear: That each country acts in its own perceived short-term interests in the belief that this will maintain or raise its economic competitiveness; that emissions will continue to rise, and wealthy nations will use their wealth and technology to achieve a degree of short-term adaptation to a rapidly deteriorating climate, allowing the developing world to take its chances.

Best bet: I expect that we shall make some progress towards managing global emissions, but I should be surprised if it will be sufficient to keep the global temperature rise below 2°C. This could be too pessimistic if a cost- and carbon-effective way of extracting greenhouse gases directly from the atmosphere were discovered.

That was what powered human history, in all its glory, until the nineteenth century brought carboniferous capitalism. England led the way a few centuries earlier as demand for shipbuilding timber prompted a shift from charcoal to coal for fuel from around the middle of the sixteenth century. But a more general switch to fossil fuels came as industrialization stretched the limits of biomass and water power. First, wood began to be replaced by coal. Then came oil and gas.

The extraction of these stored energy supplies (still the product of photosynthesis which happened many millions of years ago) made the twentieth century an era of energy consumption completely without precedent. Domestic electricity, land, sea and air transport, intensive agriculture, and all the other features of modern life were made possible by fossil fuels. Of these, oil was the favourite because it has more hydrogen in relation to carbon than coal, and so releases more energy when burnt. Natural gas is better still (four hydrogens for every carbon), but harder to use in mobile fuel tanks as it has to be liquefied under high pressure.

Coal, oil and gas together changed life for billions. We are pretty much in the position of a population of five-year-olds who have always lived on bean sprouts and oatcakes, and then find an enormous fridge full of ice-cream. How do you think their diet will change? The shift has been so huge that total energy use by our species now equals around a tenth of the energy processed by the entire biosphere of the planet. Some examples from energy analyst Vaclav Smil illustrate just how much things changed in the twentieth century.

Take agriculture. In 1900, an American farmer growing wheat on the Great Plains would have been pleased to be in charge of perhaps six well-fed horses pulling the plough. He was harnessing animals that could deliver around ten kilowatts of energy at peak output, and five kilowatts at a steady pull. After a few hours of that, farmer and horses would both need food and rest. In 2000, the same land would be worked by a farmer sitting in an air-conditioned tractor cabin, with a touch of the throttle commanding more than three hundred kilowatts, as long as there was gas in the tank.

Even those who had use of more impressive machines at the turn of the century were mere beginners by our standards. The engineer of a state-of-the-art steam locomotive in 1900 controlled about one Megawatt of power, which pulled the train at just under one hundred kilometres per hour. A century later, a Boeing 747 pilot controlled jet engines with a steady output of 45 Megawatts, flying at 900 kilometres per hour.

The twentieth century's changes were highly uneven, depending on wealth, social class and where you lived, and not everyone had access to this energy and speed. There are still societies today where people spend their days breaking rocks by muscle power. But the fact that many people had this previously undreamt of command of energy excited speculation that the upward curve might go on indefinitely. By century's end, however, it seemed that this was not the case.

Energy now: where does it come from?

In today's world, fossil fuels are still overwhelmingly dominant. According to International Energy Agency (IEA) estimates, they account for 80 percent of global energy consumption, mainly comprised of oil (34 percent), coal (25 percent) and natural gas (21 percent). The only source (apart from tidal power) not ultimately

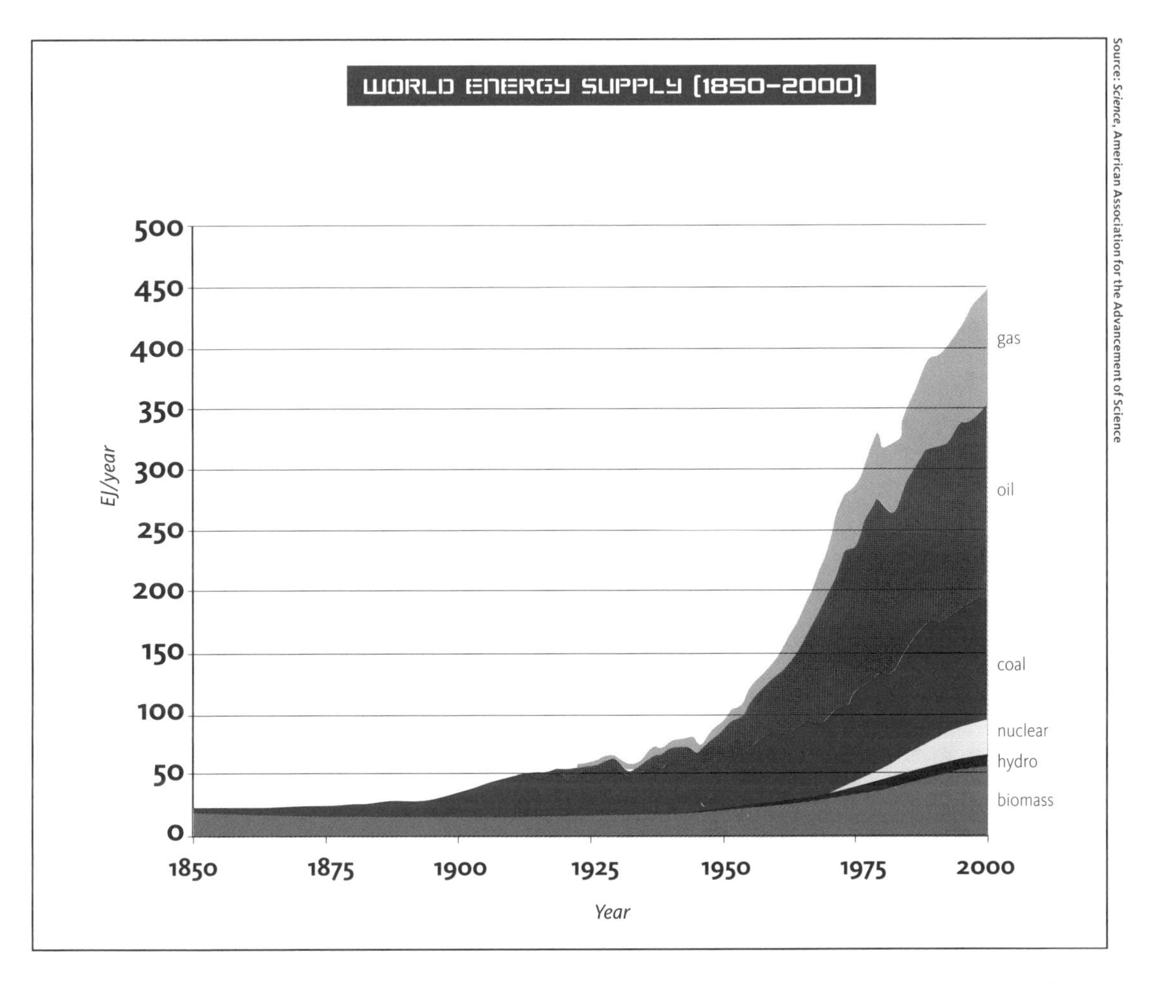

This breakdown of the energy mix shows clearly how oil and gas fuelled the vast growth in consumption in the second half of the last century.

derived from the sun, nuclear power, is a long way behind at 6.5 percent. Biomass and waste come in at 11 percent. Minor contributions, important in a few countries, come from hydroelectricity (2.2 percent) and geothermal (hot rocks), solar and wind power, which altogether only make up 0.4 percent of the total.

As the diagram on the previous page shows, coal was king at the turn of the twentieth century, but even though coal use kept growing it was soon outstripped by oil, with gas coming up fast behind. Oil – cheap, (relatively) clean and energy dense – took over transport and transformed it. Coal is now confined mainly to generating electricity and making steel and cement. Gas largely contributes to industrial and residential heating and generating electricity. The other significant contributor, biomass, is mainly used for what it has always been used for, heating and cooking, although this may be about to change. The numbers for biomass are the haziest, as much is simply gathered by those who use it.

Where are the energy sources and stores? Some, like hydropower and geothermal energy, are fixed. Nuclear power stations are more flexible as uranium is not too hard to move around, but who is allowed to get hold of it is reliably controversial. Nuclear stations are generally built to serve a national grid, with

Grangemouth Oil Refinery, Scotland, is one of nine in the UK. It has the capacity to process 210,000 barrels of crude oil daily.

some trading of electricity at the boundaries between neighbouring countries. The most flexible of the current main sources are fossil fuels, of course, which are portable and tradeable – and the means through which they're transported, using bulk carriers, supertankers, trains and ever longer pipelines, is one of the world's larger industries in its own right. This is why a country like Japan, for example, which barely registers as an oil producer, can still run one of the world's largest fleets of cars and trucks.

Japan is far from alone in using far more oil than it can extract. There are big mismatches between where oil lies and where it is used. Imports will get bigger as fast-developing countries like China and India increase oil consumption. And most oil is now controlled by state-owned corporations in the major oil-producing countries, rather than by the well-known multinational oil companies. According to the IEA, meeting world demand for oil in the coming decades depends mainly upon three countries being able and willing to increase their supplies – Saudi Arabia, Iran and Iraq. Who gets access to whose oil is thus likely to be an even larger political and economic issue in the coming decades than it has been for the past half-century.

PEAK OIL: HOW SOON?

The tensions over oil that are bound to accompany increasing demand will be heightened by the possibility of a choke on supply. This is tricky to gauge, since no one really knows how much oil there might be under the Earth, and even the best information about known reserves tends to be either a state or commercial secret. Plus, past forecasts of energy use have nearly always been wrong.

However, recent spikes in the price of oil (which topped $130 a barrel in the spring of 2008, then fell as

PREDICTIONS FILE

Mark Lynas

Writer and climate campaigner, author of *Six Degrees: Our Future on a Hotter Planet* (2007)

Highest hope: Human civilization will evolve a system of international governance (preferably a reasonably democratic one) that will allow our species to live within the boundaries allowed by the physical limits of the Earth. An obvious boundary is the need to bring carbon dioxide concentrations in the atmosphere back down to safe levels, probably fewer than 350 parts per million. But we also need to focus on other crucial aspects of planetary ecology, in particular biodiversity loss, nitrogen pollution, land use, water use and toxic accumulation.

Worst fear: Today's dominant paradigm of economic growth at all costs will stymie attempts at ecological rationality, and some globalized equivalent of the US Republican Party will call the shots indefinitely. Humanity will remain in denial about the limits to growth, and civilizational collapse – driven by ecological collapse – will become a serious possibility, perhaps even a likelihood.

Best bet: I'm sure it'll be some kind of fudge; a hybrid course between the two outcomes. We'll notch up a few successes and come up with some strong environmental treaties, but fall rather short on implementation and have to face up to adapting to somewhere between two and four degrees of warming by the end of the century. That may not sound like a hell of a lot, but it would transform the planet's surface and oceans, and push the temperature up to higher than it has been for millions of years.

the global recession took hold later that year) has led proponents of the "peak oil" theory to claim that their predictions are coming true. In their view, the rate of recovery from any given oil field follows a similar bell-shaped curve, rising to a peak, then declining over roughly the same period as the build-up as the oil gets harder to extract. Similar curves can be constructed by combining yields from many fields, as geologist M. King Hubbert did in the late 1950s, predicting that US oil extraction would reach a peak around 1970. The fact that he was more or less right about the US has reinforced predictions that an even more inclusive curve could accurately describe global oil production.

Those who conjure up such curves usually put the peak some time in the first decades of this century, in which case it has already happened. What then? A number of pundits predict that the result will be the collapse of civilization – some even seem quite eager to see it. James Howard Kunstler's *The Long Emergency* (2005) is a fair example of this kind of thinking:

> "The world oil production peak represents an unprecedented economic crisis that will wreak havoc on national economies, topple governments, alter national boundaries, provoke military strife and challenge the continuation of civilized life."

On the other hand, in *Energy at the Crossroads* (2005) Vaclav Smil points out:

> "The timing of oil's global demise depends not only on the unknown quantity of ultimately recoverable crude oil resources (which has been, so far, repeatedly underestimated) but also on the future demand … that is determined by a complex interplay of energy substitutions, technical advances, government policies, and environmental considerations."

That complexity means that other near-certainties have mixed implications. Easily recoverable oil is definitely finite. It will soon (possibly very soon) get scarcer and more expensive. Some of the possible alternatives, such as exploiting Canada's oil shales, drilling for oil in the Arctic National Wildlife Refuge or producing liquid fuels from coal, would have nasty environmental consequences even if they became economically viable.

Oil production will probably see a gentle decline after the peak, rather than a sharp cutoff. But as these trends play out, any shortfall in oil supply could quickly lead to real pain. Bringing other sources of energy supply to market is a slow process. As California Institute of Technology professor David Goodstein points out in *Out of Gas* (2004), one of the more level-headed expositions of the oil prospect, it does not take a big shift in demand or supply to create a large problem. If demand for oil grows at a few percent every year, as it has in the past, and supply declines at just a couple of percent a year after the peak, that produces a five percent gap per year. Sounds manageable, perhaps, with a little care and better efficiency and conservation, in year one. But demand might go on increasing, and supply will continue to decline. With just a five percent annual gap, that would mean finding substitutes for half the oil we're currently using in just ten years time – more than ten billion barrels of the stuff. That's an immensely tall order.

IF NOT OIL, WHAT ELSE?

In the short term, the substitution is likely to be with other fossil fuels, in spite of the attractions of renewable sources. The strongest case here comes from oil industry experts like Peter Odell. In *Why Carbon*

IMAGINING DISASTER

When the world lived in fear of nuclear war in the 1950s and 1960s, novelists turned out post-holocaust tales by the hundreds. They have not responded quite so readily to the threat of climate change, but more stories which feature extremes of heat, cold or wet are beginning to hit the bookstores.

British writer Maggie Gee foresees a watery future for London in *The Flood* (2004). The rich cope well enough on the high ground; the poor, in their high rises, are cut off from the old life. Gee's interest in extremes extends to an alternative climatic disaster in the memorably bleak *The Ice People* (1999). Ice and floods also grip Washington, DC, in Kim Stanley Robinson's trilogy about near-future climate change (*Forty Signs of Rain*, 2004; *Fifty Degrees Below* and *Sixty Days and Counting*, both 2007). It has a bit of everything – melting ice sheets, the Gulf Stream shutting down, geoengineering schemes, and a lot of Buddhism and Emersonian philosophy thrown in. More didactic than Robinson's monumental *Mars* trilogy, its focus on the politics of the National Science Foundation does not grip every reader.

A more exotic science-fiction tale featuring climate change is Ian McDonald's satisfying, panoramic portrait of a future monsoonless India, *River of the Gods* (2004). Robinson does want to save the world from climate change, though, so his books are at least an antidote to Michael Crichton's *State of Fear* (2004), which adds a tendentious view of climate science as a green fanatics' conspiracy to his customary flat characterization and twisty plot. London continues to be a popular target for flooding, with a watery future for the British capital also portrayed in Will Self's *The Book of Dave* (2006), which is strongly reminiscent of Russell Hoban's post-holocaust novel *Riddley Walker* (1980). Stephen Baxter's excellent *Flood* (2008) is also a detailed exploration of the effects of rising sea levels (caused by climate change and a fictional ocean under the Earth's crust escaping upwards), though his disaster narrative ranges around the globe. Scary stuff, though life – of a kind – goes on. The book does what the tedious *Waterworld* (1995) tried to do in cinemas: create a believable picture of a new world. Cinema remains way behind novels in depicting plausible stories of climate change, as the risible *The Day After Tomorrow* (2004) confirms.

Notable climatic-based tales from US literary novelists are still harder to find, though. Cormac McCarthy's compelling survival tale *The Road* (2007, turned into a film in 2009) has been tagged a post-global warming novel, yet his unspecified global disaster involves not just warming, but incineration. A more complex tale of twenty-first-century unpleasantness, interwoven with twentieth-century environmental activism, is the excellent *A Friend of the Earth* (2000), by T.C. Boyle. The ambiguous love story alternates between the 1990s and 2025, and depicts the later period as one of desperate decline, beset by violent weather and new diseases. The image of a mega-wealthy rock star's tatty menagerie as the last and best hope for preserving a few of the planet's more exotic species is somehow more potent than the total disasters depicted in other books.

Fuels Will Dominate the 21st Century's Energy Markets (2004), he puts the oil peak as far away as 2050, and sees major investments in coal and gas catering for the increased demand from developing economies like India and China. His predictions are on the firm side:

> "Over the twenty-first century as a whole, a total of some 1660 Gigatons oil equivalent of carbon energy will be produced and used, compared with a cumulative total in the twentieth century of just under 500 Gigatons."

This sounds like a huge expansion, but as he points out, it's lower than the Intergovernmental Panel on Climate Change's baseline scenario. The IPCC suggests that business as usual could see global energy consumption increasing fivefold, with a smaller proportion coming from relatively low-carbon natural gas, and a higher share from high-carbon coal. And even Odell reckons renewables might increase to more than forty percent of global energy supply by 2100.

There's another significant difference between Odell and other forecasters. If we do manage to prolong our reliance on carbon from under the ground, what effect will this have on the environment? Odell doubts that climate change is a serious problem. If it is, he assumes that there will be cheap methods of taking the carbon out of the atmosphere and burying it again. He may well be wrong on both counts.

Alternative energy sources, like increasing use of solar power or running cars on fuel cells powered by electrically generated hydrogen, could be more benign options than putting the environment at risk by continuing to plunder our oil supply. But the vast capital investment in existing energy systems makes them hard to shift. Countries hoping to increase their economies a lot more, or becoming more concerned about energy security (India, China and the US, to name the most important ones), happen to have lots of coal – the worst fossil fuel from the point of view of climate change. The chances are they will use it. So current trends are likely to make a poor situation worse, and in a matter of decades.

Peak oil is going to impinge on energy markets in ways that governments and consumers need to anticipate, but they've been reluctant to do so, to say the least. But stronger, more urgent constraints are imposed by global climate change. An impending oil shortage requires more investment in energy research and development, and finding substitute power for machines, such as cars and planes, which depend on highly portable, energy-dense fuels. But even if there was infinite oil on offer, its effect on climate change ensures that we would need to be doing those things anyway.

Certainty and uncertainty about climate change

If peak oil is uncertain, climate change is not. But there's no consensus on global warming, with outlying views from both "deniers" and "doomsayers". It's a mistake, however, to set either of these against mainstream scientific opinion, as if simply stating two views reflects some kind of balance. The latest report on the physical science of climate from the authoritative IPCC (published in 2007 and available in full at www.ipcc.ch) involved more than 550 authors. The two further weighty volumes that accompanied it involved similar numbers of experts, and equally long, drawn-out processes of review and discussion. Deniers suggest that the IPCC fosters "groupthink" (see p.46), but this makes a travesty of the argument and appraisal that goes into an effort like this. In fact, the elaborate and slow-moving committee discussions of the IPCC tend toward the conservative, probably underestimating future climate change and its effects. However, IPCC reports are still the best we

HERE IS THE CLIMATE FORECAST

A degree or two rise in average global temperature doesn't sound like something to get that worked up about – which is just as well, as that rise is already very likely to happen whatever we do next, as the effects of the already accumulated greenhouse gases work through the global system. But increasing evidence indicates that higher increases are on the way. In 2009, the UK Meteorological Office concluded that average warming of 4°C was likely by the end of the century if emissions continue on their current path. Such an increase could happen in the 2070s or even, in what they called the "plausible worst case", by 2060.

That average conceals much bigger likely changes. Their models suggest increases of up to fifteen degrees in the Arctic, and western and southern Africa would register up to ten degrees of warming. Those regions would also get perhaps twenty percent less rainfall, and Central America, the Mediterranean and parts of coastal Australia would probably become drier. On the other hand, India might see twenty percent more rainfall, increasing flood risks.

Some few benefits do come along with the severe costs. According to a UK team at the University of Reading, a world dealing with the 4°C average increase in 2080 could see fifteen percent of the global population (a billion people) at increased risk of water shortages, and fifteen percent of land currently suitable for farming becoming too hot or dry for raising crops. On the other hand, an extra twenty percent of land that is currently too cold might become suitable for agriculture. But the uncertainties – also affecting crucial factors such as sea-level rise, forest fires and the reliability of monsoon rains – strongly suggest a world in which the basics of food, water and dry land to live on are harder to manage, and impossible to obtain for large numbers of people in the worst affected areas.

A local resident walks on drought-cracked ground in Las Canoas Lake, Nicaragua, in April 2010. The lack of rain was one of the many consequences of El Niño.

GAIA SAVES THE DAY? PROBABLY NOT...

One of the vital insights of "Earth systems science" is that the great cycles of the planet – the weathering of rocks, plant growth and decay, rainfall and runoff, ocean currents, atmospheric circulation, and so on – are all interlinked and can influence each other. This wondrously complex interaction of physical, chemical and biological changes means that there are lots of feedback loops at work, most of them poorly understood.

One view is that they help stabilize an environment hospitable to life, keeping temperature, the concentration of atmospheric oxygen and the amount of salt in the oceans at levels that suit most living organisms. This theory maintains that the whole Earth, or Gaia, is in some sense alive, and responds to change by damping it down in ways which preserve the biosphere. Whether you believe in Gaia or not, it is at least theoretically possible that a rise in average global temperature could trigger feedbacks that would reduce temperature again. Negative feedbacks could include:

- Plants absorbing more CO_2 as they grow faster, stimulated by the increased concentration of the gas in the atmosphere.
- Warming could dry out high-altitude layers of the atmosphere, which would weaken greenhouse effects in the upper reaches of the troposphere.
- Models suggest the upper troposphere warms more than the surface of the Earth, which would increase the radiation of heat back into space.
- A warmer ocean could increase growth of phytoplankton. These generate the gas dimethyl sulphoxide, which is thought to pass into the atmosphere and help stimulate formation of water droplets, and hence clouds, which could have a cooling effect through reflection.

The list of possible positive feedbacks is longer:

- Melting polar ice exchanges a white surface for darker water, which absorbs more radiation instead of reflecting it back into space.
- Water vapour, a greenhouse gas, in the atmosphere could increase as rising temperatures boost evaporation from the oceans. On the other hand, it might make more clouds. Cloud effects are among the most uncertain of all.
- As the permafrost in Siberia warms up, there could be a vast release of the methane currently locked up in the frozen soil. This is probably the effect most likely to lead to runaway warming: the hotter it gets, the faster the methane release becomes, and the hotter it gets... According to climate change expert Joseph Romm, it's "the most dangerous amplifying feedback in the entire carbon cycle".
- As the oceans absorb more CO_2, the likely increases in temperature and acidity near the surface will reduce their ability to take in more of the gas. This is a combination of physical, chemical and biological feedbacks – all positive.
- Increasing temperatures can accelerate decay of organic matter in the soil, which releases more CO_2. Also, rising temperatures are thought likely to lead to net emission of CO_2 from tropical forests, instead of net absorption.
- Meltwater from ice may penetrate right to the bottom of the icecaps and act as a lubricant, so the whole lot slides into the sea. This would not affect temperature, but the sea level would rise faster than expected.

Even though none of these effects are certain, the preponderance of positive feedbacks suggests it might be good to try to increase the negative feedback that we have some control over – human-caused CO_2 additions to the atmosphere. Leaving it to Gaia looks likely to be less effective.

have to go on when considering the possible futures of the climate, with the proviso that the last report is now somewhat out of date, and evidence uncovered since then points to even worse outcomes.

The strongest conclusion of the Fourth IPCC Assessment Report is that "warming of the climate system is unequivocal". It estimates that at least half of this warming is almost certainly due to increases in greenhouse gas concentrations caused by human industry and agriculture. The certainty isn't total because the panel only felt able to put the probability at ninety percent, but due to accumulating evidence, this is a good deal stronger than the claims made in the previous (2001) report. The room for doubt is getting very small.

That confidence applies to the past. The future, as usual, is considerably hazier. The IPCC considers various scenarios, with its "worst case" in some respects being economic growth based on continued reliance on fossil fuels. If that plays out, their best estimate is that there will be a global average surface warming of 4°C by the end of this century.

At the lower levels of global warming that are now more or less inevitable, a couple of degrees centigrade say, there remains room for doubt about the likelihood and severity of flood or drought, more intense storms, effects on crop yields, sea-level rise, and (a more recently emerging concern) acidification of the oceans as warmer water absorbs additional carbon dioxide. But if the worst scenarios play out and the world warms up by 4°C or even more, then violent weather, drowned coastlines, agricultural blight and heavy species loss are all much more likely.

Climate models are simplified representations of an incredibly complex system. They can produce possible global averages, but offer little indication of short-term variations or regional or local differences. It looks likely that the poles will warm more than the equator, and land more than oceans. Beyond that, the models are not usually much help. But while there will be winners as well as losers in a slowly warming world, the diversity and range of the potential adverse effects is much greater than the possible benefits (see box on p.127). It seems extremely unlikely that any good will outweigh the bad.

Even more alarmingly, it's not certain the warming will be slow. In a system which is not completely understood, and in which sunlight, air, oceans, clouds, rocks, soil, ice and life all interact, there may well be feedback mechanisms which we cannot measure well, or simply do not know about. Some of these could be negative feedbacks, which could actually be beneficial for the environment, as the term refers to increases in temperature or CO_2 causing effects that would tend to stabilize things. Less reassuringly, some might be positive feedbacks, in which case they would tend to trigger runaway global warming, and make any action to reduce greenhouse gases likely to be too little, too late. Among the plausible feedbacks so far identified, the majority are positive (see box opposite).

Remember, too, that the last IPCC report was in 2007, and reviewed the science from a couple of years before that. Much of the large volume of research since then makes the positive feedbacks look more likely, and most climate scientists are now more concerned than they were when that report was published. Expect the next IPCC assessment (not due until 2014) to be considerably more alarming. We do not know if climate change will lead to some kind of ultimate disaster – the kind novelists and screenwriters like to depict (see box on p.125) – but it is a stronger candidate than any other single factor.

ACT NOW OR LATER?

Assume global warming is a reality. What should we do? Aside from climate change denial, two positions suggest that the answer is "nothing". One holds the belief that disaster is already certain. The other, less discouragingly, suggests that it would be better (and ultimately cheaper) to wait and see, and deal with negative effects of climate change as they arise.

It's hard to find anyone who upholds the "do nothing because we are already doomed" position consistently. Gaia prophet James Lovelock comes

The environmentalist James Lovelock devised the influential Gaia theory, He takes a pessimistic view of climate change and mankind's ability to prevent it, believing that nuclear power generation is essential to the future of our energy supply.

close in his book *The Revenge of Gaia* (2006), arguing that the climate is unstable, a little warming will tip it into a much hotter state, and that the warming that is already unavoidable will be enough to do that. So is there any point in trying to stop it? Well, he himself describes the book as a "wake-up call", and he has advocated both an increase in nuclear power and experiments in "geoengineering" (see Chapter 8), so it seems he thinks there is some hope after all.

On the other hand, there are those who argue that it will be OK to deal with the negative effects of climate change when they actually become apparent. Some lean towards denial, including ex-UK Treasury chief Nigel Lawson in his book *An Appeal to Reason* (2008). But Lawson, who has a better grasp of economics than climate science, also argues that the best course for overall human welfare is to allow economies to expand as they will, and then spend money on mitigating the damage caused by climate change. This produces better outcomes for more people, in his view, than reducing growth now, even a little, by diverting resources into alternative energy or conservation.

This position, also argued by statistician Bjørn Lomborg in *Cool It* (2007), comes with a moral appeal that reducing poverty and ill health should have top priority. It assumes, however, that resources that might be used for reducing carbon dioxide would be used for direct relief of poverty instead. Evidence that this is likely seems entirely lacking. It also assumes atmospheric warming will continue at a gradual pace, and not trigger any unpleasant positive feedbacks, at least not soon.

The opposite view is found in Sir Nicholas Stern's 2006 report to the British government, which attempted to put reasonable numbers on the costs of doing nothing about climate change and compare them with the expenditures which might be needed in advance to avoid the worst effects. This is even more of an exercise in informed guesswork than the IPCC's interpretation of climate models, and it has been applauded and criticized on both technical and philosophical grounds (see box overleaf).

The report is useful, perhaps, for indicating a plausible range for the kinds of costs which might be incurred in either case. According to Stern's arithmetic, it is cheaper to act now. The costs are controversial, and he has revised them since the original report in order to meet more stringent targets for CO_2 emissions.

In 2006, Stern and his team reckoned it would take one percent of global GDP to stop carbon dioxide rising above 450–550 parts per million (ppm) in the atmosphere and avoid the worst effects of climate

PREDICTIONS FILE

Bill McKibben

Writer and environmentalist, author of *Earth: Making a Life on a Tough New Planet* (2010)

Highest hope: We somehow keep global warming below the point that it wrecks huge swathes of the planet, and that in doing so we are forced to build both strong local communities and tight global ties that bring the better side of human nature to the fore.

Worst fear: We fail to preserve any kind of climatic stability on the planet.

Best bet: Since we've already managed to set the Arctic and the Himalayan glaciers melting, failure seems like a pretty good wager, but we're doing everything we can at 350.org to change those odds.

THE STERN REPORT

In 2006, a weighty report on the economic impact of climate change attracted worldwide attention. Pulled together by a former economic adviser to the World Bank and the UK Treasury, Sir Nicholas (now Lord) Stern, it was full of impressive numbers. It calculated, for instance, that a two to three degrees Celsius rise in temperatures could reduce global economic output by three percent, and a five degree rise by ten percent.

On the other hand, stabilizing emissions over the next twenty years, and then reducing them, thus avoiding the worst temperature increases, would cost one percent of gross domestic product. The conclusion seemed clear. It would be much cheaper to tackle the problem now than later, and the cost was affordable. But like the climate models used by the IPCC, these suggestions about what will happen in the future are estimates based on assumptions.

Stern was upfront about this, arguing that one key piece of the calculation should be set on ethical grounds. This was the discount rate, which economists use to express the idea that future goods are worth less than those enjoyed in the present. This is not the same as discounting for inflation, which you also have to do to measure future costs against current benefits. The reasoning is that, as economic growth continues, people in the future will, on average, be richer than those alive today. That means that having more goods will be "worth" less to them than it is to us.

But there is another, more obviously ethical element to the argument around discounting. Some have argued that benefits to those in the future are worth less *because* they live in the future. The alternative, which is closer to Stern's position, is that we should assume good things are worth as much to other people whenever they may be alive.

The result? A discount rate of 1.4 percent, from which his conclusions follow. Other economists, who take their cues from the supposedly neutral adjustments of future value made by the money markets, favour much higher discount rates of five or six percent. This makes the costs borne by those paying for climate mitigation in the near future higher, and the benefits accruing to those living a hundred years or so ahead much lower, calling Stern's conclusions into question.

change. That would spare people in twenty years' time costs amounting to between five and twenty percent of GDP. Two years later he was calling for a target below 500 ppm, and suggested that this would cost two percent of current GDP. The cost-benefit calculation still works, but this kind of adjustment does make the whole thing look more like the cost estimates for building an Olympic stadium than it did before.

Whatever expert recommendations are followed, effectively fighting climate change will take unprecedented global cooperation, which throws up its own enormous challenges. While some hail the mere creation of international climate change summits as major progress, both governments and citizens remain frustrated by the failure of these meetings to create long-lasting, binding agreements with tangible results. The 2009 United Nations Climate Change conference at Copenhagen was few people's idea of an unqualified success. It did produce a document acknowledging the peril of climate change and recommending keeping temperature increases to 2°C or less, but even this relatively mild accord was not legally binding, or close to unanimously supported.

Meanwhile, amid the hardening scientific consensus that global climate change is real and will bring serious problems, there remains room for disagreement about how bad those effects will be, how much we ought to care, and when to try to do something about it. There is at least as much room for disagreement about what to try and do.

FURTHER EXPLORATION

Brian Dawson and Matt Spanagle The Complete Guide to Climate Change (2008) Covers the range of the issue, from environmental science to politics, economics and the prospects for renewables.

Robert Henson The Rough Guide to Climate Change (2008) A thorough and accessible review of the science and the social and environmental issues posed by human-created global warming.

Marek Kohn Turned Out Nice: How the British Isles Will Change as the World Heats Up (2010) A tour of one of the relatively well-favoured countries (off the European mainland) as it might be in 2100. Kohn offers a detailed picture of life with a modified energy mix and other climate change-related adaptations.

Mark Lynas Six Degrees: Our Future on a Hotter Planet (2007) The result of a lengthy trawl through the scientific literature, summarizing the potential effects of rises in average temperature from one through to six degrees. Thoroughly alarming reading.

Oliver Morton Eating the Sun: How Plants Power the Planet (2008) How photosynthesis shaped ecology in the past, and how understanding it better might save us all in the future.

Vaclav Smil Energy (2006) Widely informed exposition of how energy underpins everything we do by the Canadian expert on energy and ecosystems. His many other books on the topic are also rewarding.

Nicholas Stern A Blueprint for a Safer Planet (2009) Published in the US as *The Global Deal*, this economist's take on the problem does not claim to have all the answers, but it does claim to have the right framework for thinking about them, and that tackling global poverty and climate change can and must go together.

www.climatecentral.org For well-written analyses of science and policy.

www.climatedebatedaily.com Provides jumping-off points for almost anyone who has anything to say about climate, from the core of mainstream science to their most severe (and frequently unreliable) critics.

climateprogress.org Up-to-the-minute news and informed, expert discussion on climate change. US physicist, climate expert and political commentator Joseph Romm was called the "web's most influential climate change blogger" by *Time* magazine.

scienceofdoom.com Numerous websites discuss the science of climate change, many in a style that generates more heat than light, but this highly recommended site contributes handsomely to clarity. It examines everyone's claims, from the IPCC to the critics, carefully and dispassionately, though it's not yet comprehensive (such diligence is time-consuming).

8

energy: a new regime?

energy: a new regime?

Energy is the hardest part of the environment problem; environment is the hardest part of the energy problem.

John Holdren, Harvard University

Whatever the costs now, the plausible risks of releasing all the carbon dioxide now locked up in fossil fuels are so serious that something must be done – and it must be done soon. Actually, make that many things, as there's no single fix. But can we cope with the move to a new energy regime that serves a population of nine or ten billion? There's plenty of analysis suggesting that it is possible, but there's no guarantee, as the task is immense, both technologically and politically.

THE SCALE OF THE PROBLEM

As well as a surplus of greenhouse gas emissions, the world has a rising stack of reports, position statements and plans which try and show how to reprofile energy production and use. They need to start from a realistic measure of what ought to be done. That is not straightforward, because there is room for disagreement about what level of greenhouse gas would be tolerable in terms of global climate change. What is tolerable will depend on where you live and what your society can spend on adapting to any changes, as well as where nasty feedbacks kick in which make climate change worse.

Although it isn't the only greenhouse gas causing concern, carbon dioxide is the main problem. The level of atmospheric carbon dioxide is currently around 385 parts per million (ppm), up from 350 in the middle of the last century and well under 300 in the pre-coal-burning era. If we carry on as we are, the IPCC estimates that levels could reach 450–550 ppm by 2050. There are many climate scientists and activists who believe the target for stabilization should be 350 ppm to avoid a high risk of runaway global warming. It is anybody's guess,

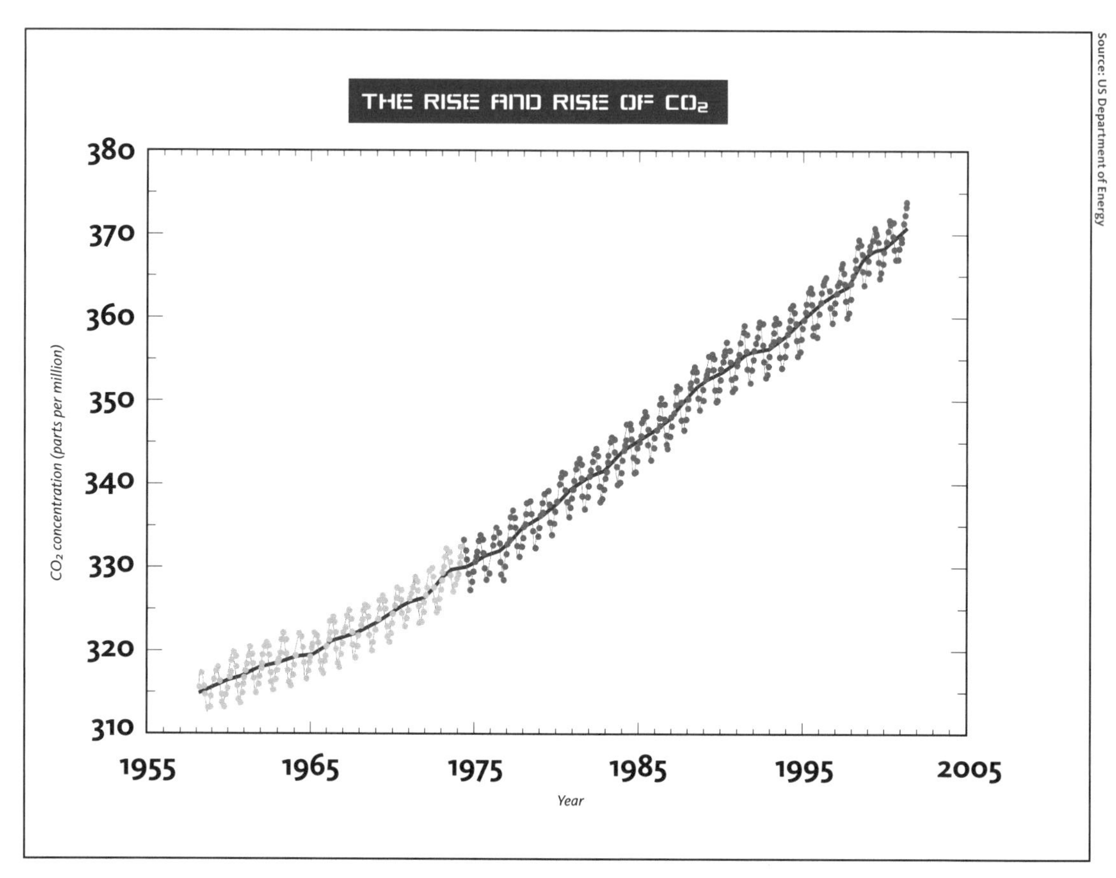

The zig-zag shows seasonal fluctuation in carbon-dioxide levels in the atmosphere, but the peak is higher each year.

frankly, if that is achievable. By way of illustration, consider adjusting the target to 500 ppm – a bit on the high side, but certainly a more accessible goal. Perhaps if a path can be laid there, we can discuss how to take it further.

A useful analysis of what's involved in this sort of reduction was done in 2008 by the McKinsey Global Institute, which specializes in economics research. They expressed the aim a bit differently, because they wanted to emphasize an extra requirement – an energy supply which maintains economic growth for the benefit of the poorer regions of the planet. The simple way to express both objectives at once is in terms of "carbon productivity". That's the amount one unit of "equivalent carbon dioxide" (a calculation that takes into account CO_2 and other greenhouse gases, including methane) contributes to GDP in dollars. The institute's overall estimate is that to curb emissions and achieve growth, carbon productivity has to increase tenfold, from about $740 per tonne of equivalent carbon dioxide now to $7300 by 2050.

Without the productivity increase, meeting even this less severe carbon emission target requires such a big drop in energy use that it wouldn't happen, short of complete social collapse. Stabilizing at 500 ppm means overall emissions need to come down to a level equal to just 6kg of equivalent carbon dioxide a day. If you are using energy generated the way it is now, and using it as effectively in 2050, the institute calculates this would allow you either to drive your car for forty kilometres, have the air conditioning on for a day, buy a couple of new T-shirts or eat two modest meals.

The productivity boost needed to avoid this kind of extreme lifestyle change is hugely ambitious, but not completely outlandish. Labour productivity in the industrial revolution probably went up tenfold – but that took 120 years, not 40. So we'd better get on with it. How you may ask? Well, the McKinsey analysis, in common with many others, emphasizes that there isn't just one thing that will help, but lots of things. They fall under five broad headings, some of which cover a host of possibilities. They are:

- Increasing energy efficiency
- Switching to low or zero carbon energy sources
- Speeding up development of new technology
- Changing behaviour, both among businesses and domestic energy users
- Using natural carbon sinks, especially forests

Combining these, the report suggests, can get the job done. Progress, however, depends on fixing energy prices and pollution penalties to produce incentives to move in the right direction, and on international agreements about targets and financing investment. As well as improving existing technology in the nuts and bolts sense of power plants and energy converters, it will need clever social technology to set up a framework in which rapid deployment of the right technologies actually happens. The rest of this chapter will look at each of these strategies in turn (except forests, which are considered in Chapter 11) and contemplate the best ways forward for both energy and the environment.

USING ENERGY EFFICIENTLY

The first priority is increasing efficiency. That will happen, if only because energy is getting more expensive. Cheap fossil fuels have encouraged profligate use, from gas-guzzling cars to overheated homes. Still, the energy intensity (amount of energy used per unit of gross domestic product – the inverse of carbon productivity) of most advanced economies has declined substantially. But there is still plenty of scope for technical improvements in efficiency in many areas. A possible snag is that these do not necessarily lead to reductions in energy use. More efficient street lighting often goes with more brightly lit streets, cars that use less fuel end up making longer journeys, and so on.

Transport may be the sector where this trade-off proves most crucial. Planes, ships, cars, trucks and quite a lot of trains depend on liquid fuel. Aside from ships, which still move most of the world's long-distance cargo, they are all pretty inefficient. Cars are

FUTURE TRANSPORT

Transport is a hard nut to crack for those plotting ways to cut carbon emissions. And long-distance travel, mostly by air, is the worst of it. George Monbiot, in his book *Heat* (2006), describes ways of reorganizing most aspects of British life to achieve what he regards as acceptable carbon emissions. Like most such plans, his would work mainly by increasing efficiency and switching to renewable sources of energy. But he gives up on air travel, pointing out that a passenger on a full airliner accounts for about the same carbon emissions per passenger mile as one sitting in a saloon car with four people on board. But the car travellers are unlikely to be going more than a few tens or hundreds of miles on any one trip – and might be induced to switch to riding trains or buses, or driving vehicles with different power units. The air passengers may well expect to be conveyed for thousands of miles, and come back a couple of weeks later. Add into consideration that planes increase global warming in other ways too, and that most governments project big increases in passenger numbers, and the problem looks pretty bad.

Nor are developments in aviation likely to help. Supersonic aircraft, should they make a comeback, fly higher and have worse effects than current commercial airliners, as they produce water vapour in the stratosphere. Planes last a long time, and even for replacements to existing fleets, there is no alternative in sight to the high-temperature gas turbine engines airliners currently use. There will probably be modest efficiency gains in the next couple of decades, but not nearly enough to offset any significant rise in air travel.

The main alternative strategy to building new runways at airports – sure to be an ever-stronger magnet for protests – is encouraging long-haul travellers to switch to high-speed trains. How? Simple: make them go faster, and build more and better-connected lines that can carry fast trains. The technology is mostly in place. Trains which can cruise at 200mph are already in service, and the time difference between train and plane for journeys of several hundred miles is small, usually vanishing when airport-to-city transfers are taken into account.

There are big plans for more high-speed rail development in Europe (already pretty well connected south of the English Channel), China (home of the 200 mph train) and the US. In January 2010, US president Obama, the well-known socialist, announced an $8 billion investment, a fairly modest effort which covers mainly north–south links such as those in California and between cities on or near the East Coast of the country. California is contributing another $10 billion of its own to link San Francisco and Los Angeles. Anyone wanting to travel coast to coast will still need to fly for the foreseeable future, but the new lines should tempt some people out of their cars.

Can trains go faster still? Some conventional trains in France have topped 300 mph, but the laws of physics still operate at ground level, and at that speed fuel consumption per passenger mile on the train is no better than for the person on board a 747. So a conventional train needs to keep its speed down to be environmentally friendlier than taking the plane, unless it uses electricity from renewable sources of course.

The more exotic approach is the "maglev" (magnetic levitation) train, which hovers just above its track and travels friction-free, but does not evade air resistance. And maglev tracks are far more expensive. The world's fastest train line in regular use, which takes people at 250 mph from Pudon International Airport into Shanghai, covers just 30 km (20 miles) and cost $1.2 billion dol-

The world's first maglev (magnetic levitation) train runs from Shanghai International airport to Pudong, China. It is capable of a top operational speed of 431 kilometres per hour.

lars in 2004. China plans to spend another $300 billion on high-speed rail corridors over the next twenty years, but they will be for conventional, German-designed, Chinese-built trains. The first country with railways, the UK, has conducted lots of studies in recent years into new high-speed rail links, but no government has yet committed the investment to build any of them. Latest plans at the time of writing envisage a modest run from London to Birmingham, with construction starting in 2017, and being completed in 2027.

PREDICTIONS FILE

Alex Evans

Fellow at the Center on International Cooperation (CIC) at New York University, and editor of the blog Global Dashboard: Notes from the future

Highest hope: By 2050, policy-makers will long since have faced up to the fact that we won't solve climate change without a safe global emissions budget, shared out between the world's people on an equal per capita basis. That would not only stabilize greenhouse gas levels, but also effectively replace development aid with something much more valuable to developing countries – property rights.

Worst fear: Policy-makers figure that it's easier to tackle thorny questions of fairness like this later rather than sooner, meaning that they're only willing to talk seriously about sharing out access to the world's atmosphere once most of the emissions budget has effectively been used up – setting the stage for a political train crash, with scant prospect of a serious global deal on climate.

Best bet: We'll recognize that sustainability depends on equity, but only under dire duress, with climate impacts howling in around us. Humanity only tends to face up to the need for greater cooperation when all other options have been exhausted. But I wish we could accelerate the process in this case – the world's poor people and poor countries will pay an increasingly awful price the longer everyone else ponders the interesting question of the political philosophy of a zero carbon world.

the worst offenders, and the ones likely to increase in numbers as people in China and India use their extra income to claim the kind of personal mobility already enjoyed by the world's richer countries. More efficient internal combustion engines can be built, up to a point. Electric power is more efficient at driving vehicles, but if the expanding auto fleets of the fast-developing economies use electric motors, the electricity still has to come from somewhere. As for planes, any prospect of using even part-renewable energy for flying depends on much cheaper biofuels that do not use up land needed for food production.

Even when all the efficiency gains have been made, the numbers may come out as they did for the United States in the last two decades of the twentieth century, well after the first "oil shock" of the early 1970s (OPEC's 1973 oil embargo, which led to both supply disruption and skyrocketing petrol prices) encouraged interest in energy conservation. The overall energy intensity of the US economy fell by just over a third in twenty years. But the population went up by more than a fifth, and the average GDP per capita by more than a half. As a result, total energy use increased by a quarter.

Efforts to increase efficiency are still worthwhile. Some of the more comprehensive plans for achieving a workable new energy mix rely heavily on efficiency. Physicist and technology guru Amory Lovins' Rocky Mountain Institute has long argued that efficient use of both oil and gas can easily be doubled through technologies already available, such as ultralight vehicle design using new materials.

Lovins and co find examples like this wherever they look. According to them, there is always profit to be made from pursuing efficiency, so a genuine market, without hefty government subsidies for fossil fuels and nuclear power stations, would quickly realize enormous gains. For this to work, governments would have to be persuaded to allow energy prices to rise, which would encourage investment in renewable sources as well. To get the money actually spent on efficiency fixes, it will also be necessary to ensure that those who meet the upfront costs, like landlords, can get their money back, rather than just allowing their tenants to enjoy lower heating costs, for example.

SWITCHING ENERGY SOURCES

The next option is to remix energy supply. There is no lack of possibilities for meeting future energy needs, but that's both good and bad. Good, because some of the technologies on offer will contribute to the new energy mix. Bad, because there are so many options it is difficult to choose – and especially to persuade investors, public or private, to commit on the scale needed to make a difference. When expanding industrial economies made the switch to oil, the choice was simple. The stuff was (then) easy to recover, cheap and astonishingly rich in stored energy. The combination is still hard to beat, and makes it harder to get support for technologies whose energy will probably be less convenient and more expensive, at least initially.

The further complication, again good and bad, is that energy is endlessly interconvertible. Physics says that, strictly speaking, we never actually produce energy. Nor do we use it up. All we do is divert energy flows to do a bit of work, and ultimately help increase entropy by turning concentrated energy into less concentrated forms (see p.29). The diversion can be fairly direct, as when someone fires up a gas boiler to heat their home. Or it can be very indirect, as when coal is burned to raise steam to power a turbine that drives a dynamo and supplies electricity for the drill you are using to make a hole in the wall. This is less efficient.

Interconverting energy goes along with interconverting fuels. We can make liquid fuel from coal. Anything that generates electricity might be used to manufacture hydrogen by passing an electric current through water. Hydrogen – a fuel, not an energy source – can then power cars via a fuel cell, which turns the chemical energy it stores back into electricity. Whether any of this is efficient and cheap enough to make a difference to the global energy economy is hard to fathom from the small experiments that are all we have to go on at the moment.

There are, of course, enthusiasts for every individual option, but the world will clearly use a range of energy technologies, old and new. We know what the menu of possibilities is for the next hundred years. Aside from fossil fuels, there is nuclear power, and wind, solar, biomass, tidal, hydro and geothermal power is there for the using. All have strengths and weaknesses, proponents and sceptics. The competition between them plays out day by day in the newspapers, and there are plenty of primers which give the basic background and likely capacities of each one. A good place to begin is Robert Evans' brief, clear and dispassionate *Fuelling Our Future* (2007).

There will be more experiments, and probably larger programmes, using all of them in the coming decades. How far they go will depend on political commitments, economics, and some things which are a combination of the two, like carbon pricing and

BIG SOLAR: DESERTEC

There is plenty of solar energy coming our way all the time, but more in some places than others. Deserts are attractive sites for solar conversion plants because they get intense sun nearly every day. But they are not where most people live. The Desertec project deals with this disparity by combining plans for solar farms in the Middle East and North Africa with new grid lines to take the power northward to Europe.

A consortium of twelve energy companies and banks, led by the German insurance giant Munich Re (when insurers are worried about climate change you know it is serious), signed up to join the Desertec Industrial Initiative in 2009. They plan to organize the finance and industrial capacity to build a network of solar thermal power plants big enough to supply fifteen percent of Europe's electricity by 2050. Some of the energy would find use in the producer countries, especially for expanding desalination plants for drinking water.

The scale of their ambition can be gauged by the fact that the project relies on the creation of a single market for electricity which extends beyond the European Union into the Middle East and North Africa (MENA). The current best estimates suggest that the upgraded transmission grid needed would cost €45 billion. If that was built over 10 years, the 30 countries who might be involved would each need to spend €150 million a year for the decade. Politically, the kind of complications – or perhaps advantages – which could be involved are captured in the promoters' suggestion that "for humanitarian reasons, a solar power plant and drinking water plant on Egyptian territory for the benefit of the Gaza Strip would be useful as a pilot project."

Similar plans have been published in outline for other regions, notably the US, but this is the scheme that has moved furthest towards reality. There is a long way to go, but it could be a model for exploiting deserts in other parts of the world, including places like China, which is well endowed with deserts and committed to big energy investments in the coming decades.

Rows of parabolic trough solar collectors operating in the California desert, heating synthetic oil to produce steam for electrical power.

trading. But there is a further level of uncertainty because there are technical developments possible in each area. Pushing the technology might turn any of these uncertain prospects into a sure-fire winner in the right location, but we have to wait and see which one it is. On the other hand, investment decisions have to be made very soon if there is any chance of changing energy supply on a large scale much before mid-century. How to decide?

The best hope is for a mix of schemes which press forward in particular areas, and encourage others to follow suit. Some will be small, or involve large-scale use of small-scale technologies like rooftop solar panels. Others, which may be more likely to make a real difference, will be huge – and require determined investors convinced that governments will support their projects with subsidies, regulated tariffs or massive improvements in electricity grids.

For example, there are massive schemes on the drawing board to supply energy-hungry regions with energy from their sun-drenched neighbours. The Desertec scheme envisages building huge arrays of solar collectors in the deserts of the Middle East and North Africa, using them to raise steam for electric power, and then transmitting the electricity throughout Europe. It is a highly centralized option, more like the large-scale generation we are used to than the distributed power production (which generates energy from many small sources) envisioned in some plans for more resilient energy systems that follow new models. What's encouraging is that it's big enough to make a difference to a large, energy-hungry region, and has attracted serious interest from companies with the resources to build it (see box opposite).

As well as plans designed to show what one technology can do, there are plenty of explorations of mixed strategies. These are usually developments of a now classic paper published in the journal *Science* in 2004, in which two Princeton University scientists proposed a set of "stabilization wedges". Each wedge is a slice of carbon emissions that could be accounted for by, say, making more fuel-efficient cars or using one renewable energy source. Add them all up, they suggested, and it is possible to stabilize atmospheric carbon levels over the next half-century using current technologies. You can gauge whether a particular technology makes a serious contribution by seeing how it could be scaled up to make a single "wedge" – equivalent to reducing carbon emissions by a billion tonnes a year. You need seven of them to keep emissions at current levels. You can read all the details at cmi.princeton.edu/wedges, which also has a nice video presentation of how it all works.

WHAT WORKS?

When it comes to the details of how such wedges can be delivered, the key to assessing effectiveness (as UK government energy adviser David MacKay argues) is to insist on real numbers. These need not be exact, but should be realistic approximations of things like the carbon emissions, conversion efficiencies and costs for the entire process. So for nuclear power, you need to factor in mining and enriching uranium fuel, treating and storing radioactive waste, and decommissioning expenses. For wind, there's the cost of manufacturing turbines, and the estimates for theoretical and actual energy output on a real site, percentage of days with no wind, downtime for maintenance and so on.

Then you need to work out how many steps or conversions there are before you get what you want. This can be a rough indicator of efficiency

– the fewer the better. So heating domestic hot water using a solar panel or mirror collector is as direct as you can get. Using solar panels to generate electricity which then splits water by electrolysis, collecting, liquefying and shipping the hydrogen produced, and then using it in a fuel cell to make electricity to power a motor which pushes your car along has rather more steps.

Most of the energy-producing technologies under serious discussion already exist, but are they fully proved, or do calculations of performance assume some improvement beyond their current state-of-the-art efficiency? Is this to come through economies of scale won via mass production, incremental improvements in materials and design, or some big breakthrough not yet achieved? If the technology needs scaling up, how big is the scale-up, and what sort of time is allowed?

These questions are a start, but choices will also depend on values and world-views. For example, some technologies are necessarily centralized and grid- based, as current versions of nuclear power seem to be. Others can be either centralized or dispersed, as with solar power and, to an extent, wind power. Some need a big new supporting system and infrastructure, like the use of hydrogen for transport fuel. So would offshore wind farms, and those massive solar arrays in desert regions.

CARBON DIOXIDE CAPTURE AND STORAGE

It would be lovely to believe in carbon capture and storage (CCS) as a get-out-of-jail-free card for dealing with climate change. Then we could burn fossil fuels – releasing the carbon they contain as carbon dioxide – and lock it all away again where it can do no harm.

In principle, it can be done. Whether it can be done economically, efficiently, or securely is another matter. Chemistry and physics will allow you to grab the CO_2 out of the waste stream of a power station, for instance, at a price. Then you have to put it somewhere. It can be pumped into aquifers or into some oil or gas fields, where it is already used in a few places to increase yields.

There are obstacles to the large-scale use of CCS, though. It makes power stations less efficient, so the energy produced is more expensive twice over (once for extra costs for building or, worse, retrofitting the equipment, and once for running it). Getting the captured carbon to suitable storage sites would need another entire gas-shipping and piping grid. It is estimated that making a real impression on the problem – storing, say, six billion tonnes of CO_2 a year – would mean building from scratch a new infrastructure around twice as big as that already operating for natural gas.

There is increasing interest in CCS from governments building new coal-fired power stations, but the real projects are pitifully small. New coal-burners are likely to be built well before carbon capture can help clean them up. A massive 2007 study from MIT on *The Future of Coal* reckoned that commercial CCS would not be working until 2030, at best.

Going beyond a few large demonstration CCS plants will probably take a decade or two beyond that, and colossal investment. And such investment is only likely if required by regulators, or promoted by carbon markets with the right pricing. Both of those need government action, and intergovernmental agreement.

If you believe, as many renewable energy enthusiasts do, that the climate crunch is the cue to reorganize society along different lines, you will opt for decentralized, small-scale local production. If you are happier with things as they are, this will matter less. Similar considerations apply to capturing and storing the carbon from fossil fuel use so it cannot do any damage – the other set of technologies widely discussed in the debate about dealing with climate change (see box opposite).

WHIZZIER ENERGY TECHNOLOGIES

Speeding up development of new technology is a matter of boosting spending on research and development and hoping for the best. But it should certainly be part of the plan. At the moment, there are things we know will work, but plenty of doubt about whether they will work well enough, or be deployed in time. Existing routes to carbon capture are a good example. So it would be good to explore other options, even if they're wildcards or fallbacks.

GET WIND POWER WHERE THERE'S THE MOST WIND – ALOFT

Wind power is at the other end of the scale from nuclear fusion (see below). The process is pretty simple, and requires little in the way of fancy technology. Materials for turbines can be improved, which is one reason they have grown more powerful – wind turbines used to be on offer in the kilowatt range, but now Megawatt turbines, and even higher-wattage models, are available to order. There are problems – the wind may not blow strongly enough (or too strongly) and cause the turbine to go off line; the windiest places may be a long way from where the output is needed; and some people do not like to look at wind farms or object to turbine noise.

Conceptual image of an airborne wind turbine being developed to harness high-altitude winds by Joby Energy in the US.

But these aren't the kind of problems technological developments will affect much, unless the option to send them airborne takes off. There are already designs for tethered wind turbines that would float one thousand feet up in the air, where there ought to be enough wind to drive the power unit any time.

PREDICTIONS FILE

Bill McGuire

Professor of Geophysical Hazards and director of the Aon Benfield Hazard Research Centre at University College London, and author of *Seven Years to Save the Planet* (2008)

Highest hope: A transformation in the way we think and behave has provided the impetus for deep and rapid cuts in greenhouse gas emissions. A sustainable world, with a shrinking population, powered by sun, wind and water, thrives as atmospheric carbon levels fall and the global temperature rise stabilizes.

Worst fear: Through a combination of stupidity, short-sightedness and plain greed, we will have bequeathed to our children and their children a ruined planet, sweltering beneath a carbon-soaked atmosphere, plundered of its resources and shorn of entire ecosystems.

Best bet: Unless we come to our senses soon and take drastic and immediate action to do what the science says is needed to tackle dangerous climate change, my best bet is for a mid-century world defined by environmental degradation, economic breakdown and social chaos.

”

The tether houses a cable to transmit electricity back to a ground station. One model due to go into production in late 2008 had a projected output of ten kilowatts – about three electric kettles' worth. So you would need an awful lot of them, and you would not want to see them around your local airport. The manufacturer hopes they will find use in rural areas, mainly in developing countries.

But that is only the beginning, according to some wind power visionaries. Go up fifteen or twenty thousand feet into the jet stream, and there is real wind. So far, the balloons, kites or whatever devices tap this theoretically very large source of energy only exist on paper, and on websites like www.jobyenergy.com and www.skywindpower.com. But they seem at least as likely to work as fusion power.

LIQUID FUELS FROM SYNTHETIC ORGANISMS

We already know that "first-generation" biofuels are generally a bad deal. Ethanol from sugar cane, for example, makes some sense, but ethanol from corn no sense at all in terms of energy inputs and outputs. Second-generation biofuels, whether alternative crops which grow on marginal (difficult to cultivate) land or carefully nurtured algae, ought to do better. Algae are a particularly attractive option as some strains use photosynthesis to make hydrocarbons not too different from fuel oil, although they are difficult to work with on an industrial scale.

But could there be a third generation of biofuels? US genome pioneer Craig Venter thinks so. He is building artificial chromosomes to modify the metabolism of microorganisms that already make simple organic chemicals, including octane, a basic hydrocarbon. Others are exploring related options, by working to modify the composition of trees, for example. But Venter is one of the main proponents of focusing on the possibilities of microorganisms.

Venter claimed early in 2008 that his team would have such organisms ready to do his bidding within eighteen months, calling the products fourth-generation fuels. They will consume CO_2 and excrete octane, using sunlight to power photosynthesis.

WHAT ABOUT FUSION?

One possible massive technical fix to our energy problems might be fusion power, which exploits the kind of nuclear reactions normally found in the heart of stars. This is theoretically attractive, but couldn't be any more technically demanding, requiring incredibly hot material (a "plasma" of bare atomic nuclei and electrons) to be confined in some way. Experimental fusion reactors use a shaped magnetic field to confine the plasma and keep it away from the walls of the reactor vessel. As the plasma has to be heated to one hundred million degrees Celsius or so for fusion to occur, the first hurdle is getting more energy out than you put in.

The next big fusion experiment is set to be the multinational ITER project (once the International Thermonuclear Experimental Reactor, but now just ITER in case people are put off by the "nuclear" bit), which was first discussed in the mid-1980s. Construction at a site in France was due to start in 2008, with plasma igniting around 2016. However, new estimates in 2008 increased the cost by almost a third and pushed the first fusion back by three years. After that comes a twenty-year investigation into the best way of managing the fusion reactions, which will last for a few minutes at a time. If the ten billion-plus Euros the project will cost turn out to be well spent, there could be a commercial follow-up. But realistically, fusion power won't appear until well into the second half of the twenty-first century.

Venter, never short on ambition, says his goal is to replace the petrochemical industry, and that he will make the methods freely available. It remains to be seen whether the promise of cheap fuel is enough to secure public acceptance of a (very) large-scale process based on genetically modified organisms, even ones equipped with a "suicide gene" so they cannot survive in the wild. If so, biotechnology could be the most promising addition to the energy options on the horizon. Venter's project is progressing – he announced the programming of a bacterium with a completely synthesized genome (albeit one copied from another closely related species) in 2010. More significantly, perhaps, his company has also launched a long-term programme to develop new algae biofuels with ecologists' least favourite oil company, Exxon Mobil. (Visit www.syntheticgenomics.com to keep track of their progress.)

SOLAR POWER FROM A REAL DESERT – THE MOON

The moon is awash with solar energy. So why not collect it and beam it down to Earth? The technical details were worked out in the 1980s, and David Criswell of the University of Houston reckons a few dozen lunar power bases converting solar energy to microwaves could transmit enough energy to replace fossil fuels on Earth. Terrestrial antennae would pick up the microwaves and convert the energy into electricity. The snag with this scheme is the current lack of moon bases of any kind.

CARBON CAPTURE OUT OF THE AIR, UP IN THE AIR

Most efforts at carbon capture try to take CO_2 out of concentrated streams from power stations. Great idea, but what about dealing with the atmosphere as

PREDICTIONS FILE

John Krebs

Zoologist and principal of Jesus College, Oxford

Highest hope: International agreement on controlling greenhouse gas emissions will be reached, temperature rise will be kept to about 2°C and new technologies will provide cheap, green energy.

Worst fear: No international agreement, global temperature rise of greater than 4°C with serious consequences for people in many parts of the world.

Best bet: No quick fix, but slow movement to an agreement as (i) the adverse consequences of climate change become more apparent and (ii) new technologies for green energy become affordable and available. My generation lived way beyond their means in terms of environmental capacity and sadly my grandchild will not have things so good. Our duty now is to do what little we can to help future generations.

a whole? That's where the greenhouse gas is, and if we could remove it from the air at will, we might not even have to ship it to a disposal site.

The concentration of CO_2 in the atmosphere is much lower than it is in a power station chimney, so the job is harder, but not necessarily impossible. At least one company has prototype technology, using a process (some details of which remain secret) for extracting carbon dioxide from the passing air. It's described in science writer Robert Kunzig and climate guru Wallace Broecker's *Fixing Climate* (2008). But there are serious snags with this scheme, according to David MacKay, Chief Scientific Adviser to the UK's Department of Energy and Climate Change. Taking CO_2 out of the air and compressing it into liquid form takes energy. The energy required to neutralize current European CO_2 output this way is roughly the same as the continent's entire current energy production. Still, if such schemes can be made more efficient, they could prove attractive.

GEOENGINEERING: THE LAST RESORT?

If alternative energy sources fall short and we're still merrily burning coal, gas and (while it lasts) oil, what then? One technical fix, carbon capture and storage, would be needed on a colossal scale. That prompts speculation about other methods for removing carbon from the air, or otherwise cooling the planet. This set of possibilities, dubbed geoengineering or even "eco-hacking", has often been raised, only to be dismissed as hubris. But more recently, some well-informed scientists have suggested it's worth investigation.

Some of their ideas do sound a bit desperate, though. One of the first serious suggestions came from Nobel Laureate Paul Crutzen, who proposed in 2006 (in the journal *Climate Science*) that releasing masses of sulphur dioxide aerosols into the stratosphere could reflect enough sunlight to make a difference. Big volcanic eruptions sometimes do just that, and temporary global cooling follows. Modelling suggests humans could manage to put enough sulphur aloft to achieve the desired effect for a few billion dollars a year. That would be only a small percentage of the sulphur that already finds its way into the atmosphere from human-created industrial pollution – we really need less such sulphur just at much higher altitudes.

There are a lot of snags, though. We'd have to continually pour it into the stratosphere, since the aerosol washes to Earth in a couple of years. And

GEOENGINEERING, OR JUST GOOD ARCHITECTURE?

Amid the exotic proposals for space mirrors and ocean fertilization are a few geoengineering possibilities which sound less like science fiction. A 2009 report from the UK Institute of Mechanical Engineers singled out three as options to buy time for governments to get their act together on more direct routes to carbon reduction.

Artificial "trees" absorb Co_2 on chemically active membranes when air passes through them. One hundred thousand trees, each costing $20,000 and absorbing ten tonnes of carbon dioxide a day, might capture all the carbon now emitted from cars, trucks, trains and domestic gas heating. How much we'd benefit would depend on the larger infrastructure needed for carbon capture and storage in power plants. These devices wouldn't look anything like trees, incidentally, but would be a new kind of installation, perhaps resembling air-conditioning units the size of a cargo container. Will people campaign to get these, or would they get bogged down in planning disputes, as many wind farms have?

We could also fit strips of algae to building walls, let them do their carbon-absorbing stuff, then remove them and use them to make biofuel. Since this doesn't use any extra land, it could give "green" housing a new meaning, though being first in the neighbourhood might involve a lot of explaining. A less slimy alternative to coating your house with algae would be using paints and other materials which reflect more sunlight than those we use now. This would be a lot less effective than the other two options, but is simpler, and builders in hot countries have been doing it for millennia.

The report stressed that these ideas should only be seen as part of a wider strategy that includes reducing CO_2 emissions and guarding against the effects of global warming. But the institute also argued that "if certain geoengineering techniques require research and testing, we should not wait until it's too late for them to have lasting effect".

Perhaps it's also time to coin a less grand term for ideas like this than geoengineering. If something as simple as painting your house is part of it, then we can all have a go.

the higher levels of carbon in the atmosphere remain undiminished, tending to acidify the oceans – an under-publicized but potentially devastating effect of rising CO_2 levels. Other possible side effects include droughts and damage to the ozone layer.

Some of those drawbacks might be avoided by another approach to blocking a portion of solar radiation – putting some kind of reflective shield into orbit. This would be far more costly, and take much longer to put into operation. A scheme worked up by Roger Angell of the University of Arizona in 2006, for example, would cost trillions of dollars.

If working from above does not do the trick, "ecohacking" from below also has its advocates. One of the older ideas still supported by some experts – including Gaia theorist James Lovelock – is to dump tonnes of iron particles into the ocean. This would boost growth of the plankton that soak up carbon dioxide, which then sink to the ocean floor when they die. Or you could pump cool water from two hundred metres deep in the ocean up to the surface, delivering nutrients that would also aid plankton growth. Another idea would increase cloud cover by using a sea-water spray to encourage clouds to form,

Artificial trees (as pictured below the wind turbines in this conceptual artwork by the Institution of Mechanical Engineers) could remove as much as ten tonnes of CO_2 a day. Indeed, if just 100,000 were built, this would be sufficient to capture the whole of the UK's current emissions from transport pollution. Such "mechanical forests" could be planted in areas such as the M25 or the North Sea.

which would need global cooperation to agree on where the clouds should be increased.

Advocates say at least some of these ideas are worth pursuing on an experimental basis in case we need to do something drastic. Critics, though, reckon that could encourage neglect of policies and technologies which, if implemented now, could still avoid the worst effects of global warming.

Changing people's behaviour

The drastic specifications for some of the wilder geo-engineering schemes emphasize the huge problem we're likely to face if greenhouse emissions can't be reduced. Perhaps they will encourage not only debates about technology and international politics, but also efforts to change individual behaviour.

There are any number of places to start, since there are so many small decisions affecting how much energy any one individual uses, especially in highly developed Western societies. Persuading people to "do the right thing" for its own sake will help, although trends in public opinion in some countries are discouraging. In the US, for example, pollsters at the Pew Research Center reported that the proportion of Americans who believed global warming was real actually fell from 77 percent to 57 percent between 2007 and 2009. Even fewer – only just over a third – believed that human activity was responsible for warming. This is not a very promising background for discussing how to get people to reduce their energy consumption. They will respond to price, of course, so peak oil may save us yet. Before then, there are subtler things that may make some difference.

The McKinsey report, for instance, stresses getting supermarkets to stop offering huge boxes or bottles of laundry detergent, which are eye-catching since they take up more shelf space. Take out the filler, and you end up with a smaller product that's easier to ship. Get Walmart, which sells more than 250 million packs of liquid detergent a year, to agree to only sell the more concentrated kind, and the carbon savings will justify the effort. Make sure people know that modern detergents work at cooler temperatures, and everyone can save energy on their washing as well. This still assumes each household has a domestic washing machine and a weekly or daily wash, but such measures could add up.

The question is how many would be as painless as this one, and how people will get the information. For many consumer goods, the carbon savings won't be large enough to make much difference in what people buy. Adding yet another layer of information on product labels will help those already motivated to make a difference, but have little influence on the rest. How far people go will probably depend on new technology to help nudge them in the right direction. Trials show that real-time energy monitors in the home help reduce consumption. At best, making the numbers go down becomes a kind of game. But a plane flight or two cancels out all the benefits of such domestic economy.

What will actually happen?

Since producing energy underpins everything we do, it's exceptionally hard to predict how future energy supply will be managed. It depends on the difference between how new technologies perform on paper and in practice, investment conditions, political trends and consumer behaviour. Concerns about security, food production, water supply and climate change will also have a big influence on how it's handled.

To try to make sense of the big picture, futurologists looking at energy futures have turned to that old favourite technique, scenario-sketching (see Chapter 3), with the Shell oil company taking the lead. The Shell Energy Scenarios for 2050, unveiled in 2008, are worth a close look. They reduce the factors that will affect energy futures to just two broad scenarios, highlighting crucial aspects of the decisions everyone has to make in the next few decades. The basic message is that there is no way to avoid big changes. We can expect "an era of revolutionary transitions and considerable turbulence", the paper states. Faced with the combined problems of impending climate change and peak oil, there are no ideal answers. And things will start to become serious a lot sooner than 2050, as the scenarios project demand outrunning supply of "easily accessible" oil and gas by 2015. (This estimate from a major oil company somehow sounds more serious than the apocalyptic warnings of the prophets of peak oil.)

The two storylines ("Scramble" and "Blueprints") involve the same range of technologies, though not deployed in the same order or at the same rate. The main differences are the speed and direction of political and regulatory change. The single most important factor behind those differences is the degree of international trust and cooperation.

In the first scenario, Scramble, in pursuit of immediate energy security national governments hasten to get a grip on supply any way they can, whether by agreements with resource-rich allies or developing their own resources. Coal, heavier hydrocarbons like those in oil shales, and biofuel figure heavily in these plans. The rapid growth of coal-fired electricity generation triggers widespread protests, but these are contained at first. There is plenty of talk about climate change, but little action until political pressures grow too strong to resist. Until then, governments assume that their constituents will not put up with significant change in their consumption habits (in the developed world) or any delay in economic expansion (in the rapidly developing countries).

The result is a long-delayed lurch towards less environmentally damaging policies. Governments belatedly impose efficiency measures and – in some hard-pressed countries – hefty price rises and draconian transport restrictions. Some face political unrest as a consequence, followed by serious investment in expanding renewable energy forms such as solar, wind and second-generation biofuels. Carbon dioxide emissions stabilize for a while in the face of a worldwide recession in the 2020s, then bounce back as growth resumes. Although the energy mix changes, by 2050 the concentration of atmospheric CO_2 is on a path to a long-term level "well above" 550 ppm. That is where climate scientists expect some of the worst effects of global change to kick in. As a result, "an increasing fraction of economic activity and innovation is ultimately directed towards preparing for the impact of climate change".

In contrast, in Blueprints, planning at many levels avoids Scramble's more ruinous consequences. Instead of a devil-take-the-hindmost approach to grabbing whatever fossil fuel supplies they can, countries follow the lead of the most farsighted individual cities or regions which are trying to fashion new approaches to energy security. New coalitions of interests promote wider alliances, which lead to the emergence of a critical mass of relatively early parallel responses to supply, demand and climate stresses. These responses are built around a CO_2 pricing mechanism using a carbon emissions trading scheme. This begins in the EU and is progressively adopted by other countries, including the US and, later, China. This proves the key to sparking enough investment to build new industries emerging around clean alternative and renewable fuels, and carbon capture and storage.

Multiple blueprints coalesce into a global regime for managing CO_2, but only as a result of long and intense political and bureaucratic efforts, sustained by governments whose people are persuaded this can help achieve energy security and sustainable growth. If it all works, the transition to a new energy and fuel mix arrives sooner, bringing with it more efficient use. The result is an early reduction in atmospheric emissions, and a good chance of keeping it at a long-term sustainable level below 550 ppm.

There is more detail in both scenarios, which are well worth perusing. Blueprints is especially valuable for its expansive view of how a combination of grass-roots initiatives and action by enlightened non-government players (such as, ahem, giant corporations) slowly help demonstrate that the right moves can be feasible at reasonable cost. To many eyes, Scramble's storyline may look both more plausible and, in its way, still optimistic; at least it doesn't lead to large-scale war, for example. But even in summary, both scenarios show how complex the paths to their different clusters of outcomes would be.

Blueprints conveys a strong message that the keys to a better (not ideal) energy future lie in economic and social innovation, not just in clever technology. The particular mechanism that helps mobilize efforts in the Blueprints world, carbon pricing and trading, is controversial. But it's vital to discuss such mechanisms, and try and figure out which ones might work, rather than just focus on getting a share of an inadequate global supply of oil and gas.

PREDICTIONS FILE

Jonathan Jones

Senior scientist, Sainsbury Laboratory, Norwich

Highest hope: Pragmatism prevails over ideology as we respond to the challenges of the twenty-first century. It will not be easy for humans to cope with the perfect storm of climate change, shortages of water, energy and food, and overpopulation. Theological arguments against, for example, GM crops or nuclear power do not help; we need to be able to use every tool in the toolbox. I recommend everyone reads *Whole Earth Discipline: An Ecopragmatist Manifesto* by Stewart Brand, *Tomorrow's Table* by Pam Ronald and Raoul Adamchak and *Sustainable Energy – Without the Hot Air* by David Mackay.

Worst fear: Political failure to ensure reduction in CO_2 emissions results in catastrophic climate change (4°C or higher).

Best bet: We will get things right eventually, after we have exhausted all the alternatives. I'm a big fan of the Desertec project to generate electricity in the Sahara using solar thermal energy and transmit it via a supergrid to Europe and North Africa; this is the kind of supranational project that gives me hope. And GM crops? Eventually everyone will see it's better to control pests and diseases with genetics rather than chemistry.

Significantly, both scenarios assume that reining in CO_2 concentration at 550 ppm would prevent calamitous global warming. The 450 ppm target now widely advocated by scientists who fear climate sensitivity is higher than the IPCC's 2007 report indicated, or the 350 ppm promoted by some activists, would be another matter altogether. The Shell team recognized that, noting:

> "Limiting the CO_2 levels to an even more challenging 450 ppm – as scientists now recommend – might deliver a world where climate change effects could be more moderate and would involve a virtual worldwide scrapping of the current approach to electricity generation and mobility. It would require a zero-emission power sector by 2050 and a near zero-emission transport sector in the same time period, with remaining energy-related emissions limited to aviation and the production of cement and metals."

While this chapter was being drafted, Shell's teams were working on framing a third scenario, which produced the lower target via a path which looked vaguely realistic. But some months later, a senior Shell planner indicated to the author that the effort had been dropped. That is discouraging, but it's clear that moving towards even lower emissions would be more likely under a Blueprints-style scenario. Shell certainly agree, judging by the commentary from the company's chief executive published with the scenarios. In a departure from their normal policy, he suggested one scenario was clearly preferable. For all its difficulties, he reckoned, Blueprints was the one to try and make a reality.

CLIMATE CHANGE AND FUTURE CONSCIOUSNESS

More than anything else, energy and climate force us to think about issues that are connected across vastly different timescales. Fossil fuels took hundreds of millions of years to form, yet once we started using oil we got through about half the easy-to-fetch stuff in just a century. We have detailed climate records as far back as eight hundred thousand years ago from ice cores, and will get even earlier ones as researchers drill deeper. But we're learning that there can be abrupt changes in climate in just tens of years, not hundreds or even thousands.

Indications are that the next twenty to thirty years will be more or less the same whatever we do owing to the colossal thermal inertia of the oceans, which means that heat they have already absorbed will go on slowly warming the atmosphere, come what may. On the other hand, we are told that we have a couple of decades at best to slow CO_2 emissions enough to keep the levels in the atmosphere below thresholds which could trigger real climate instability. But the biggest effects of climate change, particularly melting polar ice, could still take centuries or even millenia to play out. That's also the kind of timescale involved if we expand nuclear power generation, and need to deal with radioactive waste.

It will take at least twenty or thirty years, probably more, to make big changes. But carbon dioxide will stay in the atmosphere for a century or more. Meanwhile, the politicians dealing with all this are still bound to electoral cycles of a few years, and the citizens they answer to are paying bills every month.

The uncertainties of all of these timescales are perplexing. Emphasizing space, rather than time, may be a more helpful way to understand the problem. When a molecule of carbon dioxide or methane forms, it becomes part of a global system. No matter where it comes from, if it stays in the atmosphere, its effects can ultimately be felt worldwide. Globalization isn't just about the ever increasing interconnection of a world economic system which began hooking up in the fifteenth century. It's also about how the rise of automobile use in Germany, a rush to build new coal-fired power stations in China, and forest clearances in Brazil can combine to produce changes felt everywhere. Although some regions will have the resources to avoid the worst consequences, there will be no escape from the basic effects.

This is not completely new. We have known for a long while that some industrial chemicals find their way to the most remote places and leave their residues in animal tissues, for example. The depletion of the upper atmosphere ozone layer which caused alarm in the 1980s was an early example of global environmental change caused by enthusiastic use of modern technology. But the effects of greenhouse gases are likely to be more pervasive, longer lasting and more worrying.

The message of climate change, then, is that when it comes to securing future energy supplies, we are all in this together. As usual, that can be read pessimistically or optimistically. The downside is that it makes the problem hugely complex and hard to get to grips with. Any single country, any single household even, which acts in ways which seem prudent can see its efforts cancelled out by others who are less cautious, or more desperate – or both.

The upside could be that the global impact of climate change invites negotiation to avoid the worst. We have so far managed not to slide into disaster when we knew there was one big thing we must avoid doing which was guaranteed to be calamitous (global nuclear war). We have also, it appears, managed to avert a serious problem when there was a moderately bad thing happening which was the result of doing lots of small, but not absolutely essential things (depleting the ozone layer). Now, the imperative is to reduce the size of a problem which comes from things which absolutely everyone does, or would like to do. And there are almost infinitely many trade-offs in deciding who actually gets to do what, how, and at what cost, in the coming decades.

GLOBAL GOVERNANCE

Since it's a global problem, dealing with energy and climate change requires global governance. That doesn't mean fantasizing about a single body of government with the capacity to dictate policy for everyone and the power to enforce it, as science-fiction writer H.G. Wells hoped in vain. It does mean that local decision makers have to try to develop a global view.

But even effective local policy can only go so far, and there might be formidable obstacles to international cooperation. Curbing carbon emissions, for example, really demands international agreements, and one or two big countries can block them. Using, say, sulphur aerosols to reduce solar radiation could in principle be done by one country alone, or just a few nations. But if this cuts sunlight for crops in countries not involved in the scheme, other governments might regard this as a hostile act. In that respect, geoengineering would probably turn out to be as politically complex as carbon trading and emissions limits.

LOOKING AFTER THE FUTURE

The disconnections between eras and regions that make climate change difficult to handle politically also create unusual ethical difficulties. The effects that travel long distances – with big emitters harming the low emitters, who are also the poor – can be dealt with to some extent by arguing that global change puts all of us in the same boat. The rich will certainly

PREDICTIONS FILE

Gregory Benford

Physicist, professor at University of California, Irvine, and science-fiction writer

Highest hope: We'll shake off our inertia and grapple with the large-scale problems (environment, resources, etc), while still keeping our sense of wonder – especially by sending a manned expedition to Mars, to see if life persists there beneath the ground. That would be the great emblematic event of this century, just as the moon landing was in the twentieth.

Worst fear: We'll dither long enough to suffer large climate change – that could happen within two decades.

Best bet: We'll learn to reflect sunlight, make cleaner energy (go nukes!) and even offset the growing acidity in our oceans – thus averting calamity. Chances of doing this? I'd say, 50/50. I'm an optimist.

be in a better position to protect themselves from the worst effects of severe climate change, however.

The inequalities between generations are trickier. The benefits of emissions are immediate, while the costs may appear much later. Whichever generation you happen to be in, it is in some way rational to carry on burning carbon, whatever other generations might do. But is it moral? How do we weigh future generations' interests against our own?

There's even some debate as to whether we should be doing anything at all, no matter where or when. For example, critics of geoengineering tend to believe, reasonably enough, that we know so little about the complex interactions of Earth systems that interfering with them any further would entail unknown risks. On the other hand, humans collectively are already interfering with global systems rather effectively, even though the effects tend to be in a direction that is probably harmful. On that basis, you can argue that we are already committed to "planet management". The choice is just whether to do it well or badly.

FURTHER EXPLORATION

Tim Flannery Now or Never (2009) A short but passionate rundown of reasons why urgent action is needed to curtail climate change and promote sustainability, investigating some unusual strategies such as spreading Internet access to villagers in tropical forests.

Chris Goodall Ten Technologies to Save the Planet (2008) An excellent book, reviewing the range of current options soberly but, ultimately, encouragingly.

Robert Kunzig and Wallace Broecker Fixing Climate (2008) This examines several imaginative strategies for combating the worst effects of climate change, and includes fifty years of research into how humans have helped cause those disruptions.

David J.C. MacKay Sustainable Energy: Without the Hot Air (2008) An extended demonstration, by a physics professor, of why you need numbers to assess energy sources. The numbers offer convincing evidence that, for instance, relying entirely on current renewable energy technology in a small country like the UK would mean industrializing the countryside. Also available as a download at www.withoutair.com.

350.org A site with a mission to bring atmospheric carbon dioxide levels down to 350 ppm, with a blog on energy and climate change issues that's of value even for those who find that target unrealistic.

9

water

As the world economy grows, so will its thirst.

Ban Ki-Moon, UN Secretary-General, 2009

After CO_2, good old H_2O is the chemical formula which best symbolizes the global stresses and strains of a twenty-first-century world. The prospect, we hear, is of dry wells, droughts and water wars as thirsty people scramble for the stuff of life. Water, writes futurist James Martin in *The Meaning of the Twenty-First Century* (2006), is "the most critical resource of all". There is some truth in this, but the problem is not quite as daunting as the complex of issues linked to climate change and energy use.

WATER, WATER, EVERYWHERE?

There are alternative energy sources to oil, but there is no possible substitute for the world's most essential resource. On the other hand, the globe is mostly painted blue – a simple sign that there is no shortage of water on planet Earth. That is one of the reasons our planet has been so hospitable to life. Most water is in the briny oceans, though. Even around three quarters of the fresh water that falls out of the atmosphere as rain goes straight back into the ocean. That still leaves 110–120 million cubic kilometres of water a year, falling as rain, sleet, hail or snow on land. As the water cycle we learn in school implies, water is a renewable resource. The snag is that it does not fall in the right places at the right times. And the mismatch between the distribution of people and water is hard to fix because water's too heavy, and therefore costly, to transport long distances. Restaurant diners in the West can pay for air-freighted mineral water if they really want, and there have been occasional crazy-sounding schemes like towing icebergs across the seas. But the main water needs – for domestic use, energy generation, industry and agriculture – have to be met locally. Local in this context means from the nearest river basin or, in some countries, from ancient – and largely non-renewable – ground water stored in aquifers. So certain places are always going to have to be a lot more careful in their use of water than others. Before reviewing them, though, some overall numbers are useful. Where does all that water go?

WATER IN PROPORTION

Water does not stand still, and its composition changes easily, so it's not the easiest stuff to measure. As usual, a UN outfit pieces together the best worldwide data, which is reviewed in detail by the World Resources Institute and can be retrieved from its free online database at earthtrends.wri.org.

There are two ways of looking at the key numbers. One starts with all the water in the world – briny or fresh. In that big picture, 97.5 percent of Earth's water sloshes around in the oceans. Of the 2.5 percent left that is fresh, glaciers (including the icecaps) harbour 69 percent, and ground water 30 percent, with a little bit of permafrost. That leaves a mere 0.4 percent of the total fresh water on the surface of the land and in the atmosphere. But that 0.4 percent embraces every river, lake and stream, wetlands, soil moisture and water vapour in the ever-shifting clouds. A minute fraction of that, just 0.8 percent, is actually inside plants, animals or people at any one time.

A more dynamic picture comes into view when we ask where all the water goes in the annual cycle. More than half (61.5 percent) of the roughly 110,000 cubic kilometres of rain that falls on land ends up back in the air after evaporation from plants via the soil – evapotranspiration is the experts' word. That is the water used by the world's forests, savannahs and other uncultivated land which supports wild animal and plant life, and by rain-fed crops. That leaves around 39 percent, or 43,500 cubic kilometres, for replenishing rivers, lakes, wetlands and ground water.

Human use of that portion consumes between four and five thousand cubic kilometres each year – up from perhaps five hundred cubic kilometres in 1900. Most of that (seventy percent) is for irrigation. Urban water drinkers and toilet flushers, factories and power plants account for the rest, but much of it gets used several times, and then finds its way back into rivers or lakes.

In fact, for the most part water is not used up. It is contaminated, and its quality degraded, in various ways, but most of the water that humans divert gets released again somewhere. The UN estimates that agricultural and livestock use returns 30–40 percent to the environment, industry 80–90 percent, power stations (which mainly just make the water warmer) 95–98 percent and domestic and city users 75–85 percent. This means reusing water is always an option, if it can be cleaned up. As an added benefit, it doesn't have to be transported, as waste water is already at the point of use.

Here lies the main difference between the problems surrounding water use and the problems associated with climate change and energy. Carbon dioxide

PREDICTIONS FILE

Ijeoma F. Uchegbu

Professor of Pharmaceutical Nanoscience, University of London

Highest hope: That while income inequalities may not be totally abolished, every person will have enough food to eat and water to drink.

Worst fear: Sub-Saharan nations and certain nations in the Middle East will find it increasingly difficult to secure water supplies.

Best bet: Water wars will have replaced oil wars.

in the air is globally cumulative. Water use need not be. The reduction in water consumption in the US by ten percent between 1980 and 1995, even with a forty million increase in population, shows it's possible to level it off.

Even so, the prospect is for large increases in human water withdrawals from the annual fresh water flows, driven by population growth and efforts to boost food production (discussed in the next chapter). And there is a lot less margin for growth than the basic figures suggest. There may be more than forty thousand cubic kilometres of renewable fresh water runoff, but only around twelve thousand cubic kilometres is actually in places where it is usable. Much of the rest is inaccessible, or must be left in rivers and lakes so they can be used for other things, including maintaining ecosystems. On that basis, total withdrawals are already at least one third of the available yearly supply. If water use simply goes up at the same rate as population, that proportion would rise to more than half of the available supply over the next few decades. Greater affluence, especially as it leads to eating more meat, would increase it still further.

WATER CRISES – PRESENT AND FUTURE

The world has unequal access to fresh water. If you take what is classified by statisticians as "accessible runoff", Asia – with 60 percent of the global population – has just 35 percent of the annual fresh water supply. South America, by contrast, has 5 percent of the population and 25 percent of the water. Uneven access to water is also important regionally. Some large countries which have good fresh water supplies at first glance, have large water-scarce areas within their borders. As discussed below, they include China, India and the US.

That uneven supply has to be used for drinking water, sanitation, agriculture and industry without wrecking the environment. Its success at doing so is distinctly mixed. Clean drinking water – a couple of litres per person per day is all it takes – seems absolutely basic. But supplying clean water, and keeping it clean, requires massive investment in sanitation and water treatment as soon as people throng together. According to official UN estimates, there are still 1.1 billion people in the world who have no reliable source of clean drinking water and more than twice that number who have no access to sanitation. As the UN also points out, "no access" is a euphemism for having to dump faeces wherever you can, in a ditch, in the field, in the street, in your backyard or (if you can) someone else's.

The real numbers who would see the "flush and forget" sanitation installed in the vast majority of homes, schools and workplaces in wealthy countries as a life-changing development are probably a good deal larger. Official figures include returns from governments which exaggerate the delivery of clean water, especially in large cities. Reducing the number of people without sanitation while population rises is a big challenge to planners.

Then there is water's key role in food supply (discussed in Chapter 10). Agricultural production has kept pace with population increase so far, partly thanks to more and better irrigation. The irrigated area of the globe doubled between 1950 and the first decade of this century, as population grew from 2.5 to 6.5 billion. Water withdrawals grew threefold over the same period. Similar increases to meet the needs of the next two or three billion people would be hard to find. Meeting the demand will be harder still as

Water may not always fall in the right place – but that doesn't mean the wealthy can't play golf in the desert, as demonstrated at the Canyon Gate Country Club, Las Vegas.

diets change. At the extremes, producing a pound of beef for the table uses one hundred times as much water as growing a pound of wheat.

Then there are the inevitable crises that will appear in some areas which have used new drilling and pumping technology to boost their use of water stored in rocks. In some places, this is renewed by rainfall seeping downward, but many aquifers are part of the water supply, which resembles oil in how such aquifers store "fossil water" which has been under the ground for a long time. When it's gone, it's gone.

The list of countries that are taking water out of their aquifers faster than it can be replenished includes Algeria, Egypt, Israel, Jordan, Libya, Pakistan and Saudi Arabia. Within countries, large areas of India and North China are dependent on ground water which is disappearing fast. This is where millions who are likely to be short of water in the coming decades live. There are also aquifer problems ahead for many cities, and not just in these areas. It is estimated that more than half of cities in Europe with more than one hundred thousand people are using ground water faster than it can be replaced. But the largest problems are again in the developing world. Aquifers have already fallen by between ten and fifty metres in Mexico City, Beijing, Madras and Shanghai.

What about climate change?

Climate change, as usual, is likely to bring bad news for the (water) poor. As emphasized in the last chapter, climate models' power to predict local outcomes is limited. However, the IPCC expects that the general effect of rising global temperatures will be more rainfall in the moist tropics and at high latitudes, and less for regions in the mid-latitudes and the already semi-arid low latitudes. But even if there is more rain, it may come less often, making water supply harder to manage. Flash floods interspersed with droughts are likely to afflict larger areas of the globe. There are also poorly understood effects on soil moisture, and some evidence that loss of moisture in a dry spring helps increase temperatures in summer, as appears to have happened in the European heat wave of 2003.

Warming temperatures are already having noticeable effects on one key influence on water supply in some regions – glaciers. The most important areas under threat are the massive glaciers of the Himalayas and Tibet, which store water from winter snowfall and release it during the warmer dry season into seven great rivers that supply the water needs of two billion people in China and Asia. Seventy percent of the summer flow of the Ganges, for example, is estimated to be glacial meltwater. It helps compensate for India's overwhelmingly seasonal rains. Half the annual rainfall normally comes in just fifteen days, and ninety percent of the river flow in one third of the year.

On the ground measurements are difficult – the glaciers are in the Himalayas, and there are thousands of them – but satellite readings and some land-based surveys indicate thinning of many glaciers and, more controversially, increasing rates of loss. The IPCC's claim that they might all vanish by 2035 is now acknowledged to be a mistake, but there are serious signs of big changes before the end of the century. As both India and China already suffer from depletion of ground water this is a bleak prospect.

Not a drop to waste

All the indications are, then, that the mismatch between population concentrations and plentiful water supplies will widen in the next few decades. The amount of water per person is the best indicator. The UN defines areas with "water stress" as those offering fewer than 1700 cubic metres of water per person per year for all uses. Scarcity is set at less than one thousand cubic metres, and "absolute scarcity" at less than five hundred. By these measures, the most elaborate assessment yet made, in the UN's 2006 Human Development Report, indicates that by 2025:

- More than three billion people will live in water-stressed countries
- Fourteen countries will pass from water stress to water scarcity

▶ The proportion of the population in Sub-Saharan Africa living in water-stressed areas will rise from just above 30 percent to 85 percent

▶ More than 90 percent of the population of the Middle East and North Africa will be in water-scarce countries

▶ Large areas of India and China will enter water stress

This is where business as usual is likely to take us. However, one key feature of business as usual offers hope for avoiding the worst outcomes. In most of the world, most of the time, water is used staggeringly

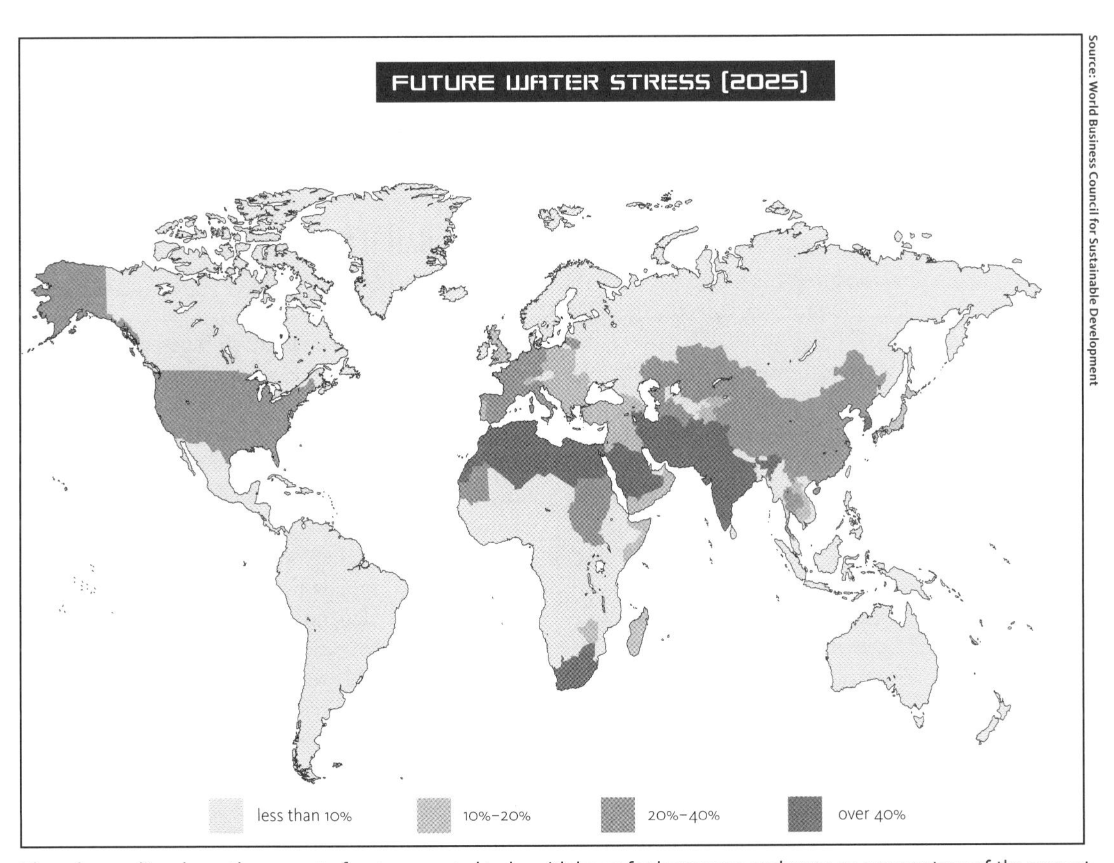

The colour coding shows the amount of water expected to be withdrawn for human use each year, as a percentage of the amount available in each region.

inefficiently. Although this is hard to quantify – in many cases we do not even know how inefficient the use is, or where the water goes – most forecasts argue that on this basis the problems, though immense, are manageable, at least in principle. Unlike securing an energy supply that does not wreck the climate, improving the water situation need not entail completely reengineering our civilization. It does not even need much in the way of new technology (though that might help – see below). It is mainly a management problem, if a rather gargantuan and costly one.

For example, the Sri Lanka-based International Water Management Institute reckons that water problems need not stop the world from being fed. In spite of all the pressures and worries, in 2007 its "Comprehensive Assessment of Water Management in Agriculture" concluded that the world has enough fresh water to produce food for everyone for the next fifty years.

They also acknowledge, however, that only big changes in planning and management can make this happen. Better use of technology is needed – though some of it is very old technology, like collecting and storing rainwater for local use. But the main requirements for more efficient and productive use of water in agriculture are political, social and cultural. Unfortunately, that does not make them more likely to be addressed. As the World Wide Fund for Nature (known as the World Wildlife Fund in North America) points out, the political effects of water scarcity are usually gradual and regional, and governments are rarely confronted with an obvious crisis.

The even more basic provision of clean drinking water mainly needs the magic ingredient, money, to ensure progress. The UN reckons $10 billion a year would enable us to meet the Millennium Development Goal, which calls for a seemingly modest halving of the proportion of the population who do not have clean drinking water or proper sanitation. Current investment aided by wealthy countries is only running at $5 billion a year, though. The drinking water part of the plan looks like it's working, but better sanitation for the mass of the rural and urban poor does not seem to have gripped politicians or aid donors. As a UN Development Programme report put it in 2006, water and sanitation have seen "a surplus of conference activity, and a deficit of action". (The health consequences of that are taken up in Chapter 12.)

RELIEVING WATER PRESSURE

What might be done in those parts of the world where water will not meet projected demand? One immediately available option is to adopt old technologies for the local collection of rainwater, especially runoff from roofs – expect to see more water butts in the future. More experimental possibilities include water harvesting from coastal fogs, already in use in Chile. Then there are a fair number of options for eking out supplies by smarter use of the water available. Some depend on making initial use more efficient. Drip irrigation systems are designed to deliver water directly to plant roots, and can be combined with computer control to optimize water use. Lower-tech versions are already increasing yields in parts of Africa.

Advances in technologies for the reuse of water offer other possibilities. Disinfection is vital when such water is contaminated with sewage, which happens in both controlled and uncontrolled ways, most obviously when we flush our toilets. (Development guru Lester Brown sees a big future for water-free composting toilets, but flushing systems are still by far the most widely used.) There is also much to do in removing chemical contaminants from industrial waste water. There is not much point recovering water after use if it is too toxic to drink or even to use on crops.

All of this is technically feasible. But applying it to vast amounts of water, cheaply and without stretching energy budgets, is harder. Improvements in membrane technology are one area where much is expected. Passing water through a thin layer of the right material can be the most efficient way of removing many contaminants. They may be needed to deal with new kinds of chemical pollution, such as hormones from mass-use contraceptive pills and substances such as plasticizers which mimic hormones, as well as older pollutants. Genetically modified micro-organisms are the other main area of research likely to yield improvements in treating waste water although, as in other fields searching for technological solutions to serious problems, putting them into widespread use may be controversial.

There will also be efforts to develop cheap treatment systems which can be used on a small scale near people's homes to complement the large-scale water treatment plants habitually installed in the past. If you can remove all the muck, bacteria, viruses and parasites from your local turbid water supply a little at a time at specific small locations, you at least get enough to drink without risking your health every time you take a swig to quench your thirst. A "stand alone" system can supply water for a school, for example, as India is doing in thirty thousand rural schools. Make such a system even smaller, a "mobile phone for water", and it becomes more flexible still.

The LifeStraw: new technology filters out bacteria and delivers safe drinking water in the developing world.

The logical limit to this kind of micro-sanitation already exists in the form of the "LifeStraw," which was developed by a Swiss company to purify water for one person as they drink it.

These filtration approaches, large-scale and small-scale, are likely to improve through nanotechnology, which studies ultra-small structures (see Chapter 5). There are a host of projects investigating the best ways to make nanomembranes for water treatment, and they do not necessarily involve new or exotic materials. Many clays and zeolite minerals already have pores as small as newly made nanomaterials, and should be useful in water filtration. Brazil, India and South Africa are all investors in nanotech water projects for local use.

DESALINATION

According to the science journal *Nature*, 75 large-scale desalination plants were under development around the world in 2008. Most of them are in the countries reliant on aquifers which will run short soon. Turning salt water fresh uses a lot of energy, and costs on average three and a half times as much as using naturally available fresh water. But the cost is still a minute fraction of what consumers in the developed world will pay for a bottle of mineral water – between one dollar and fifty cents per cubic metre at the moment.

The most basic, and oldest, method of desalination is simply to heat up salt water and catch the evaporations, which are salt-free. That takes a lot of energy, so desalination schemes are now shifting to a new process called reverse osmosis. This uses clever physical chemistry, forcing the water through a semi-permeable membrane (which lets small molecules through, though not large ones) under pressure, leaving the salt behind.

Further ahead, there are hopes of using carbon nanotubes – effectively very, very small pipes – which would theoretically be more efficient than membranes. And there are experiments with more complex osmotic techniques which are likely to offer further gains if they can be scaled up. Biotechnology may help here, too. Some microorganisms absorb sodium, and can be used to purify seawater. Others can help deal with organisms that clog up water treatment plants.

Coastal desalination raises concerns about where all the salt goes, and about marine life getting sucked into intake pipes. But the key constraint will be energy supply. Solar or wind power are often good options at or near promising sites for desalination, though.

Finally, drinking water can be guaranteed, usually as a last resort, by getting the salt out of seawater – a process known as desalination (see box).

WILL THERE BE WATER WARS?

The idea that the twenty-first century will see water wars in the same way as the twentieth century saw oil wars seems plausible. Even if you do not buy the notion that the wars were "about" oil, the number of conflicts involving countries with large quantities of the stuff and those that have too little to sustain their consumption habits is impressive.

As water is spread around unevenly, and is a life-or-death resource, the argument goes, it will be fought over more and more. But the idea of water wars owes more to journalists' addiction to alliteration than to the facts. Water is needed in quantities that are too large to ship round the world, and so is used near the source. When those sources are rivers that flow through many countries, the evidence is that geographic practicalities are more likely to induce cooperation than conflict.

According to Aaron T. Wolf of Oregon State University, director of the Transboundary Freshwater Dispute Database, no nations have gone to war specifically over water resources for millennia – even though language suggests water as a perennial source of conflict: "rivals" originally referred to those using the same river. Wolf and his colleagues reckon that international water disputes are normally resolved peacefully, even between countries that are enemies. Their database shows that outbreaks of cooperation between countries sharing river basins outnumbered conflicts by more than two to one in the second half of the twentieth century. There were only 37 interstate disputes over water which turned violent. No less than thirty of them involved Israel and its Arab neighbours in the arid territories of the Middle East – but while supplying unlimited water would make life easier there, it probably would not ensure peace breaks out in that region.

This does not mean that violent disputes over water use do not happen elsewhere, but they are more often within countries. Dams, hydropower schemes and transfer of water rights have often led to protests and sometimes more. The so-called California Water Wars between Los Angeles and the Owens Valley, whose Sierra Nevada mountain range runoff was diverted to the big city to fuel LA's explosive growth, even inspired Roman Polanski's popular 1974 movie *Chinatown* (though that's a highly fictionalized account of what actually took place).

Wolf concedes that increasing pressure on water supply will also increase the chances of international conflict. But he argues that, since water is so essential, it can also be a key stimulus to cooperation. He urges more study of long-term collaborative management of river basins which has survived outbreaks of war – such as discussion of the Mekong River among Cambodia, Laos, Thailand and Vietnam, and the Indus River Commission's role in bringing India and Pakistan together over the issue – to help understand how to uphold water peace.

Journalist Wendy Barnaby agrees. Her efforts to prepare a book on water wars at her publisher's behest foundered on the absence of any actual wars to describe, or much in the way of evidence that they will occur. She argues, along with King's College London geographer Tony Allen, that countries deal with lack of water not by grabbing someone else's nearby, but by importing things which water provides, especially food. This idea, first termed embedded water but then popularized as "virtual water", has been an important additional factor in the calculation of regional water needs and assets. Israel, Jordan and Egypt, for example, all lack access to water. But grain imports into the Middle East now account for more virtual water than the real flow of the Nile. As Barnaby puts it: "Countries do not go to war over water. They solve their water shortages through trade and international agreements".

THE REAL WATER WAR

Journalists aside, one of the loudest warnings of water wars came from a vice president of the World Bank in 1995. This is ironic, as World Bank and International Monetary Fund-induced "reform" of water supply companies is one of the sources of intra-state conflict over water, especially in cities in developing countries.

Such cities see real water wars arising from disputes over water ownership. Market reforms, privatizing water supply, treatment and distribution are designed to put an economic price on water in the interests of sustainability. On the other hand, water is not a normal commodity – there is rarely competition for supply – and companies in the private sector are less likely than public utilities to safeguard the interests of poor consumers, the vast majority of the population in most countries where such reforms are taking place. The UN and others may argue that access to water supply must be improved for everyone, but the companies which finance water supply upgrades might not be as idealistic.

In cities, such companies may maximize profits by improving supply to the monied minority who can already buy clean water, neglecting the rest of the population. In rural areas, the focus is on supplying "high value users" who are growing cash crops, rather than subsistence farmers.

The World Bank argued that public sector providers waste water, viewing private investment as the key to expanding services. However, the market-led policy it adopted in the 1990s came in for a lot of criticism from NGOs (non-governmental organizations) and advocacy groups. In 2007, this culminated with 130 groups from 48 countries calling on governments to withdraw support for the bank's policy on water infrastructure.

On the other hand, while idealists are understandably keen to establish water as a human right, non-market arrangements or public sector schemes will not necessarily be attractive to donors. According to a massive review by Transparency International in 2008, corruption (defined by the organization as "the abuse of entrusted power for private gain") in developing countries increases the cost of connecting a household to the water supply by thirty percent, on average (see www.transparency.org).

As the argument continues, the poor often remain caught between no water and expensive water. People who live in Jakarta, Lima, Nairobi and Manila currently spend more on water than if they lived in New York, London or Rome. A safe prediction is that carefully crafted case-by-case initiatives, which mix public and private funding, are the best hope for improving water supply.

PREDICTIONS FILE

William Calvin

Neurophysiologist, University of Washington and popular science writer, author of *Global Fever: How to Treat Climate Change* (2008)

Highest hope: We will succeed in drawing down the excess CO_2 before encountering a point of no return from an abrupt climate shift.

Worst fear: A serious jolt – say, a doubling in global drought areas within several years – could cause catastrophic crop failures and food riots, creating global waves of climate refugees with the attendant famine, pestilence, war and genocide.

Best bet: The tragedy of the commons will come close to sinking civilization.

From water crisis to sustainable water: three scenarios

Scenarios focusing on water issues have to take into account trends in almost every human activity – agriculture, industry, migration to cities, and so on. They nonetheless need to focus on the most important features of water consumption and influences on water use without expanding their purview to the entire state of the (future) world.

A set of scenario sketches published jointly in 2002 by the International Food Policy Research Institute and the International Water Management Institute as "Global Water Outlook to 2025" remains a useful guide. Their scenarios were based on an elaborate model of world agricultural production and markets, combined with an analysis of what determines water consumption. Although it says little about climate change, or the more recently pressing issues around biofuels or the effect of oil price spikes on food security, the food–water links it reviews are still important to consider.

Villagers on Pate Island, in Kenya's Lamu archipelago, drawing briny drinking water from holes in the sand; fresh water is hard to come by on these low-lying islands.

The authors divide their possible water futures into three main scenarios. Under "business as usual", current policies continue as they are. There are gradual efficiency gains in water use, some improvements in technology and increases in crop yields, but not much research to speed them up. Costs of supplying water to homes and factories around the world go up, which encourages more careful use, but there isn't enough investment to deliver drinkable water or proper sanitation to the billions who still need them.

Water use climbs as population increases, diets continue shifting to meat, cities grow and industry expands. Environmental use of water generally takes second place when there are conflicts over priorities. Food production increases, both from irrigated land and through rain-fed agriculture, and grain exports from high-yield countries make water use more efficient. However, some countries, in this future as now, find it impossible to pay for imports without significant aid, which may or may not be forthcoming. This business-as-usual world could carry on until the end of the period (2025). On the other hand, it could crumble into crisis, with "a moderate worsening of many of the current trends in water and food policy and investment". The second scenario, "water crisis", sees a succession of local shortages building into a wider crisis. As this story plays out, it produces the worst results on all fronts. Global water consumption ends up higher than it is under business as usual, but much more is wasted. Farmers withdraw more water for irrigation, but use it inefficiently. Food production is ultimately lower and prices higher than in scenario one, and there is little progress on improving water supply to the urban and rural poor.

These bad things can be avoided, the authors' calculations suggest, under a third scenario: "sustainable water". This features more investment in crop research and widespread reform of water management. The analysis indicates it's possible to progress toward a seemingly ideal world in which there is a serious increase in the amount of water allocated to environmental services. In addition, all urban homes have piped water, sparking increased water demand in cities. But there's still enough water left to keep food production high enough to feed everyone in the swelling population.

The scenario highlights three crucial changes necessary to make this future possible. Water prices are allowed to rise gradually but noticeably (with suitable provision for the poor) to induce efficiency in agriculture and industry. There is a shift to sustainable use of ground water, with efficiency gains allowing more cautious use of aquifers. Finally, there is a big effort to increase the productivity of rain-fed agriculture. This is crucial because irrigation cannot increase very much under this scenario, and rain still waters most of the world's crops. It is assumed that the crops that are grown are scientifically bred to achieve higher yields, but also that there is wide use of effective water management on farms. As agriculture uses such a large portion of the world's water, that seems the key element of almost any scenario for improving future management of the most precious resource.

FURTHER EXPLORATION

International Water Management Institute Water for Food, Water for Life: A Comprehensive Assessment of Water Management (2007) The weighty report focuses on the critical links between water and food, and includes scenarios considering them both. Downloadable from www.iwmi.cgiar.org/Assessment.

Meridian Institute Nanotechnology, Water and Development (2006) This paper from an international think-tank offers evidence that developing countries are already exploring nanotech solutions fitted to local needs. Downloadable from www.merid.org/nano/waterpaper.

Fred Pearce When the Rivers Run Dry (2006) Best of a bunch of books on world water, present and future, by a globetrotting environmental journalist.

Rick Smolan and Jennifer Erwitt Blue Planet Run (2007) A picture-heavy coffee-table book focusing on the human stories, images and consequences of the struggle to provide safe drinking water to the world, though the text lays out some of the basic issues.

United Nations Water in a Changing World (2009) An extremely comprehensive UN review of world data, current trends and future issues in water supply and demand. This mega-size report is available from www.unesco.org/water/wwap/wwdr/wwdr3.

World Economic Forum The Bubble is Close to Bursting (2009) Another twenty-year prospect, emphasizing limitations of future supply and possible conflicts over water sources. Updates from www.weforum.org.

10 food supply

food supply

Food isn't "just another commodity"; it is the foundation of personal wellbeing and is inextricably interwoven into a nation's culture, character and land use.

Jonathan Porritt

The doomsday predictions of the 1960s and 70s that megafamines would take hold before the end of the twentieth century were wrong. A "green revolution" saw technology applied to agriculture in ways that fostered massive increases in food production, which more than kept pace with population growth. Plenty of people still went hungry, but overall more were better fed than ever before.

FARMING THE PLANET

The expected population increase over the coming decades means another big rise in food output is essential. This time, it has to happen during a period of increasing water stress; rising costs of fuel, fertilizer and pesticides; and uncertain, but probably disruptive, effects of climate change. Can it be done? Not easily.

You would never know it if you live in one of the countries that were in the first wave of industrialization, but farming is where it's at. Employment in the wealthier countries moved away from the land over the last two hundred years. The highly mechanized agriculture of those countries now only occupies a few percent of the population, though their lobbying power can still be strong. However, the way much of the rest of the world labours has not (yet) seen the same transformation, though its population also tends to crowd into cities. Agriculture does not just use more land and water around the globe than any other activity. It employs more people too.

Recently, those people have done something remarkable. Food supply overall has just about kept pace with population growth. That has not meant an end to hunger, because access to food remains so unequal. And the increase in production has been slowing in recent decades.

Until 2007, though, the *proportion* of the world's population who are undernourished – as measured by minimum daily calorie intake – had gone down while the population as a whole went up. The drop, from around twenty percent in 1990 to sixteen percent in 2003, was not spectacular, but meant a lot fewer hungry people than there might have been.

However, in 2007 food price increases halted this trend. Even though inflation-adjusted prices are still lower than they were in 1970, recent rises have made access to adequate diets harder for many. The UN Food and Agriculture Organization (FAO) estimated in 2008 that the proportion of malnourished people had risen again to seventeen percent of the total. In absolute numbers, that meant that there were more than nine hundred million people without enough food to keep reasonably healthy and do a day's work, an increase of eighteen million since 1990. Other estimates, which take account of the fact that the poor get hungrier faster than the rich do when prices rise, put the figure higher still.

The majority of the hungriest are in Sub-Saharan Africa and South Asia. You can check the latest distribution of real food deprivation by looking up the Global Hunger Index calculated by the International Food Policy Research Institute (www.ifpri.org). The index combines data on infant mortality, underweight children and the proportion of undernourished people in a particular country. It shows the clear, and unsurprising, link between poverty and malnutrition, as well as illustrating how patchy the improvements in food supply have been in recent decades.

FOOD SUPPLY AND DEMAND

While food is a hugely complex subject closely interwoven with politics, history and culture, there is one dominant question for the next few decades. Will there be enough? Price rises for basic foodstuffs in the last few years have given the question more urgency. But doubts about the sustainability of current agriculture, and the projections for population growth, mean that the question is controversial. The way the question is framed is contentious in itself,

PREDICTIONS FILE

Hans Herren

Biologist, agronomist and president of the Millennium Institute, Arlington, Virginia

Highest hope: We will have changed to a fully sustainable agriculture that regenerates the soil, uses a large number of diverse crops and animals, is based on family farms growing food for the local communities and keeps the farmers on the farms by providing good income and lifestyle in the countryside. That food is by then a human right, not only a moral one.

Worst fear: That none of the above will come through, that industrial agriculture will continue to rape and mine the Earth to profit a few and leave the majority of people hungry and in search of water, and that in the developing countries, farmers won't have access to land, education and markets, and will therefore continue an unsustainable subsistence agriculture that won't be able to feed them, nor the people of their own country.

Best bet: There will be a strong push for industrial agriculture, because the vast multinational corporations dealing with biotech, fertilizers and seeds which control the market (having huge sums of money available to continue making false advertisements about their products' economic and environmental performance) will provide some form of aid for the adoption of their technologies, creating dependency on their products, and then cash in on the back of the farmers and urban populations. They will be aided in their plans by major private foundations that also promote technology to solve the food security problem over sound agro-ecological practices that benefit small and family farms.

as there is already "enough" food, in the sense of basic calories, to nourish the entire population. But although it is true that this indicates the problems of relieving hunger are as much economic and political as agricultural, continued increases in supply are necessary to support any feasible food policy for the rest of the century.

In terms of overall supply, optimists about future food can claim that history is on their side. Paul Erhlich's 1970s predictions of devastating famines before the end of the twentieth century were dead wrong. While nine hundred million people will not have noticed, the world as a whole is better fed now than it was half a century ago. We have had one "green revolution", optimists say, and it is time to engineer another. The first used new, high-yielding crop varieties, along with better irrigation and a big increase in fertilizers. The next will improve varieties still further, and include features like disease and pest resistance, as well as tolerance of extreme conditions.

In fact, if you look at global averages, the job might be done simply by raising yields everywhere to the best levels already achieved. The calculation works out as follows. Take the UN Food and Agriculture Organization's estimate that everyone needs 2300 calories a day. Add on a bit for safety, raising the amount to, say, three thousand calories. Then multiply by the 9 billion people likely to need feeding in 2050, and multiply again by 365 to get the total annual requirement.

Then consider farmland. On a low estimate, there are now 1.4 billion hectares being farmed (one hectare covering ten thousand square metres, or around the size of the average sports field). It turns out that you would need seven million calories per hectare every year to feed those people adequately. For an "average crop", writes Jack Hollander in *The Real Environmental Crisis* (2003), that translates to less than two tonnes per hectare. Even arid parts of Africa manage one tonne per hectare, while Europe has reached the dizzy heights of six tonnes per hectare. All this means that you can feed a world of nine billion using less than two thirds of the cropland currently under cultivation. In theory.

BEHIND THE NUMBERS

This suggests that famine is not inevitable as population increases over the next decades before, as everyone hopes, levelling off. But the figures cloak a much more complex story. Actually feeding people depends on more than mere production of calories. It means making sure they can afford to eat, can manage some kind of balanced diet, and are not missing essential vitamins. Globally, it means taking account of crops used to meet a wealthier world's growing taste for meat, demand for cash crops and, almost certainly, biofuels and industrial feedstocks. The latter, if they expand after a peak in oil production, could mean a transformation of the whole agricultural system to meet multiple needs.

On a more localized level, feeding people will involve dealing with climate change, drought, war and other calamities whose effects will vary drastically from region to region. And changing the food system entails plunging into a tangled thicket of vested interests, megacorporations, government subsidies and globalized trade regimes that make existing patterns of production, consumption and profit hard to shift unless there is a real crisis.

Another reservation – a big one – concerns the assumption that all the land stays fertile. There are losses of arable land (meaning land that can be farmed) as soil erodes, a process sped by ploughing,

FUTURE FISH

Fish and fisheries are heading inexorably toward a food crisis all of their own. Modern fishing is often a devastating combination of hunting and gathering with modern technology, from echolocation to supersize nets. The result? Depletion of the population, declining catches of mostly smaller fish and, sooner or later, complete collapse.

The crisis is still patchy. The FAO's last global review estimated that 10 percent of fish stocks were seriously depleted or recovering from depletion, another 18 percent were overexploited, and 47 percent were being captured at or very near sustainable limits. That left around 25 percent where there was room for expansion of the marine catch.

As poorer nations depend more heavily on fish as a source of protein, this sounds like a recipe for future malnutrition. At best, the prospects for increasing fish catches look dismal. At worst, as forecast by Boris Worm of Dalhousie University in a 2006 study of declining ocean biodiversity, all commercial fish and seafood species will collapse by mid-century.

The alternative is expanding fish farming, already an important global industry. In 2004, the world chomped its way through about 106 million tonnes of fish, the highest per capita consumption in history. Fish farming accounted for 43 percent of this total. However, although growing fish in ponds or pens has expanded rapidly in recent decades, it is still relatively undeveloped compared with other kinds of farming. If new ventures use the kind of intensive methods modern land-based agriculture does, they have similar drawbacks. There may be problems with disposal of waste, parasite attack and overuse of antibiotics. In addition, farmed salmon or cod feed on fishmeal, and so do not reduce the ocean catch.

Fortunately, there are more sustainable models. The future of global fish farming is likely to follow the path already taken in much of China. Accurate figures are hard to come by, but there is general agreement that aquaculture in China has expanded fast, and that the country now accounts for perhaps two thirds of global fish farming – and harvests more fish from farms than it catches from the sea. Moreover, China leads the way in fresh water farming of herbivorous fish, which are an efficient source of protein, especially as they're lower in the food chain. At best, as Lester Brown emphasizes in his book *Plan B 3.0: Mobilizing to Save Civilization* (2008), Chinese fish farms feature a clever coproduction of four different varieties of carp, which get their food from different levels in a controlled ecosystem.

Trout farm in Beijing, an example of China's world-leading fish-farming industry.

and as some of the land accumulates salt from irrigation water. Add climate change triggering expansion of deserts in some areas, and there could be a serious problem. Worst-case estimates put the proportion of arable land worldwide suffering the effects of salination at thirty percent by 2020, possibly rising to fifty percent by 2050. According to the World Resources Institute, which tends to take a gloomy view of these things, around 66 percent of agricultural land has been degraded to varying degrees over the past 50 years. It makes you wonder how food production increased so fast in the same period. But even the lower estimates of cropland loss are cause for concern that such food increases will be hard to match in future.

There might still be enough margin for increasing yields on the land that remains, even as the per capita acreage available to grow food goes down. But the averages conceal more than they reveal. The high-yielding farms of Europe and North America do furnish grain for export. But they also support an average food intake way above that three thousand-calorie minimum. How do people eat that much? Partly indirectly, by consuming meat, eggs and milk. If just a quarter of a three-thousand-calorie-per-day diet comes from animals, feeding the animals demands at least another three thousand calories a day of crop production. And some societies have gone as far as making animal products forty percent of their "average" diets. The average US citizen, for example, eats more than two hundred pounds of meat a year. If everyone else in the world ate as much animal protein, the current global grain harvest would supply less than half the population. There is also increasing pressure on the other main source of non-plant protein – fish and other seafood (see box on previous page).

From a global perspective, the supersized, industrialized burger-and-fries habits of the West help explain why some people are malnourished now, even though average production is above the level needed to supply essential calories for direct consumption. The rest are not all exactly well-nourished either. The global inequality of food distribution means lots of people are overnourished, as well as undernourished. Obesity is, ahem, a growing problem in many developing countries as well. There it tends to affect well-fed elites, rather than the lower-income groups most affected in the developed world. Tackling the health problems of overconsumption will be a big part of the story of future food, but this is taken up in Chapter 12. Persuading some of us to eat less, and especially eat less meat, would ease the supply problem, though.

If current trends continue, the World Bank estimates that overall food demand will increase by 50 percent by 2030, but that demand for meat will rise by 85 percent. This is mainly due to more meat in the diets of the increasingly affluent middle classes of China and, to a lesser extent, India. Although the mix of meats eaten differs between countries, the overall figures are clear. Meat consumption per person doubled in China between 1990 and 2010, for example, but it is still only a little over half that of the US. As economic growth in China continues, the appetite for meat is likely to grow further, outweighing any efforts to curb meat-eating in the West in response to pleas from environmentalists and animal welfare campaigners.

There are plenty of other factors affecting the debate surrounding food production and consumption. There are lots of errors in the statistics, but overall they probably underreport production. Food eaten straight out of the ground, as it were, that does not find its way to market is hard to register in any official tally. There is probably more food around now than the global bureaucracies' numbers show. On the other hand, not all food supply translates into consumption. Serious forecasts of future food supply must estimate wastage in harvesting, storage and distribution. But aside from these niceties, there are some larger constraints looming that increase uncertainty about future food security.

WATER REVISITED

Agriculture depends on water. Most cropland is rain-fed, but irrigated land is still crucial. It produces more food, but uses more water. Less than twenty percent of farmland is irrigated, but it produces forty percent of the world's food output – and, as noted in the last chapter, accounts for two thirds of freshwater use.

These numbers emphasize how the diets we choose in the future are crucial to the world's prospects for getting enough to eat. If the future nine billion or so people living on the planet are mainly vegetarian and the proportion of crops grown on irrigated land stays about the same, annual agriculture water demand might rise to around 4800 cubic kilometres. But richer, meatier diets, using crops far less efficiently, could raise the total for irrigation water to eight thousand cubic kilometres, amounting to two thirds of the planet's accessible annual runoff of fresh water. Imagine a slight increase in the proportion of agriculture depending on irrigation, and continuing expansion of water demand from industry and from ever-growing cities – recall the sanitation crisis also outlined in the last chapter – and *all* available fresh water would be needed, every year. Something has got to give.

CLIMATE CHANGE AND FOOD

Climate models are not yet good enough to help chart the overall effect of climate change on food production. Too little rain, or too much, in some areas will doubtless affect crop yields. Higher temperatures will harm some plants, but encourage others. Plants quite like CO_2 (they eat it), so raising the atmospheric portion could improve agricultural production – though there is also evidence that the nutritional quality of some crops is reduced by force-fed carbon. Making carbon dioxide easier to come by does reduce water loss caused by transpiration, and so could help stretch water resources.

This just-possible upside to gentle global warming will be swept aside in some areas by extreme weather, or inundation of cropland by salt water as sea levels rise. And in any case it only applies in the lower end of the range of currently projected temperature increases. Beyond that, it is much more likely that warming will seriously affect food supply in many parts of the world. The real trouble would then come in the second half of the twenty-first century, with the worst effects anticipated in the poorer regions of the globe. Meanwhile, there are likely to be severe localized crises from drought and as yet unknown effects on plant diseases and insect pests from increased temperatures. All of which was summarized in the Intergovernmental Panel on Climate Change's 2007 report: "Globally, the potential for food production is projected to increase with increases in local average temperature over a range of 1–3°C but above this it is projected to decrease (medium confidence)." Not a lot of help, really.

However, the International Food Policy Research Institute came up with a more daunting analysis in

2009 based on detailed modelling of crop growth under climate change. It doesn't envisage complete calamity, but does forecast a world of poorer food security and rising prices for five crops: rice, wheat, maize, soybeans and groundnuts. It estimates that child malnutrition in 2050 will be 20 percent higher – totalling some 25 million children – under climate change scenarios than it would if climate remained unchanged. The poor will be affected the worst, and South Asia and sub-Saharan Africa the regions worst off. The institute calls for $7 billion to be spent on adapting agriculture to new conditions to avoid these supply shortfalls.

Agriculture is also important not only in the discussion about the effects of climate change, but also in discussing how to slow it down. The Stern Report reckoned that farming is responsible for fourteen percent of greenhouse gas emissions. This came from producing fertilizer (38 percent) and livestock (31 percent). Cows and sheep are a continual source of methane, a potent greenhouse gas, though one that does not hang around as long as carbon dioxide.

A GREAT LEAP FORWARD?

The future of food will be played out against a background of rising population, increased meat consumption as incomes rise (especially in China), more competition for water supplies, possible land loss though degradation and sea-level rise, and less regular rainfall in some areas due to climate change. Add in the expansion of biofuels taking a larger share of some crops, and food security begins to look increasingly precarious. How, then, to ensure future supply?

Discussion about the future of food supply divides roughly between those who hope business as usual can feed a few more billions, and others who argue a radical rethink is called for. UK food policy expert Tim Lang views the two basic positions as a debate over productionism versus sustainable development.

Productionism's overriding priority is to produce more food and make it cheap enough to buy. That worked in the 1960s, when plant breeding helped develop new hybrid varieties of wheat and rice with higher yields. It could work in the future, especially if breeding is enhanced by direct genetic manipulation and embraces more resilient strains as well as the pursuit of further increased yields. If that turns out to be right, the steep increases in food prices in 2007 and 2008 will turn out to be a blip, and a longer-term trend of gradual price drops will return with new investment and new technology.

Productionism may have worked in the past, say sustainability advocates like Lang. But they also maintain it depends on increased inputs of water (through irrigation) and cheap oil (to make fertilizer). These have reached their limits, and may even have gone beyond them by contributing to soil degradation caused by displacing older mixed crop cultivation, water pollution from massive runoff of fertilizer, and resistance to pesticides induced by overgenerous application.

Those limits mean the price rises will remain in place, as they portend real shortages of key commodities. This can only be tackled, they argue, through a transformation of food production. Ideally, the charge would be led by small farmers in less developed countries, who could be encouraged to increase yields in sustainable ways that would boost production. This would begin to tackle poverty and hunger where they are most prevalent. It could also have a big effect on population movements, since it might persuade rural dwellers to stay on the land rather than migrate to the cities. This would run against the longtime historical trend which suggests that if people can possibly leave agricultural labour behind, they will. But perhaps they won't, at least not in such great numbers.

The energy crunch reinforces the sustainability argument. High energy prices increase the cost of fertilizer, make transporting food more expensive and encourage farmers to switch to biofuels. This can either happen through selling existing crops such as maize or sugar cane for fuel conversion, or by the new planting of these or more novel fuel crops. Either way, there is competition for land. The massive diversion of American corn (maize), one of the US's highest-yielding crops, toward biofuel production is a big, bad example. Current biofuels will make little impression on energy supply – even supplying ten percent of transport fuel would need at least a third, and as much as two thirds, of current US and European crops. But they can have a large impact on food supply for the poor by raising commodity prices.

FUTURE FERTILIZER

Water, seeds, sun and soil are all needed for plants to grow. But the key to the modern agricultural system is artificial fertilizer. The ultra-high-pressure, high-temperature Haber-Bosch process for "fixing" nitrogen from the air as ammonia – a vital step in fertilizer production – is just one hundred years old. And the link between fossil fuels and large, high-yielding modern farms is not just the tractors and combines which work the endless fields. It is the energy used to make the nitrogen fertilizer, almost ninety million tonnes of it a year. It allows production of perhaps forty percent of the world's dietary protein, and uses around one percent of the global energy supply. Make fertilizer more expensive, and the cost climbs steeply – as it did when oil and gas prices spiked in 2008 – and you have a real problem.

But take that fertilizer away, and you create an immediate need to feed between two and two and a half billion people. So what is the prospect for making it cheaper, or finding worthwhile substitutes? As usual, more efficient use is the first option. It's doubly desirable because nitrogen fertilizers are also a major source of pollution, leading to the release of nitrous oxide – another potent greenhouse gas – from fields and water supply contamination. Precision agriculture (see below) is one route towards greater efficiency.

PREDICTIONS FILE

Daniel Pauly

Marine biologist, Fisheries Centre, University of British Columbia, and leader of The Sea Around Us Project

Highest hope: There will be a new Enlightenment, fusing democracy and human rights to science and care for the environment, and that everywhere, religion will be pushed from the public to the private sphere, where it belongs.

Worst fear: The societal changes will not be rapid enough for our global civilization to prevent the widespread famines and genocidal wars which unchecked global warming will bring in its wake.

Best bet: We will continue to oscillate between these two options, similar to what happened in the past, and thus validate Henry Ford's statement that history is nothing but one damn thing after another.

The other options are all theoretically attractive, but not yet close to substituting for Haber-Bosch-produced ammonia. There is plenty of nitrogen in the atmosphere – there's more nitrogen there than anything else, in fact. But it's bound together in nitrogen molecules that have to be broken apart to become available for plant growth, hence the extreme conditions needed for the Haber process.

One way round this is to look for better catalysts, which would help "fix" nitrogen at normal temperatures and pressures. A few promising routes have been tried in the lab. MIT chemist and Nobel Laureate Richard Schrock hopes to develop large-scale use of a twelve-step process that uses the metal molybdenum as a catalyst. It works at room temperature and atmospheric pressure, but stops working after just a few reaction cycles. Other options under investigation are at a similar early stage – they work in the lab under carefully controlled conditions, but none look like being suitable for real-world use any time soon.

The main alternative is to try biological systems, which achieve their chemical goals more efficiently. But only a few organisms can fix atmospheric nitrogen. Modifying plants so that they can adopt the symbiotic lifestyle of legumes, which nurture nitrogen-fixing bacteria in special nodules on their roots, has been a great hope for biotechnology since the earliest days of genetic engineering. But the system is very complex, and only some of the genes involved in plants and bacteria are yet understood.

One promising avenue was opened in 2006 when the Brazilian Agricultural Research Corporation announced the gene sequencing of a bacterium that works to naturally fertilize some non-leguminous plants, including sugar cane and sweet potatoes. This does not require the root nodules that are legumes' secret weapons, as it can live between root cells, offering an easier route to enhanced fertilization. An alternative strategy is to engineer plants to over-express genes that allow them to absorb nitrogen from their roots, thus reducing their requirement for fertilizer.

At the moment, though, these are all long-term prospects, and look unlikely to help food production for decades, at least. Lower-tech approaches using mixed planting of nitrogen-fixing plants, or even trees, alongside crop plants are the preferred strategy of sustainable agriculture advocates. Such approaches are more likely to play a part in transitional schemes that seek to maintain production with fewer artificial inputs.

IS THE FUTURE ORGANIC?

For some commentators, the way forward for agriculture is clear. What can reduce fossil fuel requirements, avoid pollution from fertilizer runoff and artificial pesticides, preserve soil fertility and work in harmony with the rest of the ecosystem? The answer, they believe, is organic production. And they may be right. But is it possible to produce enough food that way for nine billion people?

Opinion on that vital question differs sharply, partly because the information to answer it convincingly doesn't really exist. Certified organic farming now takes up less than half of one percent of world agricultural land. So even if those farms are studied really closely, imagining how the agricultural system would work if it was transformed to become completely organic requires huge assumptions.

Farming remains a risky business, and the risks of big changes may be high, too. One problem is that while less intensive agriculture may be able to produce enough for everyone and reduce some environmental impacts, it does generally take up more room. An analysis of future production from Wageningen University in the Netherlands suggests that a high-input system could produce enough food for all in 2050 using just over half (55 percent) of the present global arable land area. This is in line with the world food production estimates already discussed. However, global calculation reinforces the impression that the switch to a low-input regime could be problematic, and the same analysis indicates that in that case every scrap of land is needed.

In contrast, a paper which was immediately summarized on every organic food advocate's website in 2007 suggested that scaled-up organic production might not only feed the world, but use less land than current methods. Catherine Badgley and her co-authors from the University of Michigan reviewed just under three hundred examples of organic farming and compared their yields with those achieved by

The Kulika Charitable Trust's community development programme in Uganda trains farmers in sustainable organic agriculture, which has increased both production and quality of crops.

non-organic methods. Their conclusion was that, on average, the yields were a little worse in the developed world, but a little better in the developing world.

They then used these numbers to model a global system based on organic production. Again, the conclusion was that organic production could get the job done, and "has the potential to contribute quite substantially to the global food supply, while reducing the detrimental environmental impacts of conventional agriculture". However, their critics query the reliability of some of the data. Are figures that are estimated by people running demonstration organic farms and not checked by anyone else generally accurate, for example?

More seriously, there are doubts about the assumptions used to turn the data into a model. One key factor is whether there is enough organic fertilizer. Organic farmers use manure or straw, or crop rotation with legumes that fix nitrogen. A global shift to organic production might mean, for example, that farmers had to deliver seventeen tonnes of manure per hectare to produce a six-tonne rice yield. Yet cattle manure is not freely available for fertilizer. It is an important fuel in many Asian countries. Even if there were enough to fertilize the rice as well, the labour needed to sustain yields – shovelling that seventeen tonnes a year on every hectare, every year – would defeat many farmers.

The legume option, which is assumed to be the main source of nitrogen in the model, requires an intermediate crop that is not harvested. Many rice farmers, however, already rely on another "intermediate crop" they cannot afford to lose – rice. The yield measure at the heart of the Michigan review ignores the importance of this already established double, or even triple, cropping, which relies on artificial fertilizer. So, impressive though the study (published in the journal *Renewable Agriculture and Food Systems*) looks, the big question remains open. The authors themselves were careful to point out that while they claim that organic yields could feed the world, they are not forecasting yields for any particular crop in

PREDICTIONS FILE

Louise Fresco

Agricultural scientist, former assistant director-general for agriculture at the FAO, professor at the University of Amsterdam

Highest hope: We will be able, very rapidly, to convert our economies to a sustainable and equitable mix of the large and small scale, of the global, regional and local, of greater efficiency in resource use with due attention to human values and nature. This requires collective decisions on global problems and a binding, global decision-making system, which we lack today. In all this, speed is of the essence.

Worst fear: The short-term interests of the powerful, and an unreasonable rejection of technological and social change by the majority of the population, will lead to a selfish world where the public good is an obsolete value and violence becomes the language of the powerful and powerless alike.

Best bet: I am optimistic, if only for the fact that pessimism is a luxury we cannot afford. Globalization has led to greater transparency and dialogue; some consensus is emerging on how we need to tackle our issues of sustainability and equity and no country can afford to go it alone in the long term. So I trust that the best elements will coalesce, even if slowly.

any particular region, nor saying that a global organic food system would necessarily increase food security.

Less globally, but perhaps more significantly, a recent report from the United Nations Environment Programme reviewed results of organic farming efforts in Africa. The continent is a key focus for debate because the "green revolution" has passed much of it by, and many small-scale farmers and their families are malnourished. Their farms use relatively low-input systems, and the evidence is that switching them to organic production maintains or increases yields using techniques they can afford, while protecting the long-term health of their land. The study concluded that organic agriculture could increase agricultural productivity and raise incomes with low-cost, locally available technologies, and without causing environmental damage.

A middle way?

Despite all the pressures on sustaining future supply using the current mix of methods, productionism still has its attractions. In fact, in some sense it seems essential – the population projections demand it. The big question is whether production increases can be achieved in a more sustainable way. Doing that seems likely to involve some new technology and close attention to the details of how farmers manage their land (agronomy). It's also likely to involve conserving the food we harvest, and be affected by the kind of meals everyone puts on their plate.

At the beginning of the food supply chain, there is still ample scope for improving the crops farmers grow. Researchers can aim to increase yields, or to improve traits like drought tolerance or pest or disease resistance. Routes to success include screening varieties already stored in seed banks, using modern variants of traditional breeding techniques, and – more controversially – genetic modification by directly altering plant (or animal) DNA (see p.185). As Jess Ausubel of Rockefeller University puts it, yields over the last one hundred years have increased by two percent a year on average. Population is now increasing by only one percent annually, so maintaining that sort of yield increase would ensure there was enough land to feed everyone, at least.

Realizing any benefits these make possible depends on persuading farmers to use new or modified strains, helping them learn the best way to grow them, and making sure they can get hold of enough water, fertilizer or pesticide to ensure larger harvests. It also depends on access to markets to sell new surplus crops without depressing local prices. All of this makes simple technical fixes complicated in practice, and dependent on local economic and political organization. Still, having the improved strains available in the first place is a start.

Is there a middle road between the two extremes of sticking with productionism or redesigning the whole system along more ecologically friendly lines? In practice, there has to be. Like the energy system, food production is likely to evolve gradually rather than undergo a total transformation. The outcome will be influenced by lots of different factors, including costs of inputs, competition for land, cultural preferences and so on.

For the moment, the tension between the two main approaches to future food persists. It was starkly on display in discussions around the "International Assessment of Agricultural Knowledge, Science and Technology for Development", a large-scale appraisal of how to feed the poor half of the world better in future which was published in 2008 (available at www.agassessment.org). Billed as agriculture's answer to the Intergovernmental Panel on Climate Change, the exercise, funded mainly by the United Nations and the World Bank, failed to match the IPCC in achieving consensus. The assessment emphasized the importance of "agroecological strategies" and policies for aiding the small farmers who predominate in poorer countries. It saw little role for high-tech approaches, and especially for genetically manipulated crops. But before the final reports were agreed upon, biotech companies had withdrawn from the discussion and several countries, including the US, rejected the final report.

This rather pointless exercise shows how debate fails to advance when dominated by the too-simple idea that there are clear alternatives, with organic production, for example, seen as quite separate from the use of genetically modified (GM) crops. Organic certification explicitly excludes use of GM, or any possibility of contamination with GM. This appeals to many Western consumers, but is likely bad news for long-term, global food production. In *Tomorrow's Table* (2008), Pamela Ronald and Raoul Adamchak – she a plant geneticist, he an organic farmer – argue that the best use of GM technologies is to create

FEED CHINA, AND FEED THE WORLD

One way to see how the contending ideas of how to produce enough food to meet the world's needs might play out is to look at China, where the problems already outlined will be particularly acute. The country has an expanding population, which is also getting richer and eating more meat. It faces serious water shortages in crucial agricultural areas, and is likely to suffer adverse effects from climate change as well. Many commentators see production in China falling well short of demand, with a move to larger imports which will raise prices in world markets. Lester Brown raised this alarm in his 1995 volume *Who Will Feed China?*

However, Vaclav Smil answered Brown's question in *Feeding the World* (2000), which concludes with a detailed look at China and finds that the situation is not as bad as it appears at first glance. First, there is much more land under cultivation in China than previously thought. Official estimates from land-based surveys have been superseded by data from satellites which show that there is half as much land again under cultivation. The implication is that yields are lower than initially believed, so there is more scope to increase production without reaching implausible levels of harvest per acre.

China also offers major support for one of Smil's key arguments: that reducing waste is an easy route to feeding more people. The country makes poor use of what it grows, with storage that allows grain spoilage, inefficient feeding of animals and high waste of cooked food in communal eateries. All of these are likely to be improved. Smil's conclusion was positive: "There do not seem to be any insurmountable biophysical reasons why China should not continue feeding itself during the next two generations."

If China can do it, he argued, so can the rest of the world. The Chinese, though, appear less confident the job can be done without new technology, and in 2008 the government endorsed a plan to use genetic manipulation to develop high-quality, high-yielding and pest-resistant rice and maize. This is part of a plan to increase grain output to 540 million tonnes a year by 2020, estimated to be 95 percent of the amount the country will need then to feed everyone.

plant varieties that are better suited to organic agriculture. These might include those resistant to pests, flood damage or drought.

More generally, a 2009 report from Britain's Royal Institute of International Affairs suggested that the best overall strategy will be a gradual switch from "input intensive" to "knowledge intensive" agriculture. This science-to-the-rescue approach could embrace a range of methods, from GM to precision farming using information technology geared to the needs of agronomy. Of course, this is still a slogan, and needs unpacking in terms of who has the knowledge – and increasingly, who owns it. But even patents only last twenty years, so knowledge that starts off under corporate ownership will eventually diffuse more widely.

The science, though, needs to be accompanied by other strategies, including intensive study of local cultures and farmers' needs. There are examples of how this can work in Africa, the continent that could have the worst problems. Some are at village level, but at least one involves an entire country. Malawi, a landlocked nation with thirteen million people, relies heavily on its maize crop. In 2004 it only produced 57 percent of the maize it needed, and food aid was required for 5 million of its citizens. By 2007, the maize harvest had been boosted enough to generate a surplus of more than fifty percent above the country's basic needs – and export four hundred thousand tonnes of maize to Zimbabwe, as well as actually providing food aid for neighbouring Lesotho and Swaziland. How? Better rainfall in those years helped, but the main reasons were a carefully planned distribution of fertilizer and improved maize seeds to small farmers. Such developments encourage belief that a green revolution is possible in Africa, too.

THE ROLE OF GM

Genetically manipulated crops will have a place in future agriculture, but what that place will be remains controversial. Existing GM plant strains are widely used, but not in Europe, and developed world opposition has persuaded some, but not all, less prosperous countries to treat them warily. They have also been tailored – so far – largely to suit the interests of Western corporations and, to a lesser extent, consumers.

There are many more altered crop species in the pipeline, though. Whether or not they are essential to increasing production, some of them are certainly likely to offer advantages not realized in the earlier generations of GM crops. Quite a few of these involve combinations of traits which cannot be produced through traditional breeding as they require transferring genes between species. That, of course, is exactly what makes them objectionable to some.

If such crops prove acceptable, some of the promised future benefits of GM include nutritional enhancements as well as productivity aids. They might include oil-yielding plants with better-balanced chemical composition, peanuts with improved protein content, tomatoes with more supposedly health-giving antioxidants, or garlic producing more chemicals that help lower cholesterol levels – as well as the widely reported possibility of rice strains which produce vitamin A. The latter has been criticized as a technical fix for a problem, blindness induced by vitamin deficiency, which could be solved by giving people who subsist on an all-rice diet a more varied food supply. But while such diets endure, it seems worth trying.

Then there are modifications that make crops better suited to industrial processing, such as potatoes

CALORIE CONTROL

While researchers and policy-makers toil to meet the basic caloric requirements of the future population, at least as much effort is going into saving the overfed from themselves. Food industry research continues to seek ways of inducing people to consume more fat and sugar, blended into "eatertainment". But there is also increasing work on getting processed foods to contain fewer calories, or somehow prevent the consequences of taking too many on board.

This is partly a response to public health concerns, partly a pursuit of profit. A 2008 report from Credit Suisse forecast a $1.4 trillion global market by 2012 for obesity-fighting staple foods. And a 2009 report from the partially industry-funded American Council on Science and Health waxed enthusiastic about food technology's part in strategies for tackling obesity. These include work on sugar substitutes, ways to add fibre, new chemical additives, and ingredients that produce a feeling of satiety sooner than existing products. Possibilities for that last strategy include insulin-like compounds that affect blood levels of appetite control hormones, and new fat emulsions and fibre preparations which produce feelings of fullness. (Presumably the reduction in sales would be matched by higher prices for the "enhanced" product.)

There are also more speculative possibilities for modifying metabolism, such as enzyme inhibitors which would reduce digestion of starch into sugars. The whole research area is likely to expand, directing more ingenuity to preventing the worst effects of eating too much food of the wrong kind, rather than feeding those who are going hungry.

with higher starch content. This makes them absorb less oil when they are on their way to becoming French fries. This is only one line of research aimed at helping tackle future obesity (see box above).

The paths GM might take are quite varied, but likely to be directed toward future markets in the already well-fed countries, where the money is. As Paul Roberts observes in *The End of Food* (2008):

> "While many antitransgenic activists continue to get worked up over patent protection as the main threat to poor farmers, the larger risk is that the transgenic industry will simply bypass the developing world altogether."

One key to maximizing the benefits GM might offer in other countries, as well as helping make the results more acceptable, may be to encourage development of new, modified crops by research institutes in those countries. This is likely to be non-commercial philanthropic research, exemplified by an effort to develop drought-resistant GM maize for Africa that's backed by the Bill and Melinda Gates Foundation. It's putting $40 million into the African Agricultural Technology Foundation in Kenya to help get a drought tolerance gene already identified in North America to work in varieties of maize grown in Africa. GM is not a universal solution to anyone's agricultural problems, but projects like this could make a real difference in regions where food security is already poor and is likely to get worse as the climate changes.

PRECISION FARMING

GM gets a lot of attention, but the other mainstream technical developments of the next couple of decades – in data gathering and information processing – could be just as significant for the future of agriculture. Modern, high-intensity agriculture may produce high yields, but uses its inputs – whether seed, water, fertilizer, or pesticide and herbicide sprays – profligately. One way of transforming agriculture is to use the intensive methods guided by modern possibilities for surveying and sensing. The umbrella term for this in the business is precision farming. It depends partly on the astonishing detail now available from satellite mapping of the Earth's surface. As well as finding your house on Google Earth, you can now watch how your crops are growing with a GPS receiver and a yield monitor.

Once a farmer starts looking closely at management of individual fields, or even within fields, lots of other possibilities are opened. The data from the new information systems can be used to guide application of water, fertilizer or sprays, varying the amounts over very small distances. It can be combined with maps of weeds, salinity and soil.

According to Jesse Ausubel of the Program for the Human Environment at Rockefeller University, this approach combines with genetic manipulation to create the third agricultural revolution of the last hundred years. The first was mechanical, the second chemical. The new one is tied to information. It will depend on new investment and high-tech management, and is likely to make its first impact in the intensively farmed regions of developed countries. As these are the most inefficient users of resources, the savings that result should help them cover the costs of switching to this more refined approach.

A COMPLETE RETHINK OF AGRICULTURE?

We plough the fields and scatter… and that is really the problem. From one point of view, farming has followed the wrong path ever since the original agricultural revolution ten thousand years or so ago. The grain crops we depend on are annuals. Fresh seed is planted each year in ploughed soil, plants are grown and harvested for their seed, and the rest of the plant usually discarded (though it may be ploughed back into the soil). Some of the harvest is set aside for planting for the next crop.

Today, 85 percent of the world's farmland is worked this way, and much of our agricultural effort is devoted to overcoming the disadvantages of annuals. Ploughing leads to carbon loss and soil erosion. Annual crops need lots of fertilizer, and their shallow root systems are poor managers of moisture and nutrients. Could we get round these problems? Well, many plants do already. The annual crops our ancestors adopted have perennial wild relatives. Current efforts to combat problems associated with annuals focus on wheat, which can be crossed with several different perennial species to produce fertile hybrids. Maize, rice, chickpea and millet are also potential targets for this sort of cross-fertilization.

Researchers working on perennials sketch a vision of a transformed agricultural system in which crops act as a long-term component of a stable ecosystem, not a seasonal tenant of otherwise unproductive fields. Farms would resemble natural landscapes, where perennials dominate. The perennials' root systems would make better use of water and nutrients, help reduce weeds that take hold when annuals are removed each year, store more carbon dioxide in the soil, and

KEYSTONE CROP

Wheat, maize and soya are all vital mass-acreage crops, but rice is the most important for meeting food needs in the next decades. It supplies twenty percent of calories consumed worldwide, and accounts for thirty percent of Asian diets, according to the International Rice Research Institute in the Philippines. Among the world's poorest people – those who have less than a dollar a day to live on – two-thirds rely on rice to survive. Population growth means that production needs to go up by 50 million tonnes a year beyond the 630 million tonnes harvested in 2005. How?

Using the best strains already known will help, as will proving new hybrids. Flood-resistant strains are also needed, as rice grows in paddy fields but does not like total immersion. Unfortunately, all this needs more research and development and, astonishingly, many agricultural research programmes were cut back after the first "green revolution". And it takes around 25 years for new research investment to have any appreciable effect on crop yields. There is one tantalizing longer-term prospect for rice which would definitely require genetic manipulation. The ultimate gene hack would be to convert rice from a C3 plant to a C4 plant, which would make better use of carbon. C3 refers to the fact that many important agricultural plants, including barley, wheat and potatoes, build atmospheric carbon from CO_2 into sugars via a three-carbon intermediate made by the enzyme known as RuBisCO. But quite a few plants, including maize and sugar cane, also use an alternative pathway which is more efficient at using carbon. As it uses a four-carbon intermediate in the reaction sequence, they are known as C4 plants. The two sets of reactions can operate side by side and, in principle, rice might be engineered to allow it to profit from C4 synthesis. This could make it fifty percent more efficient at photosynthesis, thus increasing crop yields effortlessly.

probably be less sensitive to climate change than annuals because they have longer growing seasons. They are also better suited to marginal land, which has low yields and is difficult to cultivate, and will become increasingly important as demand for food increases over the next decades.

Making the switch would be an enormous undertaking. Farmers would have to be convinced of the advantages of new crop varieties, and learn how to manage them in their own locality. Unlike some transformative schemes, though, it would probably not entail any radical changes in diet, as the same grains would ultimately be harvested.

So far, the development effort – which is getting underway in the US, China, Canada and Australia – is small. It does not attract commercial funding, as no annual seeds mean no annual profits for their owners. Research uses traditional breeding techniques rather than direct genetic manipulation. This is partly because the people drawn to the work tend to be suspicious of GM, and partly because the traits that allow perennial growth are complex and involve many genes and genetic controls. However, new technology for analysing complete genomes – the whole complement of genes in any species – will help tease out the details, and speed up the breeding programme. Success is a long-term prospect, though. The Land Institute in Kansas (www.landinstitute.org), which is one of the pioneers in this effort, estimates that it will be 25 to 30 years before there are usable perennial wheat varieties.

WHO NEEDS SOIL?

A super-optimist would argue that worries about shortage of cropland are, er, groundless. Look at it scientifically, and you see that plants do not actually need soil. They need things that soil provides – minerals and nutrients. Provide them, along with water, air and sunlight, and the plants will grow quite happily. This is the idea behind hydroponics, in which carefully formulated chemicals are added to water that bathes plant roots. The roots can be anchored in an inert base, or just sit in the air and interact with a fine mist of nutrient solution, a variant known as aeroponics. Soil-free cultivation is particularly popular with homegrown strawberry enthusiasts and people who want to raise a lot of cannabis plants.

So could the future of farming lie in soil-free systems? Doubtful. There is no question that hydroponics works. But it has mainly been applied in money-no-object projects, like prototype space station gardens or designs for Mars colonies and, in a

At the Kennedy Space Center's Space Life Sciences Lab, scientists experiment with plant growth under different conditions – here onions are being grown using hydroponic techniques.

more down-to-Earth fashion, Antarctic bases. People living in the United States' South Pole Research Station vary their diet of frozen and canned food with fresh produce from an automated garden. It uses hydroponics to grow lettuce, tomatoes, cucumber and peppers on site. The food's very welcome during the Antarctic winter, when it's so cold that no supplies can be flown in. But these super-salads end up costing $50 a pound (about 450 grams).

Although this is an extreme example, hydroponics operations are generally costly, mainly because of the energy for heat, light and the circulating pumps. If you are growing cannabis plants indoors, this may not matter. But legal crop plants serve a different market.

Still, there are large-scale commercial hydroponics operations in more promising environments. These are greenhouse-based, and tomatoes, peppers and

Farming in the twenty-first century: a vertical farm design proposal for Australia's Sunshine Coast by Oliver Foster of Queensland University of Technology (see www.verticalfarm.com for more designs).

lettuce are the most common crops. With sunlight freely available, the technique is much cheaper, and also uses far less water than conventional farming. Some cover hundreds of hectares and yield year-round harvests.

Visionaries see the possibility of developing the technology to build "vertical farms" (more details at www.verticalfarm.com). These would provide fresh local food for urban areas, though the lower stories would need artificial light. Professor Dickson Despommier of Columbia University reckons the technology for "farmscrapers" already exists, and just needs to be brought together. That could trigger a proliferation of design studies for farms that reach for the sky. But it would be good to see, say, a three-storey farm that can pay the bills built somewhere before vesting hopes in projects like the 57-floor sky farm proposed for downtown Toronto.

GROW YOUR OWN STEAK

There's a final technical possibility for improving agriculture, as well as eliminating the appalling animal welfare involved in the mass rearing of poultry and livestock. Along with the spaceman's meal-in-a-pill (a daft idea because it ignores the pleasures of eating), animal-free meat has long been a staple of science fiction. Early in the twentieth century, one of the first tissue cultures grown in a lab derived from chicken heart cells, and was widely reported as a "heart in a jar". Imagination soon turned this into an endless food supply, depicted in Fred Pohl and Cyril Kornbluth's satirical science-fiction novel *The Space Merchants* (1952) as "chicken little" – a vast, ever-growing slab of meat harvested and then "sliced and packed to feed people from Baffinland to Little America".

In this scenario, chicken little is unappetizing but ubiquitous. But the basic idea still appeals to some – a controlled product, efficiently grown, with no waste and no slaughterhouse. As cell biology advances, the prospect of growing meat in a vat looks more realistic. It is simply muscle, so it's not the most complex tissue. The US arm of vegetarian pressure group People for Ethical Treatment of Animals (PETA) believes in it enough to offer a million dollars to anyone who can produce lab-grown ("in vitro") chicken meat for sale by 2012.

The goal isn't to make a mere token test sample. They want a real product that's commercially available, so to win you must "produce an *in vitro* chicken-meat product that has a taste and texture indistinguishable from real chicken flesh." Then the approved product must be sold at an affordable price in at least ten US states. If cultured meat arrives in the supermarket, the environmental effects will depend on what it needs to grow. The size of the required operations won't be a problem in principle. The mycoprotein (think fungus) meat substitute Quorn is already grown in 150,000-litre fermenters.

But a chicken or steak factory would still have inputs and outputs, consuming energy and raw material, and producing waste. Novelists Pohl and Kornbluth thought of that and, impressively for 1952, had their culture fed on algae skimmed from ponds laid out in multi-storey towers and supplied with concentrated solar energy by mirrors. The actual skimming is done by workers who are more or less slaves, but sounds like the kind of thing you could easily turn over to robots.

Today, the In Vitro Meat Consortium believes that cultured muscle is a way of expanding production of the protein people most like to eat at low environmental cost. The scientific alliance, launched in 2007, outlines a four-step route to a palatable product made without animal aid.

First, they will need to culture stem cells (which multiply fast, but do not turn into specialized cells) from the right animals (cows, pigs, chickens). Then they have to be guided toward development as muscle cells that have the same nutritional value as regular meat, grown in a medium that contains no animal products, and – probably the hardest step – induced to organize themselves into the fibres which give meat its familiar chewy texture. Then, provided this can all be done at the right price, comes the stage they do not mention: persuading consumers to buy and cook an engineered biotech miracle instead of chunks hacked off a dead animal. The first cultured meat products to reach the market will be burgers, sausages, pies and other processed foods – items in which some of the content that counts as meat already bears little resemblance to actual rump steak or chicken fillet.

WHAT WILL HAPPEN NEXT?

Although food chemists and cell engineers are set to continue bringing novelties to the table, from a global viewpoint the fundamental issue remains supply of basic commodities. A useful set of scenarios was crafted by the British foreign policy think-tank Chatham House, which responded to the hefty commodity price increases of 2006–08 with a set of four projections of what might come next. Although aimed at policy-makers in the UK and European Union, they take the outlook for the world as a whole into account. According to their analysis, the main possibilities for the next five to ten years fit under these headings:

JUST A BLIP

In this future, high prices work as a market signal is supposed to, and farmers respond by growing more. The weather is cooperative, oil prices drop and stay low, and competition from biofuels eases. The result is a food surplus within five years, and stocks are rebuilt while prices fall. (Grain production did indeed increase by seven percent in 2008, but prices remained above their 2005 levels.)

FOOD INFLATION

The opposite of the first scenario, unfolding over ten years. Demand for food grows and outpaces supply, perhaps only slightly. Meat consumption grows, extreme weather harms harvests, and high energy prices boost biofuels and keep fertilizer prices high. There is still investment in increasing food production, and increasing adoption of high-tech methods, but food prices stay high, stoking inflation. But the economy adapts and the existing food system copes in the end.

FOOD IN CRISIS

A collection of worst-case factors combine forces, including new crop and animal diseases, water shortages, extreme weather, record energy prices and geopolitical conflict. The results are exhausted grain stocks and exploding prices. Governments intervene with price controls and export bans, and there are civil disturbances, wars and serious regional famines which aid cannot stem. The overall response resembles Shell's "Scramble" scenario for energy supply (see p.152), and even uses the same word to describe itself. Countries that need imports to feed everyone, notably China, tie up bilateral deals on food supply wherever they can. Those who cannot afford to do this, or move too slowly, go short.

INTO A NEW ERA

In this scenario, peak oil and climate change work together to squeeze agricultural production. Environmental regulations get stricter, and eventually there is general agreement that the old agricultural regime cannot go any further. Production slowly shifts to what this storyline terms an eco-technological approach. This embraces methods old and new including crop rotation, cover cropping (which increases soil fertility and combats weeds and pests), agroforestry, "green" fertilizers derived from waste, new plant varieties (including some that fix nitrogen), precision agriculture and more efficient water use.

There are two key uncertainties in the interactions between food, water, energy, environment and economy these scenarios are trying to reflect. The first is whether high commodity prices are a signal that long-run constraints on production are beginning to bite. The second is whether the new "eco-technological" regime can meet global crop demand. The wrong answer in either case makes crisis more likely than the other scenarios. The most hopeful suggestion is that the new approaches are established in pockets that work alongside old-style intensive agriculture, and spread as their advantages become apparent.

FURTHER EXPLORATION

Alex Evans The Feeding of the Nine Billion (2009) A report for the UK's Chatham House think-tank which begins by analysing the origins of the 2007–08 food price crises, then considers the wider issues of organizing the food system to feed the increased population of 2050. Downloadable from www.chathamhouse.org.uk.

Robert Paarlberg Food Politics: What Everyone Needs to Know (2010) Overview by US food policy academic, trying to stuff a balanced treatment of all the issues into two hundred pages. Succeeds admirably well.

Raj Patel Stuffed and Starved (2007) Another take on why current food policies allow a billion people to be malnourished in a world that produces more than enough to feed them all.

Paul Roberts The End of Food (2008) Daft title (to echo the author's previous book, *The End of Oil*), but good book. Much detail on corporate influence on how and what we eat.

The Royal Society Reaping the Benefits: Science and the Sustainable Intensification of Global Agriculture (2009) This report (downloadable from royalsociety.org/reapingthebenefits) offers a measured review of how science (not just GM) can help increase food supply through "sustainable intensification".

www.future-agricultures.org Future Agricultures' website is devoted to discussion, debate and policy analysis about agriculture in Africa – the hard case.

II
humans and
other species

humans and other species

An Armageddon is approaching at the beginning of the third millennium. But it is not the cosmic war and fiery collapse of mankind foretold in sacred scripture. It is the wreckage of the planet by an exuberantly plentiful and ingenious humanity.

E.O. Wilson

This tour of near-term futures has yet to take in the species we share the planet with, apart from the ones we choose to eat. But their fate is one of the most troubling issues of the twenty-first century. We're pretty sure that there will be fewer of them. But how many fewer, and how grievously we will feel the loss, are harder to know for certain.

EXTINCTION ALERT

Current alarm about climate change has tended to keep the decline in the varieties of living things off the front pages, though climate change will affect many other species too. But "biodiversity", the technical measure, is an abstraction which is not always easy to relate to. The word has only been around since the 1980s. There is plenty of mileage in raising alarms about the demise of individual species, preferably large, cuddly or charismatic ones. But the idea that the variety of life matters, and is being depleted, on a planetary scale is less readily appreciated.

Debates about biodiversity are also marked by a confusing mix of impulses, attitudes and impressions of its significance. Evolution is profligate, and we know innumerable past species have disappeared. Many of them went in periodic mass extinctions – there were at least five such waves over the aeons. And although it is tempting to believe that past generations observed some kind of balance of nature, the evidence suggests that the advent of humans in any particular territory has nearly always been bad news for anything large enough to eat and slow enough to catch – a story well-documented in Clive Ponting's *Green History of the World* (1992, revised and updated in 2007 as *A New Green History of the World*).

PREDICTIONS FILE

Jules Pretty OBE

Professor of Environment and Society, University of Essex, and author of *The Earth Only Endures* (2007) and *This Luminous Coast* (2011)

Highest hope: That we will have seen divergence in ways of living across the world so that the ecological footprint of humans will have been dramatically reduced. By 2050–60, human population will have stabilized, and thus it is not numbers of people that will matter, but our ways of living, and their consequent impact on the environment of this small planet.

Worst fear: That consumption patterns will have continued to converge across the whole world, and the consequent pressures on climate, water stress and energy availability will have run out of control. There will be greater poverty and hunger than today. Countries will fracture into smaller units, and conflicts over resources will escalate.

Best bet: A combination of new technologies (from solar power to robotics to food growing) combined with new institutions and ways of living (greater connectedness to nature and place) will see humans emerge from the industrial age into a new era in which we live lightly on this planet and yet still have access to modern technologies (from antibiotics to cancer treatment, and mobile phones to the Web) that connect people and keep them healthy. ”

Today, though, while hunting continues – and remains a major food source in parts of Africa – there are other, more widespread threats to non-human species. We encroach on the places where they live, pollute them, and are now altering the climate, too. The immediate effect will certainly be a less diverse natural biosphere. The further costs of that change are controversial. Do we need all those species – ecologically, aesthetically, perhaps medically? Can we preserve them artificially? Are we on the verge of creating new ones, at will? If so, the loss of the "old" biodiversity might be balanced by the advent of new, engineered varieties. For some, that would have its own fascination. And behind the debates about any immediate, practical benefits of maintaining biodiversity lies another discussion, of the place of nature in an increasingly artificial world, and of what we mean by "nature".

DON'T KNOW WHAT YOU'VE GOT 'TIL IT'S GONE

What are the facts about biodiversity loss so far? The first problem is that we do not know nearly enough, and much of the information available is contested. Biodiversity is defined to include the variety within species and types of ecosystems, as well as in kinds of creatures. There are international efforts to establish indicators of how diverse life is in particular regions, but all are based on incomplete information. These factors are all taken into account by the Global Convention on Biodiversity, which set targets in 2002 for reducing the rate of species loss by 2010 (see www.twentyten.net). The targets have had little effect so far, as their publication *Global Biodiversity Outlook* admitted in 2010. But there has at least been some progress in agreeing how to monitor species, so future targets can be assessed more scientifically.

Meanwhile, the simplest approach to assessing global biodiversity means taking inventory of all the species on Earth: a much harder job than the story of Noah makes it out to be. It includes, for example, hundreds of thousands of insects, mites and microorganisms – everywhere from the bacteria in your gut to the creatures of the deepest ocean. Even for creatures larger than microbes, the great tropical ecosystems – the most diverse patches of the planet – harbour many species that have not yet been labelled and tagged by scientists.

How many are there?

So there are two jobs that are necessary. The first is to make sure that all the species already identified are actually unique. At the moment, there are lots of different collections and databases which sometimes overlap. A megaproject to assemble documentation on the 1.8 million known species, the Encyclopedia of Life, is also at an early stage, as a glance at its website shows (www.eol.org). Then we need to know how many more there might be. There are around twice as many, according to some estimates – perhaps 3.5 million in total. But others maintain there could be as many as one hundred million species alive on Earth today.

No, there are not unknown relatives of lions and tigers lurking deep in the forest. Most of the known, and the unknown, species are insects and bacteria. There are already around 450,000 species of beetles and 200,000 species of flies in the databases, compared with just 4000 kinds of mammals and 9000 birds. A handful of soil usually has between 5000 and 10,000 species of bacteria, many of them yet to be corralled and examined.

If we do not know what we have got, can we know when it has gone? Extinction means death of a species, but it is not like the death of an individual. Establishing extinction means proving that no living specimen has been observed for a long time. There will be little room for argument if elephants go extinct, but most creatures are smaller and more retiring.

In addition, as Martin Jenkins of the United Nations Environment Programme's World Conservation Monitoring Centre points out, when we really put our minds to it, humans can rescue species from the brink of extinction, though we tend to obsess about a few at a time. This also makes it harder to assess current extinction rates, or estimate future ones.

Even estimates of species at risk can appear highly selective, once you appreciate how many species there are. The World Wide Fund for Nature (WWF, known as the World Wildlife Fund in North America) has a Living Planet Index, which uses estimated populations to classify species as endangered, that's often cited. But it's based on just 555 terrestrial species, 323 freshwater species and 267 marine species around the world. The far more abundant world of invertebrates, let alone bacteria, does not figure in WWF's alarm calls.

So why worry? The most convincing evidence that something serious is amiss derives from that basic evolutionary fact that 99 percent of all the species that ever existed are already extinct. We can use fossils to measure how long they were around. It turns out that the average length of time a species manages to hang on is roughly one million years. That meant that approximately one species in a million went extinct every year, on average, over the last billion years or so. But in the last century, the rate of bird and mammal species loss has been far faster than that – hundreds of times faster, by some calculations.

None of these methods are very precise, and different people get different results. But amid the scholarly debate, all the experts agree that the extinction rate is rising. Even the low estimates are worrisome. Ecologist and former UK Royal Society president Robert May reckons that even going by the most cautious interpretation, the data indicates a roughly tenfold increase in extinction rates in the coming centuries. Then he abandons sober language, suggesting that "we are currently on the breaking tip of a sixth great wave of extinction in the history of life on Earth, fully comparable with the Big Five in the fossil record, such as the one that extinguished the dinosaurs".

There are now only around 7500 Sumatran orangutans left in the wild, due to encroachments on their habitat by human settlements and agriculture.

WHEN THEY'RE GONE, THEY'RE GONE

As well as uncovering background extinction rates, the study of life on Earth over the past aeons also shows that species loss is a near-term prospect that will leave its mark on the future for an exceptionally long time. The most detailed analysis of fossils focuses on marine invertebrates – worms and such – which have been around for five hundred million years or so, and have suffered from earlier mass extinctions.

The good news is this shows that while extinction of species is by definition permanent, loss of biodiversity need not be. The number of distinct species is slowly restored by two consequences of calamity. Any significant drop in biodiversity is followed by a drop in extinction rates, as presumably there's less competition in the aftermath of a species cull. What's more, the record shows that high extinction rates are followed by high rates of new species appearing. Evolution has no goals or motivations as such, but it does almost look as if life is waiting for some clear space and uses it to be creative. You might think this is an argument for being more relaxed about possible loss of biodiversity over the next couple hundred years. But the bad news is that evolutionary rebound is slow. It takes between ten and forty million years. That's not long in geological terms, but in human terms it's many, many generations.

THE REAL WARNING SIGNS

It is easy to regret extinctions. Who wouldn't love to see a live dinosaur? But it is hard to convey the real extent of the losses that appear to be on the way – and hard, perhaps, to get people to really care – by

LIFE AFTER EXTINCTION?

Being extinct, for a species, is like death for the individual. You never come back. Or do you? The idea popularized by Michael Crichton's *Jurassic Park* (1990) has great romantic appeal, but we are not going to see living, breathing dinosaurs. However, there are plenty of creatures that vanished more recently, say in the last one hundred thousand years. For these, if conditions are just right, it is possible we might find remains that preserve the complete DNA sequence. If so, some biologists believe it will one day be worth trying to revive the owners of this DNA, although it is out of the question with current technology.

Science writer Henry Nicholls suggested candidates for high-tech reincarnation in a *New Scientist* feature in 2009, compiled partly on grounds of "megafaunal charisma", which means he likes them. His top ten, and the date of their last known appearance, are:

- Sabre-tooth tiger (10,000 years ago)
- Short-faced bear (11,000 years)
- Tasmanian tiger (70 years)
- Glyptodon – a kind of giant armadillo (11,000 years)
- Woolly rhinoceros (10,000 years)
- Dodo (300 years)
- Giant ground sloth (8000 years)
- Moa – an ostrich-like flightless bird (c. 500 years)
- Irish elk (7700 years)
- Giant beaver (10,000 years)

He throws in a couple more for luck: the gorilla, on the basis that it is nearly extinct, and – more ethically troubling, perhaps – Neanderthal man, who faded out around 25,000 years ago.

listing endangered species. For one thing, eliminating other mammals is more like business as usual for humans than a new problem. Humans have been wiping out large animal species wherever they have lived for the last fifty thousand years. What's more, ecosystems can survive the loss of some fairly conspicuous species. New Zealand, for example, once had 38 kinds of flightless birds. Of the nine remaining all are endangered, but the country still looks basically fine (as viewers of the *Lord of the Rings* films can testify). Could the same assessment apply to the rest of the world?

Perhaps, but current biodiversity loss goes deeper. The International Union for the Conservation of Nature's latest "Red List" names only 35 mammal species that have become extinct since 1500 AD. But another 1094 are assessed as vulnerable, endangered or critically endangered – almost a quarter of those for which there is data. The outlook for other, less cuddly creatures are in some cases even worse (see box opposite).

The destruction of some special habitats is the worst news on the biodiversity front. Two kinds of living space, coral reefs and tropical forests, harbour a large proportion of all known and, presumably, unknown species, and are under particular threat. This has long-term implications for both conservation (see below) and what we're already losing. Coral reefs, which are affected by rising sea temperatures and increasing ocean acidity, are a particular worry. In 1998, only 13 of 708 known coral species would have been classified as threatened. Now the number has risen to more than two hundred. Forests have also been extensively studied as vital habitats, and because of their importance in the carbon cycle, they deserve a separate discussion.

FROGS ON THE FRONTLINE

Of the well-studied branches on the evolutionary tree, the amphibians look set for the most severe pruning. This is worrying because they have survived the last four mass extinctions. They may be more vulnerable to some of the current environmental challenges than other kinds of creatures because they live in both air and water at different times, and may absorb chemicals easily through their skin.

According to David Wake of the University of California at Berkeley, a third of the 6300 known species of amphibians – 5600 of them frogs – are threatened with extinction. He believes the problem will get worse as climate change progresses because most of the vulnerable species are tropical and only live in small areas. The only consolation for frog-lovers is that 25 years ago there were a mere 4000 known amphibian species, so reducing the current total by a third would put the species tally roughly back where it was in the mid-1980s. But losing one in three of any kind of creature still seems like an alarming prospect, especially if it is a harbinger of what may happen to other kinds of life.

A golden poison dart frog, one of several endangered species native to Columbia.

SEEING THE WOOD FOR THE TREES

Images of teeming rainforest reduced to felled trees and bare soil regularly accompany warnings about loss of ecosystems. But the state of the world's forests is not quite as bad as that implies, because the trees can come back. There has recently been quite a lot of reforestation in the countries that earlier cleared much of their natural forest. But the new forests tend to be simpler, managed affairs, more like tree plantations – though they may still be effective at locking down carbon (see below).

For tropical forests, there is plenty of local evidence of forest clearance, and much is made of satellite photos showing depletion of forests in particular countries. But globally it is harder to establish accurate trends. The debate about tropical forests was highlighted in 2008 when University of Leeds environmental scientist Alan Grainger looked at three decades' worth of UN data, based on compilations of national inventories of forested areas of widely varying accuracy. He found that the figures did not show clear evidence of long-term decline worldwide. The finding was contested by other researchers, who pointed to satellite surveys that deliver strong evidence of forest loss. Grainger himself did not claim that his analysis suggested that tropical forest decline is not happening, only that it is hard to demonstrate convincingly, and called for a new global monitoring system.

Later that year, another team of scientists published a paper in the same US journal, the prestigious *Proceedings of the National Academy of Sciences*, which showed how it might be done. They suggest that satellite data is best, and much cheaper than ground-based surveys. It can also be checked by objective observers, whereas returns from local surveys can be affected by some degree of wishful thinking.

Their answer is to use easily available, low-resolution pictures to identify areas as small as 250 metres where forest loss is most likely to be seen. A sample of these areas can then be checked using more detailed – and more expensive – data from satellites that zoom in yet closer. The result is still a statistically defensible estimate, rather than a completely accurate measure. But it can be repeated easily, and so offers a good way to watch trends and indicate which countries or regions justify closer attention. They found that the first five years of the twenty-first century saw significant tropical forest losses, especially in Brazil and Indonesia. The figures are, in fact, lower than the most recent UN assessments, but do show serious shrinkage. Globally, they amount to a bit under two-and-a-half percent in five years. It does not sound like much, but if the trend continues, it would remove half the remaining tropical forest by the end of the century.

That conclusion is far from firm, though. A slightly earlier, controversial study, by Joseph Wright from the Smithsonian Tropical Research Institute in Panama and Helene Muller-Landau from the University of Minnesota, argues that species loss in tropical forests may be lower than this gloomy projection implies, since forests grow back. They reviewed the older UN records on population and forest cover, and suggested that declining growth rates in population, especially rural population, will allow forest to recover to some extent. There will still be clearance of primary, old-growth forest. But there could be more development of new, secondary forest than many conservationists predict. Their study was roundly criticized, but there does seem to be agreement on two counts. First, there is a possibility that the future for biodiversity in the tropical forests this century could just be bad, rather than catastrophic. And second, more research, and better data, are needed to tell.

HIDDEN DEPTHS

Forests can be surveyed from above, but to observe life in the ocean you have to get down below the surface. In general, we know a lot less about biodiversity in the oceans, especially the deep ocean. Serious efforts to rectify this are very recent. The International Census of Marine Life project began in 2000 as an effort to get a clearer picture of what dwells in the oceans, and has found thousands of new species. Its final report, due in late 2010, will review all the best data on marine species, which will be included in an online database covering perhaps 230,000 species.

The future for those species could be less uncertain than the future for those on land. Many species that are fished commercially are in jeopardy (as discussed in Chapter 10). But many others may live beyond our reach. That will not ensure they escape the effects of global warming, though. Marine species may find it easier to migrate to more comfortable latitudes when temperatures shift, as there are no barriers to their movement like the ones restricting the movement of land-based species. But the gradual acidification of ocean waters as they absorb more carbon will still catch up with them.

This will be a particular problem for creatures that grow a hard outer layer. Many plankton, for example, protect themselves with a calcium carbonate layer which is unstable and dissolves away in acidified water. They might adapt, but doing it fast enough will be tough. Geological evidence suggests that the increase in ocean acidity since the beginning of the industrial revolution is one hundred times faster than what marine organisms have experienced for at least twenty million years. The effect of rising CO_2 on the ocean in this respect is more predictable than the future of global warming, as it depends on well-understood chemistry. Raise atmospheric carbon dioxide and the oceans will absorb more, and acidify more.

There will also be severe effects on coral reefs, which are currently biodiversity hotspots. By mid-century, the acidity of the ocean will probably have risen to a level at which the reef-building varieties of coral polyps can no longer build reefs up faster than they erode.

PREDICTIONS FILE

Dennis Pamlin

Global Policy Advisor, WWF

Highest hope: Humans will evolve into a global community, moving away from nation-state perspectives with material consumption and simplistic entertainment, and focus instead on innovation, art and science. On this journey we will break down the old institutions and old ways of thinking that say humans are separate from nature and other animals, and build our society on an ethical framework.

Worst fear: We will destroy the beauty of the planet, continue to torture and kill other living beings and deny those less fortunate among us a life of creativity.

Best bet: We must move beyond the idea of guessing the exact nature of future societies and focus on principles and drivers. My best bet is that some countries, companies and interest groups will hold on to the old world (fighting over the balance between government and companies), leaving room for a new generation of entrepreneurs that dare to think big. If the entrepreneurs are not able to convince people that another world is possible, the resistance from the fossil companies/politicians will result in a world close to my worst fear. If a new network of change agents is created, and those looking beyond the industrial paradigm and the nation state within the large companies and political parties can collaborate, we will move towards my highest hope.

A NEW KIND OF MASS EXTINCTION

Previous mass extinctions seem to have been caused by untoward events like asteroid impacts or giant volcanic eruptions, perhaps causing or allied with rapid shifts in climate. There is no doubt that the current wave of extinction is caused by us. There are currently five main threats to other species. The longest-running cause is over-exploitation, currently affecting many fish species especially badly (see Chapter 10). Then comes pollution, such as fertilizer runoff and new kinds of pollution that have only recently become widespread (see box opposite). Invasive alien species, usually transported from one place to another by us, cause real problems in some areas. And climate change threatens to put many more species under stress. All of these threats are mainly or wholly due to human action. Accompanying all of them is habitat change, also brought about by people.

Loss of habitat is one of the most deep-rooted trends at work here. It covers both the outright destruction of habitats and their conversion to other uses. Extinction rates may be hard to pinpoint, but information about loss of living space is much clearer.

HORMONAL HAVOC

The effects of pollution on wildlife have been a big concern for environmentalists since Rachel Carson publicized the hazards of artificial pesticides in *Silent Spring* in 1962 (see box on p.42). Herbicides and insecticides are intended to kill at least some creatures, so it is no surprise they can be harmful in other respects as well. But more recently an unexpected threat has emerged from new chemicals that have more innocuous uses. These are a class of substances which affect production or action of hormones, and are known as endocrine disruptors.

Some are persistent organic pollutants – including some pesticides – which are extremely widespread, usually in low concentrations. Research has shown that they can affect development and reproduction in many species in higher doses. Scientific literature has alarming data on malformed reproductive organs and loss of fertility in aquatic species such as fish, frogs and alligators that have been exposed to such pollutants. There are also suspicions about other widely distributed synthetics such as bisphenol-A, used to make plastic containers and food wraps, and phthalates, softening agents found in many flexible plastic products like vinyl tubing or flooring.

Whether these are dangerous in normal use is controversial. Some studies have suggested links with asthma, but there is no clear evidence of hormone disruption in humans. But low doses could add to environmental stress on sensitive animals which are also trying to adapt to other stresses – such as climate change – and could increase the risks to some species as global warming affects their chances of survival.

Humans take up more and more of the planet, and the simplified ecosystems – farms, plantations and ranches – which we convert pristine landscape into have expanded vastly, too.

This is a new, planet-spanning way of being a dominant species. Humans have found ways of living in just about every environment Earth has to offer, and have multiplied enthusiastically as well. This puts pressure on other species not just because we make use of so much space, but simply in terms of the living room available to other forms of life.

In 2002, ecologist Eric Sanderson and colleagues published a paper in the journal *Bioscience* which pulled together a heroic assembly of satellite data and other surveys to make a global picture of just how much of the Earth *Homo sapiens* now occupies. They called this the "human footprint" on the planet, not to be confused with the ecological footprint or the carbon footprint, which are attempts to measure other kinds of impact. This was simply an exercise in mapping our use of space.

As there are few simple boundaries, this had to use a bunch of measures to form its picture of human impact. These included, fairly obviously, population density, but also took in land use, accessibility by road, rail or river, and the reach of electricity supply (the last factor being measured by determining where lights shine at night). Add these together and you get an index of human influence, which they calculated for each region on Earth.

Their results showed that there was some human influence on more than eighty percent of the land surface, rising to ninety-eight percent in regions where farmers can grow rice, wheat or maize. The maximum possible value of the index was achieved in Brownsville, Texas. The world's great cities also registered high scores, while the lowest were also in areas you would expect – forest regions of Canada and Russia, deserts, tundra and the Amazon basin.

One encouraging finding, though, is that sixty percent of the land surface charted somewhere between the lowest and highest values on the index – it was neither completely wild nor completely domesticated. In other words, "nature" coexists, for better or worse, with human-shaped landscape. This goes against the idea, popularized by US author Bill McKibben, that the current era is seeing the "end of nature". You just have to look harder to find it. The end of wilderness, though, seems a bit nearer the truth, especially wilderness that can still support life better than deserts or mountain ranges do. Any other landscape, apart from the highest peaks and most arid deserts, is likely to have been reshaped by humans.

The human footprint team has also identified the bits of the major natural ecosystems least affected by our use, termed biomes by biologists, and plotted them on a map of "the last of the wild". They suggest that this area, covering roughly seventeen percent of the Earth's land surface, may be the best place to focus conservation efforts.

FUTURE LIFE IS CITY LIFE

Amid a collection of gloom-inducing indicators, there is one trend which may be positive for the future of wildlife. As outlined in Chapter 6, the human population is shifting more and more into cities, and looks like it will continue to do so. Almost two-thirds of the world population increase since 1950 was soaked up by cities, whose population as a whole grows by a million people or so every week. The number of people living in cities – 3.2 billion of them – is now greater than the entire population of the world in 1960.

The result is that rural population has probably already peaked – at another 3.2 billion – and is projected to start falling after 2020. The rest of us will live in cities, or the new megacities, defined as those with more than ten million people. These growing urban masses will make huge demands on the land for food, water and recreation. But beyond that, whether they are content to experience wildlife in parks, zoos and on film, want the chance to see the wild in person, or simply have other priorities will be a big influence on the prospects for conservation.

WHY CARE ABOUT BIODIVERSITY?

Every extinction means the loss of a unique product of the living world, and making a case for increasing the number of extinctions would be about as easy as arguing for an increase in the human death rate. Still, some of the reasons commonly heard for preserving biodiversity may be less than convincing. Most of us like some other creatures, and even like to look after them. The ones we choose to keep as pets are probably OK, for now. But why worry about all the rest? Some argue that every species is precious. A quick look at how evolution works makes that less plausible biologically, but still leaves people able to cherish biodiversity for its own sake.

Locally, though, this is often overridden by pressing reasons for shrinking diversity to some extent. Agriculture inevitably reduces biodiversity, because our desire for more of some species and less of others is crucial to agricultural efficiency. Whether this is a bad thing depends on whether you would rather eat or observe diverse wildlife and vegetation. The two are compatible, up to a point, in less intensive agricultural regimes, but all farmers still want to change land use and fight pests.

Then there is the suggestion that all those plants, insects and bacteria we have not even discovered yet, and certainly have not taken a close look at, are stores of hidden treasure. Some may offer genetic resources for new crop hybrids, for example. More broadly, there are probably many biologically active chemicals being made by organisms whose detailed composition has yet to be examined. Some of these, conservationists argue, could be the basis for vital new medicines.

The precedents are impressive. Rainforest organisms have yielded older remedies like quinine for malaria (from the cinchona tree) and the muscle relaxant d-tubocurarine (from curare). More recently, a number of anti-cancer agents have been isolated, such as vinblastine from the rosy periwinkle plant. And soil microbes have been the source of antibiotics such as erythromycin.

It is certainly possible that more systematic "bioprospecting" could yield undiscovered riches, perhaps from herbal remedies already known to indigenous populations but not yet studied in the lab. However, plants and micro-organisms evolve defensive chemicals for their own reasons, so their efficacy in other organisms is always going to be a haphazard fit with medical requirements.

Moreover, finding new compounds from living organisms is likely to be overtaken by new strategies for generating drugs in the lab. These include "brute force" approaches in which thousands of new compounds are synthesized and then screened for activity, singly or in combination. Elaborations of this can set up an automated system to repeat in cycles that gradually improve the selection, so that scientists can "evolve" bioactive compounds faster than nature does. On the other hand, our rapidly increasing understanding of cell biology means that highly specific targets for drugs can be specified and molecules designed – or selected – to interact precisely with the target. Either route is likely to prove a more reliable method for improving future health care than waiting for miracles to emerge from the rainforest.

MORE FUN THAN NATURE?

Another argument for the preservation of biodiversity is the notion that we need wildlife for our own wellbeing – it shows us the world we came from, and are part of, as well as being richly rewarding to contemplate. Conservationists like this argument because they love nature and they hope others love it, too. The pleasure of close encounters with the natural world can support ecotourism, which contributes to efforts to preserve biodiversity in many countries. Direct encounters with wildlife do seem to offer spiritual nourishment. So when considering future efforts to maintain biodiversity, it matters rather a lot whether this feeling is itself natural. Is it one which we all have just because we are human? Or does it need to be learned, and might it thus disappear along with access to the wild?

The strongest claim for the universal need for nature is made by biologist E.O. Wilson, who dubbed it the biophilia hypothesis. We evolved, Wilson suggests, to crave close contact with other living things, and it is something modern humans retain because of the millennia during which our ancestors were in constant, intimate contact with the natural world.

Even if this is true, though, it is not obvious how it relates to biodiversity on a global scale. Our biophilic needs might be met by a stroll in the park, a walk with the dog or a few flowers on the windowsill, rather than a sight of the rainforest or a dolphin-spotting trip. Perhaps trainspotting is even as satisfying as birdwatching?

There is also some evidence that the number of people seeking stronger contact with nature than they can get near home is decreasing. Oliver Pergams of the University of Illinois showed in 2006 that visits to national parks in the US have gone down by a quarter in the last two decades, and continue to decline by around one percent a year. Two years later the trend was confirmed in a broader study that looked at other nature-based recreation in the US, and also included data from Japan and Spain. The data suggests that the time once spent visiting national parks is now spent playing video games, surfing the Internet and watching movies – videophilia is displacing biophilia, as they put it. The finding could be misleading, though, as a more extensive study published in 2009 by a group led by of Cambridge University reports that a survey of 280 protected areas in 20 countries showed a decline in visits in the US and Japan, but a general increase elsewhere.

However, other studies indicate that adults' attitudes to the environment are influenced by early experience, so city dwellers who do not visit the countryside are less likely to make experiencing biodiversity a high priority as they grow up. Humans are the most adaptable species, and perhaps they will find it easy to adapt to not seeing too much of many of the others.

At the same time, while ecotourism underpins some conservation projects, there are also more opportunities for virtual experiences of the natural world. These extend beyond interactive electronic media and zoos to new places to visit. "Dubailand", for example, a megaproject now taking shape in the desert of the oil-rich United Arab Emirates, is the most ambitious tourist destination ever created, according to its promoters. One of the six themed zones planned is Eco-Tourism World, which the publicity describes as "a series of nature and desert-based attractions integrated within their desert parkland surrounds". It will compete for the tourist dollar with Retail and Entertainment World, Attractions and Experience World, Themed Leisure World, Sports World, and a downtown area which will include Virtual Games World and Teen World. It expects to be open for business between 2015 and 2018.

The scale of the investment means that this kind of escalation of the Disneyland tradition will be limited to a few sites. But it does indicate a possible future in which unadorned nature has less appeal, at least for some affluent tourists, than its custom-designed, somewhat artificial substitute.

ECOSYSTEM SERVICES UNDER THREAT

Qualms about declining biodiversity are linked to anxieties about general environmental degradation. This is now often referred to in slightly self-centred-sounding terms as a loss of "ecosystem services" – things which natural systems provide to humans for free, and which we would be hard put to replace if they stopped. The list of services is itself diverse.

It includes providing food, fuel, fibre and building material, production of oxygen (quite important, that one), soil formation and preservation, recycling of nutrients, and water filtration and purification. You might want to include pollination, important for crop plants as well as in the wild (see box below). Other less tangible ecosystem benefits, like making people feel good, are also sometimes included, but are annoyingly hard to quantify. Efforts to measure how well the whole set of services are holding up, and to put some kind of price tag on them, are a big part of conservation discussions worldwide.

How is all this connected with biodiversity? Again, the answer is a bit unclear. Degrading ecosystems contributes to the loss of organisms and, eventually, species. And there is some evidence that the more biodiversity an ecosystem has, the more robust it is in various ways.

BYE BYE, BEES?

As well as living in the wild, honeybees are part of a reasonable-sized industry. Those nurtured by beekeepers have been the focus of much concern in recent years with the advent of "colony collapse disorder", in which the entire population of a hive – except for the grubs and the queen – essentially disappears.

This was widely seen as a larger threat to agriculture, as many crops – including most fruits and vegetables – need the bees for pollination. It is a matter of definition whether this counts as an ecosystem service, as many bees are shipped in using mobile hives when pollination is needed. But since the causes of the loss of bees were obscure, the disappearing hives were widely seen as harbingers of environmental damage.

However, a closer look at the disorder is a little less alarming. Although the term is new, bees have always been vulnerable to colony loss. Losses in the USA in the winter of 2005–06 were certainly unusually heavy, with three-quarters of the colonies in some areas disappearing. Reports of colony collapse followed from Europe and elsewhere, and it appeared there was a possibility of a worldwide bee shortage. Explanations included exposure to pesticides, GM crops, and even mobile phone radiation, as well as parasitic mites and viruses.

Bees imported from the Primorsky region of Russia. They are more resistant to disease and pests than US or European bees.

Subsequent research seems to point to the last, possibly in combination with viruses carried by mites. There have been good results in the US with imported bees from Russia that are resistant to one common mite, although investigation continues as to which bees are vulnerable to what kind of attack. Meanwhile, winter losses from commercial colonies in the US reverted to lower levels in 2008–09, so current indications are that bees are not about to disappear altogether – yet.

The data here is inconclusive, though. It is possible that less complex ecosystems could still provide much of what we want from the natural world. The deliberately simplified ecosystems of agricultural production are an extreme example, and frequently come with their own problems. But perhaps a slightly more carefully managed approach, with a little more biodiversity, would be sustainable.

PREDICTIONS FILE

Norman Myers

Professor and Visiting Fellow at Green College, Oxford, and at the James Martin Twenty-first Century School

Highest hope: The world's political leaders decide to do some leading, and alert us all to the fact – yes, the undeniable fact – that we are engaged in the opening phase of a sixth mass extinction. That humanity bestirs itself sufficiently to save the 34 "biodiversity hotspots" which cover little over two percent of Earth's land surface, yet contain the last remaining habitats of almost half of all Earth's species. Since these hotspots were first identified and defined in the late 1990s, the amount of money mobilized to safeguard them has reached $1 billion.

Worst fear: Our despoliation of the biosphere will leave an impoverished planet and a depauperized world for at least one million years (to judge from the aftermath of Earth's five previous mass extinctions).

Best bet: We shall undertake some of the activities outlined in my highest hope, while failing to recognize that extinction is irreversible – in marked contrast to all our other environmental problems.

It's not a cheering prospect. We might have to cope with biodiversity loss by moving to a world not of complete degradation, which should obviously be avoided, but of managed, low-diversity ecosystems – farms, plantations, managed forests, parks. There certainly need to be an increase in all of these. Will their overall impact be manageable? It might be.

Many ecologists are pessimistic about the prospects for preserving the richest, most biodiverse habitats for this reason. The problem, in a way, is not that disaster will necessarily follow, but that it might not. According to Robert May, reducing biological diversity will not necessarily cause ecosystem collapse:

> "It is entirely possible that we could be clever enough to live in a world that was greatly biologically impoverished in species and yet managed to deliver the natural services that we want."

May thinks this would be a dreadful outcome, comparing it to the world of the 1982 science-fiction film *Blade Runner*. While the comparison forgets the hero's final escape to a green landscape outside the grungy megacity that provides the setting for most of the action, we get the point. But if May is right, and the trend is toward a more uniform planet with fewer wild places, does that make the argument one of aesthetics, not survival?

WHAT CAN BE DONE?

It is understandable that people feel strongly about the loss of biodiversity. There can be a flavour of "repent, sinners" to prescriptions for protecting at least some of what we have left. For example, Paul Ehrlich (author of *The Population Bomb*, 1968) and his Stanford University colleague Robert Pringle suggested in 2008 that while prospects were bad, there were seven strategies which, if they were implemented, could preserve a good percentage of global biodiversity. Two of them are, it is fair to say, rather general – namely, "actions to stabilize the human population and reduce its material consumption" and, slightly more vaguely, "the fundamental transformation of human attitudes toward nature". However desirable those might be, are there more specific things we can try in the meantime? Perhaps there are. They also list: use of endowment funds and other strategies to support conservation areas; steps to make human-dominated landscapes hospitable to biodiversity; accounting for the economic cost of habitat degradation; land reclamation; and education and empowerment of people in the rural tropics.

One particular problem that endangers biodiversity – the introduction of the "wrong" species into new areas – can be tackled where there is a will. Well, that is, it can be tackled if the area in question is isolated and hence easy to protect from further invasion – in other words, an island. In the last few decades, hundreds of islands have been cleared of rats, goats, pigs, and even cats and rabbits. The results are good. Vegetation recovers and threatened species like birds and small reptiles regain their territory.

These are relatively minor efforts, though. The immediate key to slowing biodiversity loss – it seems ultra-optimistic to think of reversing it – is finding ways to pay for it.

Biodiversity strategies

International policy has set ambitious goals for alleviating the problem. In April 2002, the countries party to the International Convention on Biodiversity committed themselves to achieving a significant reduction in the rate of species loss by 2010. The target was endorsed by everyone from the World Summit on Sustainable Development to the UN General Assembly. However, the agreements appear to have had little effect. The 2010 *Global Biodiversity Outlook* conceded that none of the targets have been met, and in fact things have generally got worse. Most indicators of the state of biodiversity show negative trends, with no significant reduction in the rate of decline. More importantly, perhaps, the report concluded that although there have been some local successes, overall:

> "The five principal pressures directly driving biodiversity loss (habitat change, overexploitation, pollution, invasive alien species and climate change) are either constant or increasing in intensity."

A separate technical report on biodiversity scenarios for the twenty-first century concluded that the outlook is probably worse than previously realized. It suggested that the latest modelling reveals possible "tipping points" resulting from interactions – that had not been taken into account before – between two or more of the pressures driving species to extinction. The chance, for example, of a widespread dieback of the Amazon rainforest caused by the feedback loop of deforestation and climate change reinforcing one another has been "substantially underestimated" in previous assessments.

These reports also underline that the future of biodiversity depends on action taken in other areas. The main aims on the list for curbing the depletion of species in the coming decades, for instance, are improving agricultural efficiency worldwide and mitigating climate change. Add the fact that the biodiversity experts see tipping points occurring with a rise in average global temperature of just 2°C – and most climate scientists now expect a rise of 4°C by the end of the century – and the outlook seems poor indeed.

In the face of the Convention's admission that there has been little effective action so far, there's a lot of work going into economic analysis of the benefits of our ecosystems, to help persuade people that there is real payback if they're preserved, and potentially heavy costs if they're damaged. There is also an effort to convince residents of particular regions that their local ecosystem's inhabitants are worth more alive than dead. That's a tougher task, as it tends to demand a shift from arguments about the

FORESTS: WHERE BIODIVERSITY MEETS CLIMATE CHANGE

Humid tropical forests – what most of us call rainforests – are hotspots in more than one sense. They are the ecosystems in which many of our current concerns converge. They are a collection of biodiversity hotspots; they are under pressure from agricultural development, which is likely to grow as populations increase; and they are at the centre of concerns about global carbon cycles, since removing forests releases stored carbon. As a consequence, they are also policy hotspots. The tropical forests are on the front line of efforts to work out the indirect costs, or externalities, of exploiting natural resources – and to figure out policies which can take these costs into account in ways that will be effective on the ground.

The interconnections between key issues that play out in forests make things complicated, but may turn out to be good news for protecting the remaining biodiversity. Their role in modulating carbon dioxide in the atmosphere has focused attention on forests in a way that concerns about biodiversity alone never quite managed. Forest advocates estimate that deforestation causes around twenty percent of global carbon emissions as the trees are cut and burnt.

One of the few bright spots in the outcome of the 2009 Copenhagen climate summit (see p.132) was a boost for forest preservation as a way of combining action on climate change with the conservation of biodiversity. The countries taking part all endorsed a proposal to create a new version of a UN initiative called REDD (Reducing Emissions from Deforestation and Degradation) – REDD plus, which will channel funds from developed countries to developing countries that protect their trees.

There are various ways this might be done, including using the global carbon market that emerged in the wake of the 1997 climate summit in Kyoto. The US Union of Concerned Scientists concluded after Copenhagen that "it is now clear that all countries see REDD-plus as a fundamental tool for addressing climate change". Some results are already emerging: Brazil passed legislation in December 2009 calling for an eighty percent cut in deforestation there by 2020.

Tackling forests offers an appealing strategy for reducing carbon emissions because it is relatively cheap and gets results fast. A review for the British government in 2008, which followed up the Stern Report's global cost-benefit sums for climate change (see box on p.132), came up with numbers suggesting that $17–33 billion would have to be spent through the carbon market annually to halve forest sector emissions by 2030. But the benefits would be much larger, amounting to $3.7 trillion over the long term.

The UK-based Global Canopy Programme (www.globalcanopy.org) and Forests Now (www.forestsnow.org) are good sources for the latest developments on forest-related issues.

Illegal logging is rife in Brazil – as seen in this picture, taken near Anapu in Pará – and has led to record rates of deforestation in the Amazon rainforest.

indirect costs of biodiversity loss to actually giving people money to protect local species.

The strategy currently emphasizes biodiversity "hotspots", an idea first put forward by environmentalist Norman Myers in the late 1980s. Definitions of hotspots vary, but the idea is that there is more conservation "bang for the buck" in some places than others. Influential zoologist Edward Wilson estimates that fifty percent of known plant species and just over forty percent of mammals, birds, reptiles and amphibians are found in hotspots that take up just four percent of the Earth's land surface. He also calculates that all of them could be protected – which would involve both paying for the land and looking after the people who now depend on it – for around fifty billion dollars (which he reckoned was 0.1 percent of one year's global GDP).

Perhaps the best hope for raising such sums is the overlap between conserving the most biodiverse regions and protecting against climate change. Tropical forests are key regions in both contexts. While all ecosystem services are important, locking down carbon is the one that governments and citizens are getting ready to actually pay for. That ought to make it possible to find policies that channel some of the money committed to reducing carbon to people who live in or around the tropical forests, and to persuade them that those forests really are worth more if left alone (see box on previous page).

DOES THE WILD HAVE A FUTURE?

The interplay between food production, possible expansion of biofuels, population growth, and climate change and efforts to reduce it will be complex. One set of scenarios which capture some of the influences such factors might have on future life appeared in 2007 in Futures of the Wild, a project of the US-based Wildlife Conservation Society (www.wcs.org).

As usual, the scenarios are not predictions, but illustrations of where future trends might lead over the next ten to twenty years, according to various assumptions. Strikingly, all entail continuing rapid loss of biodiversity.

All these possible futures include the expected increases in human population, urbanization and globalization. The Society then singles out five additional key influences on the prospects for conservation. Technology tops the list, followed by four other areas to watch – energy, food and agriculture, information technology and biotechnology. Then it considers meat consumption and wildlife trade, economic development, environmental change, and energy prices and consumption.

Evaluating how the connections between all these things may play out and how they could affect wildlife and conservation, the report highlights three uncertainties as the most important:

- Will future environmental changes be slow and manageable or sudden and disruptive?
- Will cultural and societal attitudes shift toward stronger environmental protection and wildlife conservation?
- Will new technologies have a net positive or negative impact on global sustainability?

The answers may be unknown, but some of the ways they are linked can be plausibly predicted. For example, an environmental calamity or two, local rather than global, might well increase support for environmental protection and conservation. Slow change, on the other hand, passes unnoticed more easily and provides little spur to action, though it is not ruled out.

Some of the other interactions are much harder to foresee. One which comes up in several of the six scenarios this exercise offered is how attitudes to conservation could be affected by the progress of genetic engineering and synthetic

The first scenario sketches a world in which environmental change is gradual, support for conservation weak and growing slowly, and technology is doing more good than harm. However, there are more ambiguous developments. The US launches a large-scale "genome seed bank" program to sequence the genomes of all endangered organisms so future generations might be able to restore lost species. But this means the public cares less about species preservation, believing that organisms can be recreated at will by clever geneticists.

However, if biotechnology could work well enough to offer a genuine safety net for endangered species, it would be unlikely to stop there. Other scenarios suggest that genetic engineers tinker with existing organisms to create new varieties, and that zoos or reserves might be stocked with re-creations of creatures long extinct – the genetic difference between a woolly mammoth and an elephant, for instance, is unlikely to be very large. We won't see a real-life Jurassic Park, though, as dinosaurs died out too long ago for their genetic material to be preserved. But genetically enhanced pets could become all the rage.

Some are enthusiastic about the prospect of domesticating biotechnology, foreseeing a new generation of gene hackers banding together to make home-brewed pets much as dog breeders or pigeon fanciers created their own fertile subcultures in the past. According to physicist and technology visionary Freeman Dyson:

> "Do-it-yourself kits will become more available to everyone. You will be able to read and write your own genomes and produce roses and orchids and lizards and snakes or any kinds of creatures, according to your own design."

He reckons this could even extend to biotech games for children, "where you give the child some eggs and seeds and a kit for writing the genomes and seeing what comes out." It is a vision of a new kind of creativity that would reinforce connections with the natural world by redesigning it, piece by piece – though it perhaps neglects the downside of, say, the GM equivalent of the folk who dump the puppy they bought at Christmas when New Year comes.

This sounds like it's heading toward science fiction. But add in other developments such as an increase in virtual experience of the wild and a reduction of interest in the real thing, and this century seems likely to see an accelerated shift in relations between people and "natural" ecosystems. We will continue to expand urban areas, and simplify more and more of the remaining ecosystems on land to grow crops, fuels and perhaps fast-growing GM trees to soak up carbon. At the same time we will explore the history of life in virtual spaces, while creating its future in the laboratory and the marketplace.

If biodiversity continues to reduce worldwide, all this will be joined by a scramble to save creatures threatened by rapid climate change and other effects of human activity. All in all, the outlook for at least the larger kinds of other species appears poor. Their best hope, as detailed in Alan Weisman's *World Without Us* (2007), is for humanity to simply disappear. But that is just a thought experiment – isn't it?

FURTHER EXPLORATION

Eric Chivian and Aaron Bernstein **Sustaining Life: How Human Health Depends on Biodiversity (2008)** A review of the science of how humans benefit from other species, from medicines to food, and how to preserve them.

Thomas Lovejoy and Lee Hannah **Climate Change and Biodiversity (2006)** A comprehensive reference to one of the main threats to biodiversity, though it's in need of updating.

Edward Wilson **The Future of Life (2002)** One of zoologist Wilson's series of books sounding the alarm about biodiversity.

WWF **2010 and Beyond: Rising to the Biodiversity Challenge (2008)** The leading wildlife and conservation NGO's latest report (downloadable from www.wwf.org) on actions that need to be taken to protect biodiversity in the years to come: hope springs eternal...

www.cbd.int The Convention on Biodiversity's site includes its secretariat's reports on the field's current state of affairs.

12 future health

future health

Medicine of the future will be predictive and preventive, examining the unique biology of an individual to assess their probability of developing various diseases and then designing appropriate treatments, even before the onset of a disease.

Institute for Systems Biology, Seattle

In health, as in other aspects of life, some of us are incomparably better off than our ancestors. In affluent countries, the chance of living out the biblical span of three score and ten years is greater than it has ever been – and increasing numbers of people live a good deal longer than that. Spreading the benefits of modern medicine more widely ought to be possible in the coming decades. At the same time, technological advances promise further improvement, at high cost, in the richer countries, perhaps resulting in even longer lives than ever before. If that happens in just a few places, a new dimension of inequality will unfold.

THE GLOBAL HEALTH OUTLOOK: HEALTHIER

Overall, the immediate global health outlook is not too bad. Although massive global inequalities extend to health, health services and disease, these are slowly reducing. In 2006, the World Health Organization (WHO) published global projections of death and disease up to 2030 which indicate steady overall improvements.

Before examining these numbers, and how they are calculated, a word about how to read them is advisable. As health relates directly to life and death, writing about the big picture can make anyone sound like Stalin, who reputedly said that a single death is a tragedy, a million deaths a statistic. For their nearest and dearest, every one of the million is a cause for grief. Yet, with a world population of six and a half billion people and rising, a million deaths, dispersed in time and space, can be a small part of the picture. So let's endorse the not very inspired headline writer's cliché – "too many deaths from [choose your disease]". Although it invites the question "how many would be enough?", there are too many deaths from plenty of diseases that could be tackled right now if we cared to. Malaria, for example, kills a million people a year, though effective, cheap preventive measures exist. UNICEF estimates that ten million children die annually from this and other preventable causes.

Even so, there are grounds for optimism that, overall, global health can improve. The WHO generated numbers using three scenarios that they labelled baseline, optimistic and pessimistic. In truth, all are fairly optimistic. All three indicate a shift in the distribution of deaths from the young to the old. This is accompanied by a diminishing toll from infectious disease, poor nutrition, and death in childbirth (for mothers) and during or soon after birth (for babies). The basic scenario has the proportion of deaths from noncommunicable diseases rising from 59 percent in 2002 to 69 percent in 2030. Life expectancy rises everywhere. The size of the increase, but also the uncertainty in the projections, is largest where the average lifespan is now shortest – in sub-Saharan Africa and South Asia.

In other words, more parts of the world can realistically hope to make the shift already seen in the first countries to industrialize. The result will be longer lives, most of which end with one of the illnesses we associate with old age, such as heart disease or cancer. The overall figures suggest other positive and negative trends. By 2030, it looks like the biggest causes of death or ill health will be HIV/AIDS, depression – which does not appear in the mortality figures – and heart disease. Smoking-related deaths are likely to rise as more people in developing countries take up the habit.

Deaths and injuries from car accidents are also set to increase as rising incomes increase car ownership. Although trauma care can now often cope with people with very severe injuries, immediate death in a car crash still awaits perhaps two million people a year in 2030, unless there is a major change in our curious tolerance for this risk. The optimistic scenario assumes that strong economic growth in current low-income countries will cut death rates from infectious disease and reduce child mortality, but the same growth also has its downside.

The numbers depend on lots of assumptions – about economic growth rates, how many people will smoke, how many will get anti-AIDS drugs, and so on. Barring disasters, the general drift is probably correct, though. And even real disasters only produce modest changes in the overall global figures. The main error in the first attempt at this exercise, in 1996, was an underestimate of the impact of AIDS. The number of deaths from AIDS is now put at 2.8 million in 2002, with a mid-range projection of 8.3 million in 2030. But even an epidemic on this scale is not enough to make much of a dent in overall global health improvement, though the impact on some countries is devastating.

What else might happen to make things appreciably better – or worse – than these statisticians' projections? And how will we judge what counts as improvement? There are plenty of promises of new breakthroughs in treatment and prevention. But the amazing developments in health care of the twentieth century seemed to leave people "doing better, but feeling worse". Longevity is a plus, but it also means more time for illness, as medical historian Roy Porter pointed out at the end of his magisterial history of medicine and medical care, *The Greatest Benefit to Mankind* (1998).

Other tensions mark the future of medicine and health as well. Research promises great things, including the tantalizing possibility of extending life still further. On the other hand, the (relatively) rich often risk denying themselves longer lives by using their money to eat and drink themselves to death. And even the wealthiest countries are likely to have trouble affording the cost of health care as patients and providers – including those seeking commercial gain – join in raising expectations of what can be done for illnesses of all kinds. Future health care may well benefit from what will seem like miracle cures. Few of them are likely to be cheap.

NEW DISEASES, OR OLD SCOURGES RESURGENT?

While there are real hopes for improvements in health, there may also be setbacks. Economic or environmental calamity would wreck the optimistic statistical projections. But the main directly health-related setback would be a new pandemic.

The very worst fears are based on experience of past catastrophes, amplified by apocalypses of the imagination. A plague-driven die-off has been a science fiction staple since Mary Shelley wrote *The Last Man* (1826), which has been followed by a host of literary and cinematic visions of humanity succumbing to a new disease. The Black Death killed between a quarter and a half of the population of Europe in the fourteenth century, and probably affected China even more badly. More recently, the worldwide epidemic of Spanish Flu in 1918–19 caused 25–100 million deaths (we only have good estimates for the more developed countries of the time – it is now thought likely that as many as 50 million people died in India). However, a new Black Death is not really plausible.

Even if a completely new disease appeared that humans had never encountered before, which is unlikely now that we have spread and mingled over pretty much the entire globe, it would be quickly characterized and monitored. The SARS virus, for example, which first appeared in November 2002, was isolated soon afterwards, and its complete gene sequence worked out by the following April. The Black Death spread in a world in which no one had any real notion what caused it, let alone use of modern pharmaceutical and anti-viral technology. Add modern surveillance systems and the likelihood that a truly deadly disease would kill people before they had much of a chance to spread it, and the chances of containment would be good.

On the other hand, a less deadly but still infectious agent would be easily spread, especially through air travel. It would also find hosts in many newly mushrooming cities, which are often overcrowded and poverty-ridden, so we need to keep a close watch for signs of such outbreaks. The Global Viral Forecasting

PREDICTIONS FILE

James Pinkerton

Fellow at the New America Foundation, and blogger about health and health systems at seriousmedicinestrategy.blogspot.com

Highest hope: that we do what the American people want us to do, which is to fight a real war against disease.

Worst fear: that we will continue to miss the point about health care, which is the search for cures, as opposed to the search for health care-financing mechanisms.

Best bet: I am confident that the march of science will push toward new medical breakthroughs, and that the stall of the last decade or so is only temporary.

Initiative, run by Nathan Wolfe at Stanford University, is setting up an early warning system that can spot when a new virus jumps from animals to humans. It was started with modest US government funding, but got an $11 million boost in 2008 with a grant from two private-sector donors, Google.org (Google's philanthropic branch) and the Skoll Foundation. Keep tabs on their progress at www.gvfi.org.

None of this means that a new epidemic couldn't happen, just that it wouldn't match the horrors of centuries past. Most experts believe the most likely threat, by some margin, is a virulent new strain of flu emerging, probably from Southeast Asia, and being quickly spread worldwide by air travellers. As health planners warned at the outset of the 2009 outbreak of swine flu, this might result in around a fifth, a third or even half of the world's population falling ill. The fact that the mass infection governments feared did not materialize does not mean, as some commentators have since claimed, that the risk was never there, just that some critics don't understand the meaning of the word "risk".

Virulent strains of flu are quickly spread by air travellers: staff at Hong Kong's airport wear face masks during the swine flu outbreak in April 2009.

If a more contagious viral strain did appear, the economic, social and political effects would be immense, though probably short-lived.

A new strain of flu, derived from strains that now affect other species, such as birds or pigs, could cause a severe and widespread illness. But flu evolves quickly as different strains of the virus mix their genes – that's one reason why new outbreaks keep happening. It's just as likely that a strain that could readily infect humans would be more like existing human-resident types, and cause a condition which for most people, while miserable, would not be life-threatening. Another reason for such a prediction is that a dead host can no longer spread a respiratory virus. So it is in the virus's own interest, as it were, to cause a mild illness, with lots of coughs and sneezes along the way, rather than a mass killer.

So much for new viral nasties. Other future risks will come not from new diseases, but from old ones that can get around our protection – antibiotic-resistant bacteria. The early successes of antibiotics after World War II, and the wider use of vaccines, prompted happy predictions of the end of infectious disease. These were dead wrong. Bacteria are small and reproduce very fast, and so were always very likely to evolve resistance to the drugs.

The spread of multi-resistant bugs is a cause for concern, but such concern is a spur to research. Public health organizations have criticized the big drug companies for neglecting research and development aimed at new antibiotics. They may prefer to work on more profitable drugs for chronic diseases rather than medicines that cure people after a short course of use. But if old antibiotics fail, there will be handsome profits to be made if you happen to have a patent on a new one. An industry report in 2007 unearthed 370 potential antibacterial drugs in development. Many are small variations on existing drugs but some, mainly those in earlier stages of development, are completely new. The effort to control infection is not a matter of once-and-for-all success or failure, but will keep evolving, along with the bacteria.

TROPICAL DISEASES EVERYWHERE?

Could global warming lead to big reversals for public health? If tropical diseases were to become a serious threat to the affluent nations of the Northern Hemisphere, it might strengthen motivation to reduce greenhouse emissions, but this seems pretty unlikely. The most detailed scientific reviews suggest that the health impact of climate change will be greatest in the poorest regions, especially Africa. As with other effects of global warming, they are also likely to be the least able to deal with them. But beyond that, the precise effects are highly uncertain.

A hotter climate is itself bad for some people. However, as the World Health Organization's latest assessment of possible climatic effects points out with clinical detachment, "we do not know to what extent deaths during thermal extremes are in sick/frail persons who would have died anyway". In addition, increased deaths during heat waves might be accompanied by reductions in deaths from extreme cold in the winter months.

The WHO also estimates that the risk of diarrhoea in some regions could go up by twenty percent by 2030, but stresses that there is little data to go on to support this. There are also worries about infectious diseases spread by insect vectors, such as the mosquito-borne malaria and dengue fever, which can increase their range in warmer climates, although the insects need moisture as well as heat. However, the WHO suggests that temperate regions will not suddenly see the advent of, say, malaria. They will either remain outside the climate conditions that allow the mosquitoes to breed (most of Europe) or have good prevention already in place to prevent reinvasion, as in the southern states of the US.

The overall verdict is that yes, climate change is already affecting the pattern of disease, and the risks will increase. However, the current impact is small compared with other risks to health. If the worst climate scenarios materialize in the latter half of the twenty-first century, there will be graver health effects, but they are likely to be less important than other serious issues, like lack of food and water in hard-hit regions.

RESEARCH FRONTIERS

Health systems are perennial budget busters. But research, which costs a lot less, is now done on a pretty awesome scale. By far the biggest spender is the US, whose biomedical research outlay in the public and private sectors runs to nearly $100 billion a year. (This is far more than has ever been spent on energy research, incidentally.) In comparison, the WHO's Tropical Diseases Research program, one of the main efforts to tackle neglected conditions, which affect more than one billion people worldwide, spent $50 million in 2007.

Where might the big-spending research of the wealthy nations lead? Getting results in medical research depends partly on improvements in basic knowledge, but also on investigating specific conditions. In the US in 2007, the leading causes of death as written on doctors' certificates looked like this:

- Heart disease: 615,661
- Cancer: 560,187
- Stroke (cerebrovascular diseases): 133,990
- Chronic lower respiratory diseases: 129,311
- Accidents (unintentional injuries): 117,075
- Alzheimer's disease: 74,944
- Diabetes: 70,905
- Influenza/pneumonia: 52,847
- Kidney disease: 46,095
- Septicemia: 34,851

Another way of looking at it is to assess "actual" causes of death, the ones defined rather discouragingly by the US Centers for Disease Control as lifestyle and behavioural factors that contribute to death. According to their slightly older figures, the top contributions to these theoretically preventable deaths in 2000 were tobacco (435,000), poor diet and physical inactivity (400,000), alcohol consumption (85,000), microbial agents (influenza, pneumonia, etc, 75,000), toxic agents (pollutants, asbestos, etc, 55,000), motor vehicle accidents (43,000), firearms (29,000), sexual behaviour (20,000) and illicit use of drugs (17,000).

It's striking that the top three things people do (or fail to do) which spoil their chances of enjoying a spry old age contribute heavily to deaths from cancers, heart disease and strokes. They reinforce the need for researching not only cures, but also the disease prevention that would result by simply targeting the behaviour that causes the illnesses in the first place.

Life is not so simple, of course, and much research focuses on genetics (which contributes in many subtle ways to disease risks), cell biology and specific conditions that are likely to become more common. Diabetes, for example, looks set to become a growing problem in the US where more people are overweight. Alzheimer's will also become more prevalent as the population ages, unless there is a research breakthrough. Some other countries are seeing similar trends, but not all. In Russia, for example, the current population decline is due to a rise in conditions such as HIV, tuberculosis and alcoholism, particularly among men.

Still, the diseases at the top of the cause-of-death list do attract the largest share of research funding in the developed countries where the vast majority of the work takes place. There is a massive research effort into heart disease and treatment, for example,

FUTURE HEARTS

Good news: the death rate from cardiovascular disease has come down by half in the last forty years in most developed countries. Bad news: as the incidence of obesity rises, heart disease deaths are likely to go back up again.

The overall statistics hide a complex web of causes and effects. That fifty percent cut is measured from a time when deaths from coronary heart disease, in particular, were alarmingly high. And despite the impressive reduction it remains the biggest single killer of men (21 percent of them in the UK in 2004, for example) and women (15 percent) in affluent countries. Counted together, cancers have recently overtaken heart disease as a cause of death in the UK, though not quite in the US, but no individual cancer rivals heart disease.

The fall in heart disease deaths is due to successful campaigns against smoking, the introduction of a very effective class of drugs (cholesterol-lowering statins), and improvements in emergency care for heart attack victims. But these may soon hit their limits.

The determined eaters of America offer a preview of the problems that may soon face the affluent in other countries. The proportion of people classified as obese rose in the US from 16 percent to 31 percent between 1980 and 2003. As a result, *The Journal of the American Medical Association* reported in 2006, white people in the US have double the rate of diabetes and fifty percent more heart disease than their British contemporaries.

Future affairs of the heart will be shaped by these conflicting factors. Diet and exercise advice will persuade some to lower their risk of a coronary artery blockage. Taking drugs for life will suit others better, although statins' side effects are a disincentive. And there are a host of high-tech options which, in theory, could help those whose heart or arteries need repairing. Surgical techniques will continue getting better, and less risky. Cell research may make it possible to regrow portions of damaged heart muscle. Using animals as a source of transplantable hearts is still a real possibility, if concerns about possible virus hazards from the animals of choice – pigs – can be overcome. If not, there are a range of designs for sophisticated electromechanical hearts which benefit from improved understanding of the dynamics of the biological prototype.

Whichever of these developments proves most important, doctors are still likely to be hard-pressed to keep hearts beating in people who exercise little, smoke or eat large amounts of industrially produced food that's cleverly tailored to old, evolved appetites for fats and sugars. Ever-increasing incidence of obesity, diabetes and high blood pressure is the true affluenza.

and it has been more successful than most people realize. Whether that success can be sustained is harder to say (see box on previous page).

Despite our unhealthy habits, there are real hopes for research-based improvements in prevention or treatment of many conditions. How realistic are they? Let's take that question in stages, looking at particular areas of research.

STEM CELLS AS MIRACLE CURES

It is safe to predict that the next few decades will see a steady stream of research breakthroughs which are hailed as harbingers of miracle cures, only to prove disappointing. That is not to say that none will work, only that they will take longer, and be more complex to use, than initial reports suggest. The current most plausible source of such reports is stem cell research. It's plausible because knowledge of how cells work is improving very rapidly, and making it more likely that we will be able to manipulate their development.

This is promising because the different types of cells in our tissues and organs develop from less specialized ancestors. Adults retain some less specialized stem cells, which can reproduce and replenish one or a few cell types – blood cells, say, or muscle cells. Embryonic stem cells can grow into any cell type: they are totipotent. As we unravel how all this works, it ought to be possible to steer the process so that a stem cell can be induced to produce any type of cell we choose. At the moment, controversially, the research often utilizes stem cell lines derived from embryonic tissue, but better understanding will soon bypass this requirement.

There have already been successful experiments designed to make cells from adult tissues behave more like stem cells, by manipulating the cellular controls that govern the development of different types of tissue. Soon, there will be better control over differentiation, as biologists call it. And the treatments that develop from this will be based on turning on or off genes that the cells already have, rather than engineering new ones.

So getting a handle on stem cells, or even switching regular cells back into the same state, is an incredibly interesting idea for researchers trying to tackle an injured organ – whether it be a heart muscle damaged by a blocked artery, a liver scarred by alcohol, or a brain depleted of vital cells by stroke or degenerative disease.

In principle, this could be a route to curing lots of diseases which result from having too few of particular types of cells, as well as repairing damaged tissues. There are already promising results, or trials planned, for some basic but important repair jobs. They include cell injections to speed recovery of heart muscle after a heart attack and potential treatments for the aftereffects of strokes caused by blood vessel blockage (which triggers the majority of strokes – the rest are due to internal bleeding). Other work is focusing on cancer treatments, diabetes and Parkinson's disease. With all of these innovations, it may be possible to get good results by increasing one cell type within an existing organ, which already has its architecture in place. However, despite the Internet sites already offering cure-all therapies to the gullible or desperate, the practice will be complex, and real benefits will only arrive slowly.

A good simple example is diabetes. There is a real chance of fixing so-called type 1 diabetes – the kind that tends to appear early in life and leaves people needing to inject the blood sugar regulator insulin. It arises because one cell type in the pancreas fails to make the insulin hormone. So arranging for the right kind of stem cells might result in a new supply of insulin being made in the patient's own organ.

In fact, it may prove even simpler. In 2008 a Harvard University team reported that they had been able to tweak the gene regulation of the more common pancreatic cell type – so-called exocrine cells – and turn them into the rarer insulin-producing cells (in mice). This rather cleverly avoids the need to get them to turn back into pancreatic stem cells first, and is a nice example of the kind of finer control over cell types researchers will be seeking.

On the other hand, while it might liberate diabetes patients from a lifetime of hormone injections, type 1 diabetes is already reasonably well managed. So this is a case where a revolutionary treatment would have relatively little effect on health care overall. Ninety percent of diabetics have the more complex type 2 diabetes, which most often arises when tissues become insensitive to insulin, not from cell loss in the pancreas. And it is that condition, formerly known as "late onset" diabetes but now increasingly found in younger patients, too, which is becoming more common as the incidence of obesity increases.

A good complex example of how stem cells might lead to revolutionary treatments is taking place in heart research. Work on rats and mice shows how stem cells from embryonic lines, or perhaps from bone marrow, could help damaged hearts. But they have to do three things exactly right, in order. Stem cells for treatment would have to find their way to the injured region of the heart muscle, presumably after injection nearby, and make new heart muscle cells in enough numbers to make up the deficit. Then the new cells would have to join the rhythmic beat which enables the heart to do its job.

The fact that many other species – including amphibians, fish and molluscs – can already repair their own cardiac muscle encourages the belief that this might work for us. But there is more work to do on analysing the genes that control this in the different species under study, and figuring out how this can be used to coax stem cells into doing the right thing in a patient recovering from a heart attack.

So overall, stem cells, like other "miracle" cures, will produce mixed results, and become part of a

PREDICTIONS FILE

Alistair Tweed

Director, Ageing Management Ltd, UK

Highest hope: There will be enough progress in anti-ageing science to deliver significant real world results.

Worst fear: I have two. That lifespan will increase and will not only leave us with extended quantity, but reduced overall quality of life. And that any life extension techniques will be used to extend working life at the expense of leisure, as so many other technologies that promised to free us have just enabled us to work harder for longer. Employers offering only their most productive workers a year's life extension as a yearly bonus would result in a tensely competitive and rather miserable existence for the successful, and bring a whole new meaning to redundancy for the less talented.

Best bet: There will be marginal progress in life extension and people will be free to choose what to do with their extra years.

more varied, and often more effective, set of options for doctors. But they will not eliminate cell-based diseases any time soon.

RE-GROW YOUR OWN

In the ultimate vision of a stem cell-enabled future, cells – perhaps taken from the patient's own tissue – would be cultured, geed up and injected back into the body to heal, repair or even grow enough tissue mass of just the right kind to replace whole organs. But before getting too excited about our imminent ability to grow a new spleen, or a replacement kidney, look at what we need to achieve to do something a bit simpler – get some new teeth.

In principle, teeth look like an easy target. We normally grow two sets already, but only two. This is a bit inconvenient now that humans tend to live longer and eat loads of sugar. But the machinery for replacing absent teeth clearly already exists. Besides, some creatures – fittingly, in view of their reputation, sharks and crocodiles are the best known – replace old teeth throughout their lives. Perhaps all we need to do is understand how this works a bit better, and dental implant specialists can all be sent for retraining.

The basic idea resembles other stem cell treatments. Get hold of some stem cells of the right type, perhaps from embryonic tissue or adult cells from wisdom teeth; culture them in the lab; and implant them into the jaw. Sounds simple? Maybe, but the tooth bud has to "know" quite a few things to develop correctly. Is it in the upper jaw, or the lower? Is it a molar or an incisor? And it has to start and stop growing at just the right time, so that it implants properly in the jaw, pointing in the right direction, and ends up nestling comfortably next to its neighbours. Anyone who has had an impacted wisdom tooth knows why.

Or perhaps the bud just needs to be induced to grow into a generic toothlet, which can support an artificial crown. Even so, it will need to generate nerve cells, soft tissue for the pulp inside the tooth, and the crucial hard enamel. The whole thing has to be firmly anchored in the jawbone, but with openings for the nerves and a blood supply.

As with other organs, this can obviously be done. It is happening every day in normal development, when we lose and grow new teeth at a young age. But making it all happen to order will still be no mean feat. It's understandable, then, that news reports of stem cell implants in animal models that produce something resembling teeth invariably say clinical trials are five to ten years away. In medical research, "five to ten years" is code for "we don't really know".

So while teeth are an excellent prospect to be the first stem cell-derived tissue of use, it would be no surprise to wait five years and still find real results are just as far away. I would love to believe otherwise, as conversations with my dentist all too often include the words "root" and "canal", and the promise of a hefty bill, but looking after your own teeth would probably be a better bet than relying on stem cell replacements, for the next two or three decades at least. That might still mean that harvesting stem cells from baby teeth and putting them in store could turn out to be a good investment for the parent who wants to think of everything.

GENETIC MEDICINE

Genetics will produce a wealth of results relevant to medicine, though they might not be as useful as supporters of the effort to map and sequence all the human genes (the genome) believed in the 1990s. That effort was accompanied by predictions of cures for cancer, better understanding of mental illness, and the advent of a new age of "predictive medicine" in which everyone would know what to avoid doing to minimize their personal disease risk. Such hopeful forecasts focus on human genetics, but knowledge of the genetic details of other organisms will also be important. It is likely to be the best route to new and improved vaccines, for example, which could transform health conditions in some parts of the world if they're delivered where they're needed.

The catalogue of human genes available since completion of the Human Genome Project at the turn of the millennium opens the door for better understanding of how genes act, and interact, to affect the chances of disease. But the interactions, in most cases, are turning out to be more complicated than the architects of the project foresaw. We now know that few genes have simple, or large, effects on disease. Most conditions are affected by many different genes, as well as by developmental and environmental factors. And there are lots of new layers in the regulation of the interactions being uncovered. These include "epigenetics", biologists' term for how DNA is tagged to mark genes as active or inactive. They also include the use of many new kinds of the other DNA-like molecule, RNA, which carry information from one place in the cell to another.

All this will take time to unravel. This will still be "the century of biology", but cashing in all that genetic information is still a few decades off. However, the advent of detailed genetic profiles will give individuals a better idea of whether they are at high risk for particular genetically influenced conditions, even when many small genetic effects are involved. Will this lead to a new era for preventive medicine? Perhaps, but it will bring its own problems. Some people respond fatalistically to risk information, thinking it means there is little they can do. Some risk information is hard to convey responsibly to the people who might benefit. If there is a genetic signature that denotes an increased risk of schizophrenia, for instance, telling the person at risk could be a source of just the kind of stress which they find unhelpful.

More positively, genetic risk readouts for some conditions will be accompanied by improved technologies for self-monitoring. At the moment test kits for blood glucose or cholesterol come from the pharmacy, and blood pressure checks, for example, require access to a machine. But cheap, wearable technology will come into use that can monitor such things continuously and transmit the readouts to computer systems which will alert us, or our doctors, when something needs to be done. More elaborate gadgets could even include monitors that automatically call for help or administer basic treatment after a heart attack or stroke. As sensor technology and analysis of the results becomes more sophisticated, it will draw on the growing understanding of systems biology that some forecasters anticipate.

That will lead to a transformed medicine, according to Leroy Hood of the Institute for Systems Biology in Seattle, where "in the near future physicians will collect billions of bytes of information about each individual – genes, blood proteins, cells and historical data. They will use this data to assess whether your cell's biological information-handling circuits have become perturbed by disease, whether from defective genes, exposure to bad things in the environment or both". The "near future" referred to here, he says, means between five and twenty years. Physiologist Colin Blakemore agrees, suggesting that "in

IF THEY CAN PUT A MAN ON THE MOON...

Will our new, high-tech approaches finally produce a cure for the common cold? Well, they just might. Colds are caused by lots of different but related viruses. They are quite simple, as viruses go, with a genome around seven thousand DNA "letters" long, which tells infected cells to make just ten different proteins. As of early 2009 we have a complete gene readout of more than one hundred such viruses, which is enough to draw up a family tree. The tree shows which parts of the virus vary and which are constant no matter where or when the illness is found. Those constant regions ought to become targets for drugs (vaccines are not likely to work in the nose) which can destroy the virus or stop it reproducing. Presto: a cure!

What's stopping the scientists now, then? Well, it takes quite a few years, and quite a few hundred million dollars, to make a new drug and get it tested and approved. As colds are not a big deal for most people, unless they additional health problems, any new drug would have to work really well, and have no drawbacks, to get through licensing. And would the snuffling millions pay enough for a drug to ensure the company bringing it to market made a profit? Or would they just buy a cheap, ineffective over-the-counter remedy and soldier on?

In 1928, Professor Bordier from the University of Lyons, France, unveiled a device which he claimed could cure colds in a few minutes. Eighty-odd years on, a cure is theoretically possible – but don't hold your breath.

twenty or thirty years, people will have an implanted chip that will monitor a wide range of indicators of their state of health, coupled remotely to an Internet-based personal prevention diagnostic system."

In the meantime, genetic information, especially about variation between individuals, will find other uses in medicine too. It will lead to more accurate diagnosis of some conditions, including infections which will be rapidly identified through DNA analysis. (More detailed knowledge of genetics, incidentally, will also improve prospects for tackling some troublesome bacterial and viral infections – see box opposite.)

Risk factors for some common diseases will be catalogued and some people at high risk identified. In most cases, there will be a few people at high risk because they have one or two gene variants that make a big difference, others who have an increased chance of disease because they have quite a few genes that each raise their risk a little, and a larger group, probably most of us, who are not in the clear, but have just one or two of the low-risk gene variants. This pattern is now emerging for the risk of coronary heart disease, for example.

It will become clear, in fact, that everyone has some genes which are not quite all they could be. This information could end up being treated as something which is useful to know, along with how to eat well and exercise sensibly, but not – for most people – enormously important. On the other hand, if it gives a good guide to probabilities of particular conditions, it could create a large class of potential patients who are uninsurable, or who have to conceal their gene profile – perhaps legally – from insurers. In that case, maybe the higher quality of personal genetic information would increase pressure for universal medical care in countries which now rely heavily on private sector provision, cutting across controversies about "socialized" medicine.

As well as revealing indicators of disease risk, genetic variation will also be relevant to treatment when doctors prescribe drugs. The way pharmaceuticals are processed by the body affects whether they work and what side effects they may have. So knowing about individual biochemistry will be helpful in choosing one drug rather than another. The professional term for this is pharmacogenetics, and it is expected to help improve drug safety and efficacy.

Those are all ways of using the information extracted from genes, which will be easy to come by as the cost of sequencing an entire genome comes down. What about fixing the genes, though? Progress toward this long-anticipated goal is likely to be slower. Turning off genes to block their activity will get easier, with specifically synthesized stretches of so-called "antisense" DNA or RNA binding to one of the strands of the DNA double helix. That just leaves the problems of delivery to the right cells and, probably, repeating the treatment periodically.

Actually modifying genes or adding new ones is harder. We can do it fairly readily in other organisms – bacteria, plants and animals – but only by, in effect, discarding all the failures and picking one or two new specimens where the experiment went right. This would be frowned upon if applied to humans.

There is also a widespread ban on modification of germ cells (eggs or sperm) as opposed to somatic (body) cells on the grounds that we don't want reproducible modification just yet. That reluctance might be overcome if there were proven modifications that people really wanted, like immunity to a dread disease. The fact that gene therapy is one of the best hopes for a cure for AIDS (see box overleaf) increases the chances of that happening. While we wait for safe and reliable methods for gene-tweaking in humans, the possibility of making a genetic selection, for health or other reasons, will be open to more people through use of in vitro fertilization and related techniques. Here, though still controversial, setting aside embryos that are not quite up to specification is more acceptable. While the vast majority of people will continue to prefer making babies by the traditional method, assisted reproduction is now an established option, accounting for one percent of births in the US, for example, in 2005. (That's one percent of births, not one percent of pregnancies, since many IVF treatments lead to multiple births, but it still adds up to more than forty thousand babies.)

As IVF becomes cheaper, safer and more reliable, its use will be more attractive not just to couples who have difficulty conceiving, but also to people who know they are at high risk of a disease with a strong hereditary element. Preimplantation genetic diagnosis – in which a cell is removed from the very early embryo as it's developing in the lab, so its DNA can be tested – can prevent such conditions. At the moment it is in use for diseases like cystic fibrosis and to select embryos free of genes which carry a much larger risk of developing certain cancers, such as breast cancer. When a complete genome readout becomes possible as a matter of routine, exactly which embryos to select, and on what criteria, will get much harder to decide.

TACKLING AIDS AND HIV

Despite huge efforts, there is little sign of a working vaccine against HIV, the virus that causes AIDS. And while there are antiretroviral drugs (ARTs) that can keep the disease under control, you have to take three at once to combat viral resistance. They are also expensive, and often have unpleasant side effects.

Some people, however, have an inborn resistance to AIDS thanks to a mutation which prevents the virus from latching on to white blood cells. In at least one case, a bone marrow transplant from a donor who already enjoyed this protection has cured a patient infected with HIV.

That method couldn't treat thousands of people, but it also happens that white blood cells are a good target for genetic modification because they can be "harvested" from the blood, processed, and then reinfused into the patient. Inactivating the gene that makes the relevant cell-surface protein seems to work in the same way as the natural mutation does. And animal models suggest that, as hoped, this is effective even if only some white cells are treated, since they survive when they are back in the bloodstream while the unaltered ones are slowly destroyed by the virus.

This is only one of several possible ways gene therapy is being developed in the search for effective anti-AIDS measures. But the work is likely to progress rapidly because of the size of the prize and the accessibility of the cells which are the target for treatment.

ROBOT SURGERY

Of the many applications of information technology in medicine, using robots to treat people is one of the more outlandish, but is nevertheless an active research area. Wars have often helped push along medical technology, and they are now spurring efforts to achieve new levels of automation. The film and TV series *M*A*S*H* popularized cynical-but-heroic surgeons in Korea, its acronym standing for Mobile Army Surgical Hospital. Trouble was, it was never all that mobile. Studies since Korea have

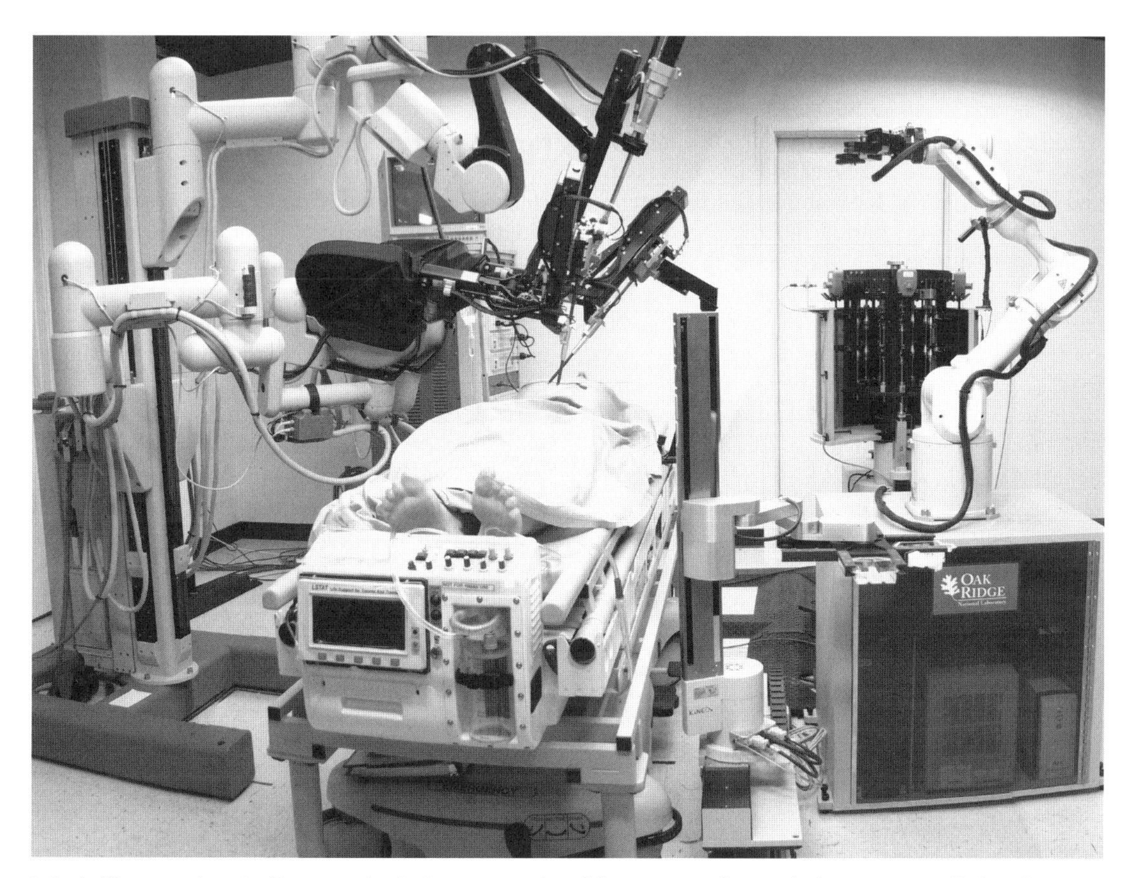

Robotic lifesavers: the US military are developing unmanned, mobile "trauma pods", in which remote-controlled machines can operate on injured soldiers on the battlefield, while the human surgeons stay out of harm's way.

shown that cutting the time before the wounded get medical help increases survival rates a lot. The solution? Buying more and faster helicopters for evacuation helps, but creates more targets. Moving doctors nearer the action gets them to the wounded faster, but puts the doctors themselves at risk too. Now, there is serious investment in robotic substitutes.

Leading the way is the US Defense Advanced Research Projects Agency (DARPA)'s Trauma Pod program to develop an unmanned, mobile operating room. It will be small, hard-wearing and carry automated surgical and diagnostic systems. As well as seeing mobile use in war zones, it is also likely to become part of disaster relief efforts, and could even be used as medical backup for accident victims in remote rural areas far from city emergency rooms.

Some of the pod's functions will work by themselves – they need to be able to find the worst injuries and control bleeding, for example. But they are also likely to use remote-controlled surgical instruments, manipulated over radio and computer links by doctors who remain safely out of harm's way. The radio contact will be maintained by unmanned robot aircraft, and should in theory make distance no object for the operators.

The technology largely exists already, but needs miniaturizing and modifying to survive on the battlefield. If it all works, DARPA's plan envisages a kind of operating theatre on a stretcher. A soldier cut down by shrapnel, say, would be lifted into the nearest trauma pod, which would scan injuries and administer antibiotics and anaesthetics as needed. Then a remote duty doctor would come online and operate surgical robot arms which could extract the shrapnel, staunch bleeding and close the worst wounds. The job would then be finished after airlift in the normal way, but with a much improved chance of patient survival.

When will this wonder technology be seen in action? The developers suggested a couple of years ago that it might be ready for deployment in 2009, but the year came and went with no sign of a real trauma pod. Existing demos show standard-sized robots moving stuff around, but not actually operating. The nearest the project has yet come to reality is a YouTube video animation, which supposedly depicts a battle scene in 2025. It also bears a striking resemblance to the casualty "cocoon" described in science fiction writer Robert Heinlein's *Starship Troopers* (1957), except in that case evacuation after initial care in the cocoon was by spaceship, not helicopter.

NANOMEDICINE?

Nanotechnology crops up in many discussions of future medicine, realistic and otherwise. This makes sense, as the basic units of life – cells – are very small, and are already chock-full of intricately evolved nanomachines which keep them alive. So being able to intervene at this level, using similarly tiny things, should be a boon to medicine.

Many applications, including most of the ones that will appear soon, will offer slightly better ways of doing things that are already part of medical practice. There will be smaller sensors for biochemical readouts and imaging, which can zoom in closer and yield data at finer resolution. Also potentially important will be new ways of delivering drugs to exactly where they can do most good, especially in tumours. "Theranostics" may combine such delivery methods with mobile sensors, enabling devices to identify a target and guide administration of the right medicine. And there will be new nanomaterials for implants and for use as scaffolds for tissue engineering.

But real nanonauts go much further. The ultimate goal, they say, is medical nanorobotics, which will be a source of cell-sized or even smaller machines that can roam around inside us and fix things that go wrong, one bit at a time. This would go a step beyond treatment to upgrading the body with miniature bionic components. Nanoenthusiast Robert Freitas is the leading advocate of this idea, and published a detailed blueprint for a robot oxygen carrier – a "respirocyte" – as long ago as 1998.

In his vision, a respirocyte is something like a red blood cell – which is basically a watery bag full of haemoglobin – except much more efficient. A tiny, diamond-hard sphere, it is packed with oxygen under high pressure, which can be loaded and unloaded from the respirocyte by "molecular sorting rotors". The rotors in turn are linked to sensors which can tell how much oxygen is around, so that they take the gas in the lungs and release it in the tissues that need it, just like the boring old red blood cell does.

The respirocyte overcomes the rather tiresome limitations of natural oxygen storage and transport, however. As it works at high pressure, it has much more oxygen on board. Cue a splendid nanofantasy: "What if you added one liter of respirocytes into your bloodstream, the maximum that could possibly be safe?" asks Freitas. "You could then hold your breath for nearly four hours if sitting quietly at the bottom of a swimming pool. Or if you were sprinting at top speed, you could run for at least fifteen minutes before you had to take a breath!"

But what if the respirocytes go a bit wrong? Will you actually explode as all that oxygen bubbles out? You will have to wait until they are actually built to know for certain. Freitas's design remains on the drawing board and, as with other nanobots, many doubt that it can ever exist or operate inside the body (see Chapter 5).

WHO WILL PAY?

Talk of the medical wonders to come is overshadowed by a simple question: will we be able to afford them? Health care costs are a big political issue everywhere, and the benefits of advanced health systems are already accompanied by large inequalities in delivery both between and often within countries.

Some factors will shrink costs. In developed countries, the incidence of chronic disease has been declining for some decades, and the rate of decline is increasing. On the other hand, the cost of treatment, when needed, keeps rising. Prevention could offset that, but early diagnosis and more successful treatment can mean that more people need treatment for longer periods of time. Future demand for health care will also be affected by people's rising expectations. Finally, there is the rise in the number of elderly people.

That last fact worries health planners more and more. According to a UK Department of Health report in 2008, "our ageing population poses a challenge to the sustainability of the NHS". The number of people in the country over 75 years of age is set to increase from 4.7 million to 8.2 million by 2030, and older people use more health resources. The average 85-year-old is 14 times more likely to be admitted to hospital than a 15- to 39-year-old. In fact, US figures indicate that the cost of health care for those 85 and over is more than 75 percent higher per person than for those aged 75–79.

Even so, according to a 2008 study by Nobel-winning economist Robert Fogel for the US National Bureau of Economic Research, changes in the average age of the population are likely to produce only a fairly modest increase in health expenditures – around 16 percent between 1999 and 2040. But he nevertheless foresees a much larger rise in health care costs, driven by increasing demand. The wealthier people are, the larger the proportion of their income they spend on health. He suggests that expenditure on health in the US is likely to increase from around 15 percent of GDP to 29 percent by 2040.

Other countries, which perhaps have more efficient systems than the US (where about 15 percent of the population lacks any kind of health insurance), spend well below 15 percent now – the European Union average is 9.5 percent. But the same assumptions might still see a doubling of their proportional spending on health care. Even if much of that goes on non-essentials, as Fogel suggests, the standard set for high-level treatment is likely to be a big problem for publicly funded health provision. Current controversies about who gets expensive drugs that give small increases in life expectancy for cancer patients, and who pays, will be rerun many times over. A fully realized nanomedicine might one day offer surveillance and control of bad cells and finally eliminate cancer. Until that happy day, the advent of more effective drugs is more likely to succeed in turning various cancers into chronic conditions somewhat like AIDS now is in countries where patients have access to therapies. That means lots more patients who can survive on expensive medication for a long time.

PREDICTIONS FILE

Aubrey de Grey

Biomedical gerontologist, chief science officer of the SENS Foundation, author of *Ending Aging* (2007)

Highest hope: We will have achieved "longevity escape velocity" worldwide. We will have developed, and made universally available, regenerative therapies that repair or obviate all the molecular and cellular "damage" that underlies ageing – not necessarily perfectly, but well enough to allow us to stay one step ahead of ageing, improving these treatments' comprehensiveness faster than the aspects we still can't fix catch up with us. (The necessary rate of improvement actually declines as more progress is made, so once we achieve LEV we will almost certainly maintain it forever.)

Worst fear: Irrational knee-jerk opposition to tampering with something so "natural" as ageing will slow progress in the necessary research, such that millions (quite possibly even billions) of lives will be lost unnecessarily to the scourge of ageing. The timeframe for achieving LEV is sensitive to scientific unknowns, of course, but also to disproportionate political and sociological fears.

Best bet: I think the chance of my optimistic scenario coming about by 2060 is actually quite high – at least eighty percent.

UPGRADING HUMANS 1.0

While funding fights continue over regular medical treatments, they will be complicated in an increasing number of areas by debates about where treatment of medical conditions ends and making people "better than well" – enhanced, in a word – begins. Is this part of the job of medical researchers? It is often hard to draw the line. The advent of brain-enhancing drugs (see below) is one of the first areas where medical developments targeting the affluent might enhance life for people who are not ill.

All the precedents indicate there will be huge demand for such products. Real but modest effects, or a hint they might be there, already sustain high sales of caffeine, ginkgo biloba extracts and energy drinks, so it's safe to assume that even remotely effective brain-enhancing drugs will attract eager customers too. Bringing the new products to market won't be straightforward, but if the "war on drugs" is anything to go by, keeping people away from drugs they want is more likely to create criminal supply routes than control access. Controversies and compromises will abound as more and more chemicals attract a non-medical market.

As well as drugs and superfoods, there will be new prosthetics, including limb replacements, sensory aids like artificial retinas and improved cochlear implants for the deaf, and perhaps organ replacements. These are less likely to move rapidly out of the medical arena. Bionic limb developers, for example, are looking to serve a large market created by the increase in diabetes. Canadian bionics company Victhom estimates that there will be 140,000 surgical lower-limb amputations in the US by 2020, an increase of nearly half from the current figure.

It is hard to imagine anyone choosing to have their limb removed, even in the unlikely event that an artificial replacement did the job better than the original. But eye and ear implants might one day offer advantages which seemed worth the trouble to some people. And if you need help because your senses are deteriorating with age, why not have them enhanced at the same time?

Hank Greeley, Professor of Law at Stanford University, highlights five issues raised by the likely advent of enhancement technologies, whether drugs or devices.

- The most obvious one is safety. Sick people may take a risk to get a benefit. Healthy people looking for a boost may take risks they don't need to, as the use of steroids in sports training shows.

- Then there is choice. Could use of enhancement come to be expected – by employers, for instance? Could there be a self-imposed or media-hyped necessity to improve your appearance?

- Concerns about fairness begin with basic access to treatment, and get more complicated if enhanced people get richer and can pass their advantages onto their kids by buying them the same treatment. It might come to be seen as just an extension of giving children a good start by paying for privileges like private schooling.

- Then there is fairness to others in a competition, whether an exam, a sports event or a beauty contest. How are results to be judged when some have used drugs or other high-tech boosters, and some have stuck with their natural ability or appearance?

- And, finally, what counts as natural anyway?

Expect to hear more arguments around these questions. But be prepared, too, to see them widely ignored if genuine enhancement becomes available. Like today's boob jobs and facial tucks, they will be sold as things that allow individuals to make their own decisions to improve their lives.

Eighty-million-dollar hand: the Proto 1 bionic arm developed at Johns Hopkins University, 2009.

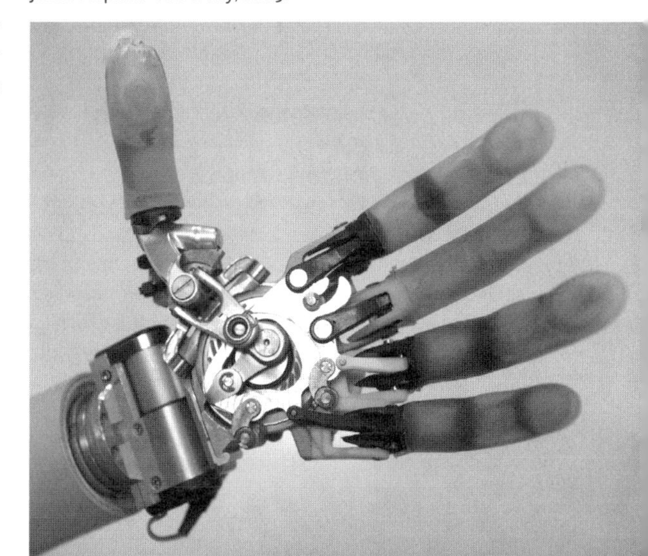

DRUGS FOR THE BRAIN

Most cultures have had their favourite neuroactive drugs for recreation, and they are still with us, legal and illegal. The last century saw a new set of drugs that work on the brain for treating illness, especially mental illness. Most were blunt instruments. Even relatively recent drugs such as the widely used selective serotonin re-uptake inhibitors (Prozac is the best known) for depression have effects which are poorly understood. There are at least a dozen different receptors for the cell signalling the molecule (or neurotransmitter) serotonin in the brain, so a drug that probably stimulates all of them at once is a pretty haphazard treatment.

This century will see more powerful brain drugs, with more specific effects, designed using much better knowledge of how some brain systems actually work. What will they do? Here are some speculations drawn from a review of drug futures by the UK government's Foresight unit, which analyses how numerous issues are likely to develop in the coming decades.

Key research targets will be neurodegenerative disorders like Parkinson's disease and Alzheimer's, both of which are increasingly important in ageing populations, along with addiction and mental health disorders. At the same time, better understanding of the brain could lead to drugs for pleasure which are less harmful and less addictive than the ones widely used in the past. Later there could be drugs to "help us learn, think faster, relax, sleep more efficiently, or even subtly alter our mood to match that of our friends".

NO MORE HANGOVERS?

A good example of where better knowledge of brain chemicals might lead is the possibility of fashioning drugs which offer the things we like about alcohol without the less popular effects, such as nausea, hangovers and liver damage.

British psychopharmacology professor David Nutt published a paper in 2006 outlining how it might be done. The action of old-fashioned ethanol is complex – it interacts with a range of brain receptors. But the nicer effects appear to be caused by how it binds to receptors for the neurotransmitter Gamma-aminobutyric acid (GABA). Only some GABA receptor types produce the relaxation and sociability triggered by a few drinks. Others are linked to the aggression, staggering about and memory loss heavy drinkers also experience. So a cleverly designed chemical which targeted just the right set of receptors, a "partial agonist" in the jargon, might offer the more desirable effects of drinking alcohol without the others.

Such a drug would ideally come with its own antidote, which might work like existing chemicals that already counteract the effects of drugs that act on GABA receptors. Nutt cites the fast-acting flumazenil, which is used to save people who have overdosed on the antidepressant benzodiazepenes. In his scenario, you might enjoy a party on the new alcohol substitute, pop the antidote to drive home safely, and awake the next day feeling as bright and alert as non-drinkers do now. There are a few social barriers once the science is done, though. As a new drug, most countries – including Nutt's native Britain – would require the alcohol substitute to be licensed under medicine control legislation and prescribed by doctors. He argues that the massive reduction in the harm alcohol causes would be a good reason to revise the drug laws.

Here a blurring of the boundary between medical treatment and enhancement is simply unavoidable. Memory enhancers would be a boon to Alzheimer's sufferers, for example, but also to students cramming for exams. Drugs for sleep disorders could help the healthy stay awake and fight fatigue.

The experts consulted for this report believed that novel drugs would not make it possible to increase brain performance across the board. Rather, in ten or fifteen years there might be a range of substances that allow us to optimize our brains for specific tasks like following a complicated argument, talking with friends, or just going to sleep after a demanding day.

MEDICINE'S ULTIMATE GOAL?

Keeping people healthy and curing the sick are goals as old as the medical profession. But for the most fortunate, these are now nearing their limit. Suppose you follow all the advice about diet and exercise, avoid accidents and infections, sidestep cancer, and enjoy a comfortable old age in command of all your faculties. Sounds good? Sure. But there will come a point when some of your body's systems just stop working. However robust your constitution and however crafty your medical regime, you will get frail and shaky, and then die.

Average life expectancy has certainly gone up. There are a lot more centenarians about than there used to be. But there are vanishingly few 110-year-olds. The record for longevity is 122, recorded by Frenchwoman Jeanne Calment in 1997. Some think postponing ageing is medical research's most important remaining goal, and some believe it will be possible quite soon. The most prominent advocate of this view is British scientist and activist Aubrey de Grey (see box overleaf). He argues that within a few decades, intensive research should be able to stop ageing altogether.

Even scientists who do not proclaim they seek life extension are led to that goal by the overall drift of medical research. As infant mortality has been successfully curtailed in many countries, age-related deterioration is the most important problem left to tackle. Almost all of the major research targets are diseases which affect the elderly, or become more likely with age. Cancer, for instance, affects elderly people far more often than the young or even middle-aged. Increased life expectancy means that diseases of old age are more widespread, and will become still more so.

Slowing the ageing process is our best chance to prevent most of the conditions which burden health care systems now and in the future. Research on specific conditions – cancer, Alzheimer's, Parkinson's, arthritis, heart disease, type 2 diabetes, osteoporosis and so on – has its place. But retarding ageing could affect all of them at once. According to a group of doctors and researchers in the *British Medical Journal* (2008):

> "The pursuit of extended healthy life through slowing ageing has the potential to yield dramatic simultaneous gains against many if not all of the diseases and disorders expressed in later life."

So what are the chances? Proponents say there are good reasons to believe that ageing is unnecessary. The logic of evolution suggests that genetic traits are selected to keep an individual alive until he or she has reproduced, and perhaps nurtured their offspring – but no longer than that. That does not mean the genes are programmed for decline, simply that there is no advantage in preventing it.

THE NEW METHUSELAH?

Aubrey de Grey, based in Cambridge, UK, is founder of the Methuselah Foundation and a tireless propagandist for Strategies for Engineered Negligible Senescence (SENS). He contends that to intervene in the ageing process, you do not need to know how all the damage to cells that accumulates as we grow older happens, just how to put it right. What is more, he reckons that we already know enough to launch applied research on doing just that.

As one example, ageing tends to accumulate extracellular junk in the form of protein tangles that the body finds it hard to get rid of. The plaques which form in Alzheimer's disease are just one form of this refuse. Fundamental research might show how they form, and help prevent them from building up. All fine, says de Grey, but meanwhile it would be more useful to find ways to remove them. Perhaps a vaccine could persuade your immune system to dispose of them, for instance.

SENS combines seven strands of research like this to achieve the ultimate goal of longer life. How much longer? Well, de Grey really does mean negligible senescence. He wants healthy adults to die only from accident or other mishaps. Their bodies would never wear out. He outlines the full strategy in *Ending Aging* (2007).

In the short term, the Foundation is concentrating on mice. The Methuselah Mouse award will be offered to the research team which produces the longest-lived laboratory mouse. As they tend to be short-lived compared with wild species, there are some tricky issues defining what counts as true life extension for the rodent, however.

There also seem to be plenty of leads to advances that might extend lifespan. One particular set of genes which control a set of metabolic switches can be tweaked to multiply average lifespan several times over in the simple nematode worm, for example. The same pathway also appears to exist in fruit flies and mice, and so could even work in people, too.

However, at the moment there are more sceptics than believers in life extension among specialists in the study of old age. They argue that complex, long-lived organisms are likely to accumulate damage of many kinds. Chemical nasties build up, mistakes in copying genes proliferate and mechanisms for correcting errors get less efficient. The eventual triumph of entropy (see p.29) can be delayed, but not denied.

Practically, this means there are only trade-offs between different kinds of damage, rather than simple ways of doing away with it altogether. For example, ageing expert Professor Tom Kirkwood of Britain's Newcastle University points out that studies of other species suggest that those which destroy cells at the smallest sign of damage may have fewer cancers. They also show earlier signs of other age-related pathologies which are due to cell loss.

Kirkwood, like many of his colleagues, supports more research into the complexities of ageing, not least because of the continued increase in life expectancy. But he dismisses any near-term jump in maximum lifespan. Athletics is a good analogy, he says. No one thinks the current world record for running a mile can never be broken, "but no one seriously expects the mile to be run in two minutes any time soon".

Add up all these views, and it seems reasonable to expect that more people will go on surviving longer in the countries which already have lengthy life expectancies. Many of them will be in better health. Real improvements in combating ageing may come in a few decades, and ageing populations will want to pay for research aimed at achieving that. If real progress is made, the distribution of the benefits will be a moral and political challenge of a new order. We now tolerate large inequalities in wealth between countries, and big disparities in average life expectancies, though elites in poorer nations can buy access to high-quality health care. But what if the already well-off were able to pay their way to living significantly longer – a couple of additional decades, say? Perhaps they would have longer for their consciences to kick in and make them work to ensure access to the technology for more people. Or perhaps it would turn out to be so expensive they will need every penny to hold on to their advantage.

Even without serious life extension, existing trends in ageing will cause some social disruption. Imagine a world in which, say, one person in a hundred lived to 110 years of age, and one in ten made it to 100 years of age. The social, political and financial consequences would be wide-ranging. It might resemble the world portrayed in Bruce Sterling's novel *Holy Fire* (1996), in which society is dominated by a bunch of ultra-wealthy and rather cautious centenarians, though the novel also features a radical life-extension technology for those who can afford it – returning a one-hundred-year-old to the physical condition of a twenty-year-old – which remains pure science fiction.

FURTHER EXPLORATION

Francis Collins **The Language of Life: DNA and the Meaning of Personalised Medicine (2010)** A vision of the future of medical care informed by knowledge of everyone's genetic risk for everything – by the former director of the US Human Genome Project and current head of the National Institutes of Health.

Aubrey de Grey **Ending Aging: The Rejuvenation Breakthroughs that Could Reverse Human Aging in our Lifetime (2007)** The longest account of Aubrey de Grey's reasons for thinking that he, and everyone else, has a shot at immortality. He thinks it's not just a cool idea, but a moral imperative.

Pete Moore **Enhancing Me (2008)** An excellent, measured review of the near-term prospects for different types of enhancement by an author in possession of an excellent hype detector.

Steven Schimpff **The Future of Medicine: Megatrends in Health Care that Will Improve Your Quality of Life (2007)** Upbeat, technology-focused and US-centric survey of possibilities in genomics, stem cells, vaccines, imaging, surgery, etc. You can also read the author's blog at www.medicalmegatrends.blogspot.com.

www.who.int The World Health Organization monitors worldwide developments and tries to orchestrate research on the most important health problems across the globe.

www.nih.gov The National Institutes of Health is the powerhouse of US medical research, allocating the government's billions for work on future diagnosis and treatment.

13
future war

future war

What is unveiling itself right now in war sounds like science fiction and therefore keeps us in denial.

Peter Singer

The twentieth century is often seen as having been horrifically violent. Will the twenty-first be just as bad? Clearly, some aspects of war have changed a good deal – not least the possibility that all-out war between nuclear powers could end civilization. Yet wars still go on around the world. How will they be fought in the decades to come? And who will be doing the fighting? There's one potentially game-changing answer: robots.

War changes even while it stays the same. Technology, training and tactics all evolve. Yet the underlying principles remain: war exists when one group of people try to impose their political will on another and are prepared to kill, and risk being killed, to do so. According to international relations expert Colin Gray, in *Another Bloody Century: Future Warfare* (2005), war is not going to go away, or even be radically transformed. Gray styles himself a realist, arguing that the best hopes for peace will always be disappointed. He believes that suggestions that the nature of war is changing significantly are overblown. How convincing are these predictions?

Specific forecasts, as Gray emphasizes, are as likely to be wrong about war as they are about any other subject. Yet military planners are professionally required to imagine the future – not least because complete weapons systems can take decades to develop and commission, and remain in service for decades more. The first B-52 bomber flew in 1952, but much updated models bombed Iraq in 1991 and, with more advanced avionics, the planes known as BUFFs will probably still be in service in 2040.

Generals often go wrong by trying to fight the last war, but military futurists continue to devote their efforts to fighting the next one, or the one after that. Worryingly, planning like this can turn into self-fulfilling prophecy. The kind of war fought, if not the outcome, will be influenced by the kind of war envisaged years earlier, and the tactics designed around the kit which was then ordered to be built.

Many of the techniques of would-be future-shapers – such as Delphi forecasts and scenario sketching – were developed in the RAND corporation's wargamers' playpen during the 1950s nuclear arms race (see Chapter 3). So today a great deal

PREDICTIONS FILE

Freeman Dyson

Physicist and observer of technology, author of *The Sun, the Genome and the Internet* (1999)

Highest hope: That some right-wing Republican president of the US will decide that our nuclear weapons are useless for winning wars, and will get rid of them in one afternoon by executive order, as Nixon did with our biological weapons, without waiting for international negotiation and Senate ratification.

Worst fear: That some gang of mischief-makers will steal a few of our nuclear weapons and explode them in such a way as to start a full-scale nuclear war.

Best bet: That a nuclear terrorist explosion will happen, but not be as bad as my worst fear, and it will produce a response not quite as good as my highest hope.

of future-gazing goes on in intelligence agencies, defence departments and associated think-tanks. In the US, it can seem like more thought is given to future war, how to fight it and what weapons could be used to do so than to future peace. But should we really take future war for granted?

KILLER PRIMATES

There is no getting away from it: however peaceable we feel personally, people are prepared to kill each other. But is this willingness to wreak violence on our fellow men and women getting worse? Many believe so, but this may just be because we hear more about it courtesy of modern mass media, and we care more. In fact, there is some basis for optimism that tolerance for violence, and violent deeds themselves, are gradually declining. According to psychologist Stephen Pinker, careful studies of the past show a decrease in all kinds of violence throughout human history. This applies to murder, torture, enslavement and mistreatment of animals for sport as well as to war. There have still been plenty of atrocities in recent memory, but the overall picture is encouraging. We just think it looks worse because we are more aware of the gruesome things which do still happen, and worry about them. The interrogators of the Inquisition would be bemused at the way today's torturers deny the practice where they would once have advertised it.

Even the war-torn twentieth century does not look so bad if you take population increase into account. If the conflicts of the last hundred years had killed the same proportion of the population that died in the wars of the average tribal society, according to Pinker, there would have been two billion deaths instead of one hundred million.

Pinker's position gets some support from the best attempt at comprehensive statistics on global conflict, though it covers a much shorter span. The Human Security Report (www.humansecurityreport.info), first published in 2005, documents a striking decline in violence – including wars, genocides and human rights abuses – since the end of the Cold War. More recent assessments by the same team have suggested that – in sharp contrast to most expert opinion – the threat from terrorism has reduced in the past few years, and that warfare's declined even in reputedly conflict-ridden regions such as sub-Saharan Africa.

The conclusion about terrorism is partly a matter of interpretation, as the report chooses to classify civilian deaths in Iraq as due to civil war rather than terrorism, and these account for the vast majority of "terrorist" deaths recorded in other studies. However, this draws attention to another aspect of recent conflicts. According to the Centre for the Study of Civil War in Oslo, civil war is now the dominant type of conflict. Most wars nowadays are within states, not between states.

Such trends are not guaranteed to continue, but the numbers do not suggest that the world is getting more dangerous. However, it is still safe to predict that there will be new wars in years to come, and that they may be fought with frightening new weaponry. As Pinker puts it, although violence may be on the wane, it is still true that the world has never before had national leaders who combine pre-modern sensibilities with modern weapons.

Modern guerrilla warfare: Mujaheddin fighters atop the thirteenth-century Herat fort in Afghanistan.

FUTURE SOURCES OF CONFLICT

From one point of view, the geopolitics of the next decades are simple. Since the end of the Cold War between the US and the former Soviet Union, the US has been the world's undisputed top nation. For some years, the US had unstoppable "hard power", or military might, as well as impressive "soft power", or economic and cultural leverage over other nations. It retains the first kind of power, even as the second is increasingly in question.

So how does a world with a single military superpower evolve? That depends partly on the

TALKIN' 'BOUT MY GENERATION...

In the late 1980s, US strategists proposed the need to plan for a fourth generation of warfare. According to them, modern war began with the introduction of the smooth bore musket, which was used by armies drilled in line and column tactics. As firepower improved, this was abandoned because the lines could slaughter each other too readily. It was superseded by fire and movement, with gradually increasing emphasis on artillery. When this eventually produced the trench-mired stalemate of World War I, it led to a third generation, with more emphasis on rapid manoeuvre and infiltration.

The fourth generation, they reckoned, would see some parts of the third carry over. But it would look different in several ways. This kind of war is a fuzzy beast. It can exist in a kind of mosaic with peaceful interludes, or areas largely free of action. There are rarely real battlefields or an identifiable front line. Who is a fighter and who a civilian can be hard to tell, and whole societies are drawn into conflict, which is political and cultural as well as physical. Big military sites like airfields, communications complexes and generals' headquarters become places to avoid, if possible, since they tend to get blasted from the air.

This sounds like a familiar recipe for how to fight an enemy that is – in old terms – militarily superior. It's a war of terrorism and insurgency against an opponent it would be foolhardy to confront directly. It's the story of Iraq and Afghanistan. That opponent, in turn, adopts its own fourth-generation tactics, and technologies to go with them. Large bases may be established, but operational units are small, linked to broad communication networks – often with robotic outposts – and continually redefined targets.

The fourth generation idea caught on due to the high-profile wars fought by the US and its allies. It is a useful label, though it has its limits. Some of the features of the fourth generation are not new. And there will still be old-style wars fought between forces armed with ancient Kalashnikovs or even machetes. But the new style does seem to fit some key conflicts. As Colonel Thomas Hammes, author of *The Sling and the Stone: On Warfare in the Twenty-first Century* (2004), put it in 2003:

> "Fourth generation warfare uses all available networks – political, economic, social, and military – to convince the enemy's political decision makers that their strategic goals are either unachievable or too costly for the perceived benefit. It is an evolved form of insurgency."

Meanwhile, some strategic thinkers are outlining the next (fifth) generation. This denotes an extension of existing trends, factoring in, for instance, access to biological weapons. If that happens, we might see small groups in a position to challenge nation-states.

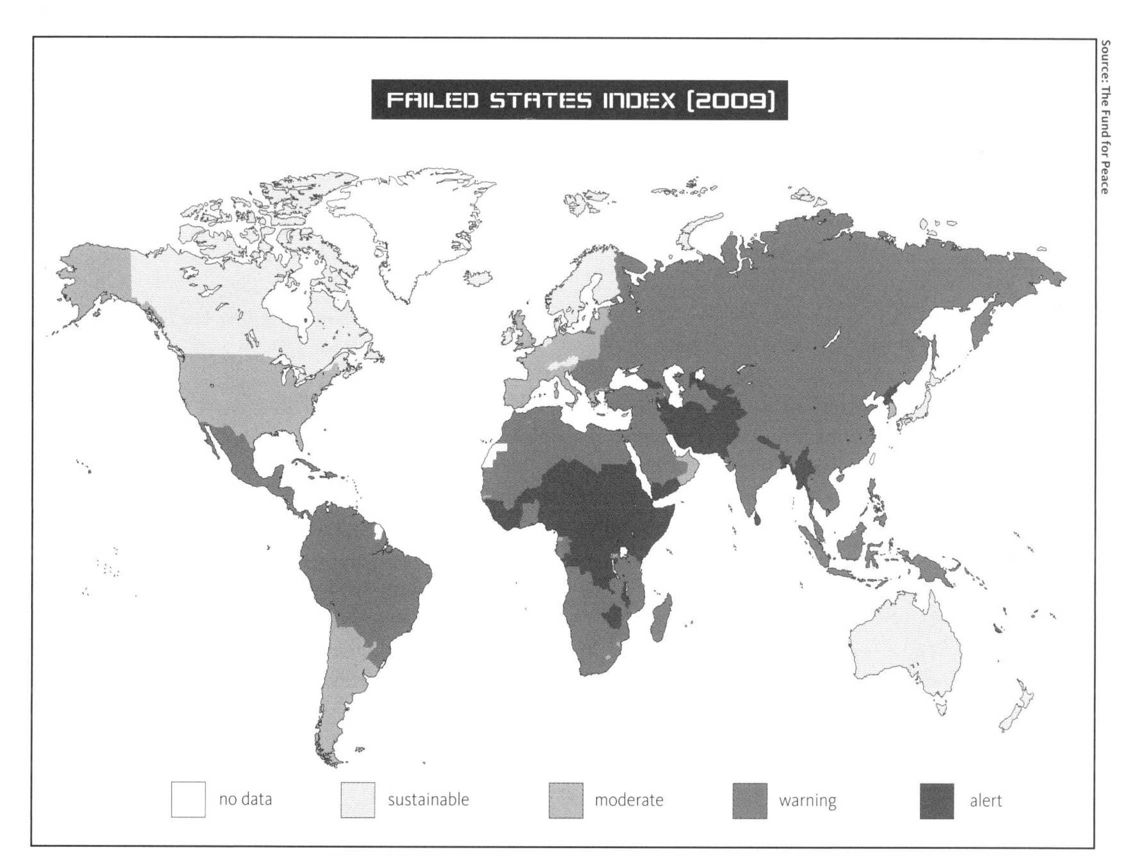

The Failed States Index combines political and economic indicators to spot likely areas of instability or conflict.

superpower, partly on others. In 2006, the US Quadrennial Defense Review Report emphasized two main strategic guidelines: the "long war" against Islamic terrorism and, less predictably perhaps, the ability to influence China by military "dissuasion". This does not mean America is preparing to invade mainland China. Indeed the 2010 review softened the language and talked of "whether and how rising powers integrate into the global system" being a defining question for the twenty-first century. But it does mean the US wants to be in a position to counter Chinese efforts to increase their sphere of influence in Asia, and possibly other parts of the globe such as Africa.

Against this backdrop of declining American hegemony and growing Chinese influence, there will also be regional conflicts. Where and when these occur will depend on too many factors to allow much in the way of informed guesses about the parties to any future wars. But access to resources, energy, minerals and perhaps water will often play a part. And climate change will influence regional tensions in ways as yet unknown, but which could be increasingly important as the century progresses.

The stresses and strains induced by climate change may also tip more countries into state failure, marked by the collapse of government and services. Such social breakdown tends to be followed by more complex internal conflict, as warlords and militias seize available weaponry and vie for supremacy, and can spread to neighbouring countries. The magazine *Foreign Policy* (www.foreignpolicy.com) publishes an updated Failed States Index each year, rating the countries most at risk of collapse.

Seven of the countries in the 2009 top ten are in Africa, and two others – Iraq and Afghanistan – are already being fought over. The final entry, Pakistan, is earning worse ratings each year, and is the only nuclear-armed state in the top few dozen.

HIGH-TECH WAR

The future, they say, is already here – it is just unevenly distributed. Nowhere is this truer than in military technology. Sure, other countries have armed forces, weapons industries and military R&D (research and development) budgets. But rather like the future of advanced health care, the future of how war is waged is being designed in the US.

The US will not be able to fight the wars of its own choosing, or necessarily prevail. But the ways those wars will be fought will be shaped by the work going on in labs and testing grounds for the US military, and the equipment designed to support the kind of wars they now foresee. Precise figures depend on how you distinguish R&D from procurement, but the annual US military budget can be reasonably estimated at $700 billion. Perhaps ten percent of that goes on R&D. The total budget is more than ten times that of the next biggest spender, China. In fact it is more than the rest of the world put together, and the R&D budget is even further ahead of other nations.

What do they do with it? Many programmes are classified, but the US is an open society and quite a lot of the effort is accessible, at least in outline. As the sole global superpower, this side of US policy-making has more significance for the future than many attempts to look forward. In weapons systems, in particular, outfits like the Defense Advanced Research Project Agency (DARPA) support the kind of "blue-skies" research which can turn predictions of future possibilities into working technologies. These often have effects on civilian technologies, too. The modern Internet, to take just one example, grew out of DARPA's concern to protect military communications by designing networks to route messages flexibly.

So what kind of wars do such programmes envisage in the coming decades? One answer, as investigative journalist Nick Turse explains, is that the conflicts may be urban. The slums, shantytowns and favelas of the developing world are where Pentagon planners see their infantry needing a new technological edge. And these twenty-first-century street fighting men and women will need twenty-first-century weapons. DARPA's project descriptions take a bit of decoding. For instance, a "Cognitive Technology Threat Warning System Program" will try to develop "soldier-portable visual threat detection devices". In English, please? Smart binoculars. Other examples of projects underway include:

- VisiBuilding: new sensors and 3-D reconstruction techniques to alert troops to where enemy troops or equipment may be hidden.
- Camouflaged Long Endurance Nano Sensors (CLENS): a system of tiny, networked sensors to monitor an area 24/7.
- Urbanscape: a project to collect data and images from a city district using unmanned aircraft and robot vehicles, and then re-create a virtual, 3-D city to help soldiers get around quickly.

These are just a few of many sensor projects. The details matter less than the level of ambition. In terms of surveillance, that ambition's evident in a US Department of Defense study which suggested that "the ability to record terabyte and larger databases will provide an omnipresent knowledge of the present and the past that can be used to rewind battle space observations in TiVo-like fashion and to run recorded time backwards to help identify and locate even low-level enemy forces. For example, after a car bomb detonates, one would have the ability to play high-resolution data backward in time to follow the vehicle back to the source, and then use that knowledge to focus collection and gain additional information by organizing and searching through archived data."

DARPA's lengthy prospectus is impressively funded proof that the military will generally adopt any technology they think will help them do their job. They may also try to speed up developments in areas they think could yield military advantage. In the US right now that means information technology and, to a lesser extent, biotechnology and nanotechnology. Other efforts focus on robotics, with airborne drones, remote-controlled or self-guided vehicles, and new generations of smart munitions under development. There is also extensive work on new materials, to create lighter armour for example, and on improvements in trauma care for soldiers injured on the battlefield (see box on p.228).

In many of these areas, the unwillingness to rule anything out means that ideas that sound like science

PREDICTIONS FILE

James Howard Kunstler

Author of *The Long Emergency: Surviving the Converging Catastrophes of the 21st Century* (2005)

Highest hope: The energy predicament suggests that we are now entering a long period of contraction. This includes the human population. War, shortages of food and fossil fuel, hardships and dislocations caused by climate change, and epidemic disease are likely to have substantial impacts. The economy that emerges from these disorders will centre on agriculture. I believe it will resemble the early nineteenth century – if we are fortunate – with a residue of some modernity, perhaps some electrification. The basic theme will be the twilight of the industrial adventure and the beginning of a new medievalism. My highest hope is that we can avoid the use of nuclear weapons in the disorders I have described above.

Worst fear: Nuclear weapon proliferation is already so out-of-hand. My worst fear is that some of these weapons will be let loose upon the world.

Best bet: A steady contraction punctuated by hopes for return to the "old normal", dashed by intermittent socio-economic and political crises and traumas.

fiction – and probably are – still get tried out at the research stage. In fact, US science-fiction writers are quite often consulted about future technologies and threats by the planners and wargamers at the Pentagon. Several sci-fi author for instance, were early proponents of the futuristic "star wars" Strategic Defense Initiative. This was based on the hope that incoming ballistic missiles could be taken out by super-fast interceptors, laser beams or some other high-tech network of surveillance and response. Tests so far have been uniformly disappointing.

TWENTY-FIRST CENTURY MILITARY KIT

A fully-equipped modern soldier, who is supposed to be fast and agile, may be lugging around 140 pounds (63.5kg) of equipment. But new materials and miniaturization could transform this bulky figure into a sleek sci-fi warrior. At least that's the ideal being pursued by MIT's Institute for Soldier Nanotechnologies. They envisage a twenty-first-century soldier going into battle wearing "a bullet-resistant jumpsuit, no thicker than ordinary spandex, that monitors health, eases injuries, communicates automatically, and reacts instantly to chemical and biological agents". Now in its second five-year programme, the institute spends around $10 million a year on developing new materials for lightweight protection, nanosensors, and novel medical and communication systems.

In its quest for the ideal battle suit, the institute is pursuing a bunch of individual projects. Find a nano-coating for waterproofing clothes and kit, and the soldier can leave a poncho behind. Design a radio which fits into a button-sized tab on the collar, and junk the hefty transmitter-receiver which has to be carried in a harness. But the larger plan is for all these things to be integrated into a suit which responds to insult or injury. Miniature sensors would monitor the wearer's body fluids, for example, and the suit might even automatically administer drugs or antidotes to chemical or biological agents.

MIT president, Dr Charles Vest, greets US servicemen wearing prototypes of future combat uniforms at the opening of the Institute for Soldier Nanotechnology, 22 May 2003.

The institute is mainly funded by the US Army's Future Force Warrior programme. This envisages a soldier equipped with an "Iron Man"-like array of gear, including:

- ▶ A hybrid machine gun/missile launcher, firing either bullets or 15mm rockets which guide themselves to a target.
- ▶ A drop-down visual display no more than an inch across which can show sensor data from night vision or thermal imagers, and pick up images from other viewpoints. These will go along with smart earpieces, which relay radio messages while blocking loud noises like firing guns or explosions.
- ▶ And if the wearer comes under fire, the suit will incorporate responsive body armour which is flexible in normal use but turns rigid fast enough to absorb the impact of a bullet (that is, very fast indeed).

Failing that, the foot soldier might be equipped with an armoured exoskeleton which would provide protection, increase strength and solve the problem of carrying so much weighty gear if the miniaturization programme fails to deliver. But here the visions of the military technologists do blend into science fiction. As with the fabled jet pack, the downfall of the invincible supersoldier speeding across the landscape in an armoured exoskeleton will be the need for power while avoiding overheating. However brilliant the technological thinking, it will still have to obey the laws of physics.

Information war

Amid all the new technologies used in modern warfare, the primary real-world developments tend to be found in information systems. Information technology is so important to war today that it overwhelms everything else, according to US defence analyst Bruce Berkowitz's *The New Face of War* (2003). There are two aspects to how such technology is deployed. One involves retrieving information from combat zones so that commanders actually know what is going on. A panoply of technologies is being brought together to do this, including satellites, robot aircraft and vehicles, and instruments carried by individual soldiers. They're bound together in networks that can be made and remade as forces move around, and linked to computers that can analyse the data.

The goal is to get as close to real-time analysis as possible, but the operational systems are also related to simulation and gaming efforts which are used to run future wars inside the computer. Military systems and computer games come together in schemes like DARPA's "Deep Green" software project, a "battle command decision support system that interleaves anticipatory planning with adaptive execution". The aim is to provide commanders with ultrafast simulation of the outcome of battlefield options, which are fed in and evaluated according to the latest information on the ground.

Whether this ambition can be realized, or such a system can ever truly dispel "the fog of war", is open to doubt. But forces equipped with the latest IT systems are certainly better informed than their predecessors, from commanders down to infantry. Less well equipped enemies will, of course, try to find ways of disrupting these systems, and better means of concealment, to reduce this advantage. They could even use cybertrickery to co-opt other weapons systems, perhaps including nuclear arms. A paper published by a disarmament think-tank in 2009 put forward the alarming possibility that terrorists looking to use nuclear weapons would find this easier than stealing them or making them themselves. This could be

UP, UP AND AWAY?

The dream of personal human flight has inspired lots of technological effort. Some of it worked out, some didn't. One project which never quite delivered was the jet pack, which became a long-standing symbol of gee-whiz futures. What happened? The details are instructive because the jet pack, aka rocket belt, is one of those science-fictional ideas which was taken up by military project planners eager to make it real, but turned out to be too good to be true.

The rocket belt was the coolest piece of gear used by Buck Rogers. The twenty-fifth-century adventurer was first featured in the sci-fi magazine *Amazing Stories* in 1928, but had a long and varied life in comic books, radio and TV serials, and even in the cinema as late as 1979. In the real world, the belt turned into a backpack and, er, a front pack. German experiments with front and rear personal rockets were designed to help soldiers leap over obstacles, but never saw action.

After World War II, German rocket technology was shipped to the US. But the rockets then in use were either too big, or too low on power, to allow engineers to design a plausible one-person rig – even if any sane person was willing to try it out. In fact, those annoying limits set by physics have never really been overcome. US Army experiments in the 1950s with state-of-the-art mini-rocket motors using hydrogen peroxide eventually produced a system which could lift a man in 1961, but for less than a minute before the fuel ran out. This limitation was ignored by the screenwriters of the James Bond movie *Thunderball* (1965), who used a jet pack to allow Bond to make a spectacular getaway.

Off the screen, a few more million dollars from the US Defense Department produced a jet-powered outfit. This was a bit like a mini-Hawker Harrier (the vertical take-off fighter plane), with one jet engine feeding thrust to adjustable control nozzles. But it still weighed 75 kilos, and fine control was tricky, to say the least. This and later projects never saw action. Even if they had worked, the idea of people flying around strapped to jet engines or mini-rockets is pretty absurd. The fact that it was pursued for so long shows the power of some technical fantasies to sustain false hopes.

"Rocket man" Peter Kedzierski at the California State Fair, Sacramento, 1964.

done either by hacking into launch systems or, perhaps more likely, fooling early warning systems and triggering retaliation against a nonexistent attack.

That is one face of the new cyberwar. The other, which is harder to define, is the possibility of attacks on IT systems that underpin key aspects of civilian life – the "critical information infrastructure" – and how to defend against such attacks. This raises crucial questions about the vulnerability of complex, interconnected societies and how to make them more secure. There are always options to protect computers from cyber attack, and defence systems are likely to be screened by special firewalls and surveillance routines. But clever design cannot circumvent a basic trade-off: the harder it is to communicate freely over a computer network, the less useful it may be.

ROBOT WARS

Robots are also high on the military research agenda in the US and other countries. They could be seen as a subset of the general push to make better use of IT. But they are worth considering separately as they raise some troubling questions about how future wars may be fought.

As Peter Singer of the Brooking Institute argues in his comprehensive book on current and future military robots, *Wired for War* (2009), it is easy to see the appeal of intelligent, mobile machines to defence planners. Robots can spare human soldiers – no more letters home to grieving parents. They ought not to get tired or scared either. And, unlike many human troops, they will shoot to kill when ordered to do so. As one high-ranking robot enthusiast in the Pentagon put it in the *New York Times* a few years ago:

> "They don't get hungry, they're not afraid. They don't forget their orders. They don't care if the guy next to them has just been shot. Will they do a better job than humans? Yes."

This kind of thinking means that robots are coming into real wars fast. A US Joint Forces Command study in 2003 suggested that robots on the battlefield would be the norm by as early as 2025. And the effort to make better, smarter ones is a major part of the $127 billion Future Combat Systems project – the biggest military contract in American history. This envisages development of a vast array of new

PREDICTIONS FILE

Parag Khanna

Director of the Global Governance Initiative and senior research fellow at the New America Foundation

Highest hope: All territorial nations of the world (there could be around four hundred by then) live in a state of equilibrium, content with their borders and allowing trade and migration across them.

Worst fear: Poorer parts of the world will be exploited for their resources to the point of evisceration. Much of Africa could be turned into a giant resource pit that is uninhabitable by humans.

Best bet: The worlds of reality and cyberspace will coexist uneasily, just as the haves and have-nots do today. There will be many different zones of the world – some well-managed, others very poorly run and dangerous. Technology will be the great divide between the winners and losers.

gear, involving networked weapons, robots, drone aircraft and computers. Soldiers will still be involved, but will be surrounded by technology. Singer emphasizes, however, that robotics does not have to involve the highest of high-tech resources, and some will be available to many countries with smaller R&D efforts than the US. He counts 43 countries with development programmes for military robotics. Even terrorists will be able to build home-brewed robots from off-the-shelf components.

For US strategists, as well as avoiding all those messy – and politically costly – casualties among deployed troops, the robots will be much more efficient. According to national security analyst John Pike (director of GlobalSecurity.org):

> "Armed robots will all be snipers. Stone-cold killers, every one of them. They will aim with inhuman precision and fire without human hesitation. They will not need bonuses to enlist or housing for their families or expensive training ranges or retirement payments. Commanders will order them onto battlefields that would mean certain death for humans, knowing that the worst to come is a trip to the shop for repairs."

This is encouraging for commanders and relatives of people in the forces, but disquieting too. Battle robots that actually work would make wars easier to wage, and thus more tempting to enter.

How big a change will this be? From one point of view, the arrival of robots on the battlefield continues the long-running trend of distancing users of weapons from the consequences. Tracking your enemy through a gunsight feels different from hand-to-hand combat with swords, and artillery, bombs and missiles all help the perpetrators of violence stay detached from their targets. Nowadays the detachment may even extend to remotely controlling devices half a world away, and going home to the family at night. War truly does become a videogame for the drone pilot or missile targeter.

But robots with real military capability create an option that could be the exception to Colin Gray's claim that war doesn't really change. Humans will gradually be removed from making the decisions. Overall strategy, presumably, will remain in the hands of politicians and generals. But the decision to open fire could be automated. That does seem like a path to a different kind of war.

The move from remote control to autonomy is still in the future. The US Future Combat Systems Program included plans for developing a collection of robotic ground vehicles, including a robotic armoured car equipped with light weapons. Although prototypes exist, the land vehicle side of the programme was cancelled by the Obama administration early in 2009. However, the US Army almost immediately began making plans for an improved version, which it hopes to have ready for use in five years or so. Such vehicles will be used for reconnaissance and surveillance, but use of weapons is still intended to be subject to human decision.

Will that last? Logic suggests not. Singer argues that the assertion of robot autonomy will come about gradually, through operational pressures. For example, the Future Combat Systems plan has two humans sitting at remote controls and tweaking the settings of ten land robots. In theory, the robots can roam around and find stuff out, but will still need human permission to fire weapons. But things happen fast in war, and experiments suggest that even supervising two robots can be one too many. Then there is the question of what happens when a robot comes under fire. Waiting to shoot back is unlikely to enhance their operational life. At some point, timescales get too tight for humans to stay in the loop.

Exactly when this may happen depends on predicting the future of robot intelligence, as discussed in earlier chapters. But it seems likely it will be possible in the second half of the century, if not the first. There could be good and bad results, and sometimes it'll be hard to say whether they are good or bad. War waged remotely, with lots of carefully filtered images from automated cameras – more like a computer game, less like a matter of blood and death? Bad thing. Robot systems which avoid risking troops and so encourage intervention in countries where less advanced armies are murdering their fellow citizens? Maybe a good thing.

The debate about the ethics of robot wars is already underway. Ron Arkin, a Georgia Tech professor and author of the reassuringly titled *Governing Lethal Behavior in Autonomous Robots* (2009), believes it is worth exploring "whether intelligent robots can behave more ethically in the battlefield than humans currently can". They would be less likely to harm a few passing civilians out of anger that a comrade has just been killed, for example.

Arkin is thinking about a near-term future which features not *Terminator*-style robot supersoldiers but semi-autonomous "battlefield assistants". These would track snipers, or clear buildings – situations in which their design will have to take into account that things happen too fast for a decision to be deferred to a remote human operator. That means such a robot needs not just weapons, but also a set of rules about whether to use them. How are people being targeted and identified? Could they be civilians, or already wounded? Are they perhaps trying to surrender? Programming decisions to any of those questions will challenge software engineers. But Arkin, while agreeing that robots will never be infallible, thinks they might get good enough to do a better job at making these kind of judgements than combat-stressed humans.

An Iraqi policeman prepares a robot to check a roadside bomb in Karbala, southern Iraq, 2009.

SPACE WAR

Space-based systems are already important in war, for those who have them. As far as we know, there has not yet been war *in* space. But at some level, space war is bound to come. If satellites are tracking your movements and directing enemy fire, there will be strong incentive to attack them. Ground stations for receiving satellite data are the obvious first targets, but space attacks will follow. On-board satellite systems may be technically highly sophisticated, but equal sophistication will not be needed to knock them out, or knock them down.

Could space weapons transform the whole picture of war, though? This is unlikely. They are most realistically seen as an extension of the historic expansion of war into the third dimension with the growth of air power. The evolution of space war will probably have four stages:

- Space to Earth (passive). Observation – tracking, spying and targeting – and communication. This stage is already well established, and ties in with information war and cyberwar.

- Earth to space (active). Weaponry which can attack satellites by jamming them, bashing into them, or destroying them with radiation. Some elements of ballistic missile defence systems are Earth to space weapons of a kind, as they are designed to intercept incoming missiles as high in their trajectory as possible.

- Space to space (active). Ground-launched weapons could attack satellites, but space-based systems would also come into play. Some would be defensive, for example designed to fend off missiles. Some might be offensive, aimed at an opponent's satellites. This is the stage which preoccupies some arms control thinkers as it represents the "weaponization" of space. Once that threshold is crossed, they argue, there is nothing to stop the move to the final stage…

- Space to Earth (active). Here there is much speculation, but no experience yet. The science-fiction scenario is that space-launched missiles or high-powered lasers could be used to rule the world. But the leading technological power, the US, can already project firepower anywhere on the planet it chooses, using planes and missiles which travel through the atmosphere. Firing them from a higher altitude would not alter anything fundamental. In particular, it would not solve the old problem that controlling territory ultimately needs ground troops, however plugged in they are to satellite networks. In addition, space weaponry may lose its advantage because it is harder to conceal than ground-based systems and less mobile, in some senses, than aircraft.

BIOWAR, AND BIOTERROR?

One feature of war in the twentieth century was marked by a new phrase: weapons of mass destruction. The first WMDs on the list, nuclear bombs, clearly warranted the description. Even so, the classification is not always so straightforward. Conventional high-explosive weaponry can rival small nuclear warheads in destructive power, as can incendiaries used in quantity to create a citywide firestorm. On the other hand, chemical and biological weapons are routinely counted as WMDs even though their small-scale use might cause relatively few casualties.

The distinction is not as pedantic as it seems. Nuclear weapons are genuinely scary, and rather numerous. International efforts to curb nuclear proliferation, and to frustrate any attempts by terrorists to get their hands on nuclear materials, will continue to have high priority. Risks from chemical and biological weapons, on the other hand, are harder to assess.

There are reasons to be fearful, especially regarding biological weapons. The science underlying biotechnology is advancing very fast, and is relatively accessible, the argument goes. So defence agencies have to worry not just about countries building biowarfare arsenals, but also about terrorists recruiting a rogue genetic engineer and designing themselves a new superweapon. And, since they might not care about their own fate, they could deploy a deadly new bug even if it was dangerous to themselves – one of the standard problems in imagining how bioagents might be used in "ordinary" warfare.

The threat of bioweapons, old and new, is certainly not negligible. But it has been judged as much higher than that in a series of commentaries on future prospects for new kinds of atrocity. In 2002 Martin Rees, president of Britain's Royal Society, bet $1000 that an accidental or deliberate release of bioengineered pathogens will kill a million people by 2020 (he hopes to lose). More recently, the 2008 report from the cumbersomely named US Commission on the Prevention of Weapons of Mass Destruction

PREDICTION, OR FICTION?

Forget fancy genetic engineering. Biological weapons could wreak havoc by old-fashioned methods. But stories about how this could happen bear close examination. For example, Lawrence Wein of the Stanford Business School told readers of the *New York Times* in 2005 that they ought to get worried about terrorists putting botulism toxin into the milk supply.

The scenario he considered "most likely" was certainly pretty scary. A terrorist would follow instructions in a 28-page manual called "Preparation of Botulism Toxin", which has been known to appear on jihadist websites. Using toxin bought from "an overseas black-market laboratory", the terrorist dumps a gallon or so of broth into a tanker collecting from a dairy farm. By the time the tanker's contents have been mixed with others in a processing plant, distributed and drunk, hundreds of thousands of people would have swallowed contaminated milk. As one millionth of a gram of toxin can be fatal, he estimated that half those exposed would die. One terrorist would have thus caused at least one hundred thousand deaths. It certainly sounds like a weapon of mass destruction.

But wait. Aside from assuming that the instructions work, the terrorist is competent to carry them out, and that a "black market" in botulism exists – for which there is no evidence – there are some other crucial uncertainties. In the paper Wein published in the US journal *Proceedings of the National Academy of Science* which details his analysis, he mentions three: the dose-response curve for the toxin, the rate of inactivation by heat treatment of the milk, and the amount the terrorists could release.

How uncertain? Well, the paper suggests each is impossible to estimate to better than "several orders of magnitude". So each one could be off by a factor of a thousand. Put all three together in a calculation, and the result could be off by a factor of a trillion. So the headline figure of one hundred thousand deaths which got the study coverage in the *New York Times* could, in fact, be zero. Suddenly the whole thing, even if it is believable, looks less scary.

PREDICTIONS FILE

Arlan Andrews

Engineer, science-fiction writer, futurist, entrepreneur

Highest hope: A future world that is democratic, with free minds, free markets and free spirits, and with all individuals educated, healthy and wealthy; all driving a civilization that is expanding throughout the solar system and beyond.

Worst fear: Technology and inept politics will facilitate dictatorships, tyrannies and theocracies in a world of Balkanized nations and tribes that are continually at war – a global Hundred Years War scenario but with nuclear, biological, chemical and nanotech weaponry.

Best bet: None of the current predictions will account for disruptive changes in technology, economics, politics, philosophy and natural events sufficiently to be of much use, other than for laughs or to regret opportunities missed. To wit, there may be contact with extraterrestrial civilizations, or practical immortality, or uncontrolled artificial intelligences, or time travel, or instantaneous transportation, or comet impacts – any one of which would create a world inconceivable to us here fifty years in its past.

Proliferation and Terrorism – a high-level Congressional enquiry – warned that it was "more likely than not" that a weapon of mass destruction would be used in a terrorist attack somewhere in the world by the end of 2013, and that this was more likely to involve a biological weapon than a nuclear one.

Their book-length report was pretty short on evidence to support that conclusion, however. As they put it, "Al-Qaeda is not known to have successfully stolen, bought or developed agents of bioterror. But scenarios of just how such an incident might occur have been developed for planning purposes."

They describe those scenarios as "chilling", and so they are. But are they realistic? On the whole, no, says veteran biowar expert Milton Leitenberg of the University of Maryland. He has strong criticisms of two widely reported wargame-style scenarios designed to test US defences against biological attack. In one, named "Dark Winter", a terrorist group released smallpox at three different locations in Oklahoma, infecting three million people across the United States within two months and causing a million deaths. Similar catastrophe resulted from "Atlantic Storm" in 2005. This time a terrorist group used a commercial dry powder dispenser to spread smallpox in six cities in the US and elsewhere.

However, in Leitenberg's analysis the assumptions about what the terrorists could deliver, not to mention how the virus spread, were rigged to produce the worst possible outcome. The scenarios glossed over the multiple problems of getting hold of disease agents in any quantity, keeping them active, putting them in some kind of aerosol, and spreading them around, all without anything going wrong to kill either the virus or the terrorists. He attributed the extreme depictions of bioweapon threats to a bidding war between US security agencies vying for responsibility over the most deadly threats. In his view, justified at length in his 2005 report "Assessing the Biological Weapons and Bioterrorism Threat" (downloadable from www.strategicstudiesinstitute.army.mil), there was no evidence that any group has gone even a few steps toward achieving the capability of the ones in the defence planners' storylines. At the time of writing, he still holds that view, in spite of the latest Congressional warnings.

While reassuring on the whole, this conclusion applies only to existing technologies. The real problems may arise when more powerful biotechnology becomes widespread. Nuclear production, for example, requires large facilities which are hard to conceal (and easy to attack). Artificial genes will soon be synthesized using desktop machines that can be operated by anyone with basic training in biology (and IT). Although the report from the Commission on Prevention calls for tight security at all biotech labs and speaks of the technology being "locked down", this almost certainly won't work. As they say themselves, the only way to rule out the harmful use of biotechnology would be to stifle beneficial

QUANTICO: SECURITY SCENARIO OR SCI-FI?

If military planners' scenarios often look like science fiction, that's because they consult science-fiction writers about their storylines. After 9/11 US security agencies in particular decided they needed all the help they could get, and that sci-fi authors might know a thing or two about possible future threats.

This new alliance feeds back into the writers' own work, as it helps them get access to technical briefings and to get a better idea of how agencies work. One product of this convergence of interests was Greg Bear's techno-thriller *Quantico* (2006). Bear, whose past novels have used a lot of detailed knowledge of molecular biology, was involved in a series of security discussions which helped inform the book – Quantico is the location of the FBI's training headquarters.

It is a good read, with lots of near-future thrills, terrorist fanatics of one kind or another (Muslim and Christian), and a complex bioweapons scenario which involves weaponized anthrax that has supposedly been tailored to infect specific ethnic groups. However, this turns out to be a red herring, designed to put agents off the trail of the real terror weapon, a yeast strain engineered to induce a new prion disease – a progressive condition that affects the nervous system, a bit like Creutzfeldt–Jakob disease but with more forgetfulness.

The whole thing fits together alarmingly well while you're reading it, and the author cites his encounters with the military and security agencies as evidence for its credibility. But look again at how the terrorists' weapon is made, and you find elements which seem, well, a tad unlikely – notably the autistic boy genius who happens to live in a disused winery which he adapts for home biotech, while teaching himself biohacking skills which would do the US Army's Fort Detrick medical research facility proud. He then runs into a disillusioned former agent with just the right contacts to link the boy's creation into terrorist networks.

Bear has an afterword which indicates that he wants his fiction to be read as a plausible warning, but has left out technical details which he is privy to in case Bad People read the book. The result is a caution which probably ought to be appended to most of the "official" bioterror scenarios that make headlines when they are discussed by security agencies: "The biological weapons and processes are possible, but not in the way I have described them."

www.quanticothebook.com

applications as well. People are going to want some of the products of new-generation biotechnologies badly enough to make controlling them next to impossible.

This "dual use" problem will keep bioweapons on the security agenda. Even if the chances of synthesizing smallpox or "improving" some other pathogen are small, the possible consequences are so dire that governments are bound to take them seriously. While making a new pathogen is still not the same as making a weapon – the production and delivery problems are much the same as those facing would-be spreaders of anthrax or smallpox – the risk of one getting out and triggering panic is real. They would be effective weapons of mass disruption, at the least.

On the other hand, dual use works, as it were, both ways. By the time would-be attackers have mastered the synthesis of enhanced organisms which have all the qualities needed to spread infection, evade the immune system and cause harm, our understanding of how to ward off such infections will also have improved, courtesy of the same biotechnology. As with newly emerging diseases (see p.217), the technological future may ultimately give cause for optimism that the problem can be contained.

FURTHER EXPLORATION

Colin Gray Another Bloody Century: Future Warfare (2005) Analysis of military policy by an international relations scholar who argues that, when it comes to war, the past remains the best guide to the future.

Patrick Lin et al Autonomous Military Robotics: Risk, Ethics and Design (2008) A report to the US Office of Naval Research, available from www.ethics.calpoly.edu.

John Robb Brave New War (2007) Slightly overwrought look at future terrorist capabilities, arguing the need to decentralize essential systems to reduce vulnerability. The author keeps track of developments at www.globalguerrillas.typepad.com.

Peter Singer Wired for War (2009) Review of the development of robots for fighting wars by US analyst (not to be confused with the moral philosopher).

Nick Turse The Complex: How the Military Invades Our Everyday Lives (2008) A wide-ranging review of the Pentagon's reach into every aspect of life in the US. Wars to come are covered in part 7.

www.au.af.mil/au/awc/awcgate The US Air Force Air University maintains an Internet gateway to an extensive library of documents on war, strategy, technology and futures.

www.darpa.mil The US Defense Advanced Research Projects Agency (DARPA) publishes details of what its current projects aim to do, which adds up to a vast speculative catalogue of future weaponry.

www.hsrgroup.org The UN-supported Human Security Report logs trends on armed conflict and "other forms of organized violence".

www.wired.com/dangerroom A well-sourced blog on military affairs, from a US point of view, with on emphasis on new technology.

14
dealing with disaster

dealing with disaster

You can only reasonably prepare for disasters that leave your world largely intact.

Bruce Schneier

Will the future bring more and bigger disasters? We already know we could bring about the ultimate disaster ourselves using nuclear weapons. However, our fascination with imagined calamities makes it hard to gauge the chances of other world-changing events, from extreme weather induced by climate change to asteroid impact. Disaster planning is a growth industry, but may be more a product of our hope to control an unpredictable world than a realistic endeavour.

This guide has already looked at some of the possibilities for large-scale setbacks for humanity brought about by war, climate change, disease, or a combination of these. But disasters come in all shapes and sizes, with journalists quick to use the term for anything involving more than a handful of people (see box opposite). Most are a mix of natural and human-made factors. If there are no buildings, for instance then an earthquake on land may cause little damage.

Disasters also vary in both scale and probability. We know there will be earthquakes, hurricanes and erupting volcanoes in the future, and we can roughly estimate their frequency. But the possibility of disasters that may have little precedent are even more troubling. At the extreme, there might be disasters which wipe out all humanity, but that can obviously only happen once. Others worth pondering fall into a class of "high consequence – low probability" events, which are especially hard for governments and policy-makers to deal with. Is there anything we can do about them?

WHAT COUNTS AS A DISASTER?

Disasters are in the news all the time, and disastrous they surely are for those involved. Most, though, are limited in scope, especially in comparison to the ultimate disaster – there will be no reports on the *real* end of the world.

Some last longer than others. An air crash is awful, but is quickly over and done with, and soon fades out of the news. A nuclear accident may have effects which last for generations. There is still dispute about how many people died, and may still die, as a result of the nuclear reactor meltdown at Chernobyl in 1986 – estimates range from a few thousand to nearly one hundred thousand, depending on wildly varying interpretations of cancer rates.

Even bigger, shorter-term disasters can also have uncertain consequences. A massive serial dam collapse after typhoon-driven rains in China in 1975 was a state secret for the next thirty years, so accurate numbers for the victims are hard to come by. The best estimates seem to be that 26,000 people were swept to their deaths by the flood wave, but another 145,000 succumbed to hunger and disease in the vast area that was inundated.

That is truly a disaster. But the baseline definition of disaster starts low. The UN-funded Centre for Research on Epidemiology of Disasters (CRED) at the University of Louvain in Belgium maintains a global database of disasters since 1900. It records events in which ten or more people were killed, one hundred people or more were affected, a state of emergency was declared, or there was a call for international assistance.

While the body count can seem like a morbid obsession, it's actually a good indicator of the scale of a disaster, and it's usually easier to calculate than financial factors. Other measures, like property damage, make it look as though disasters in rich, developed countries are worse than those that happen in poorer ones. In terms of deaths, the opposite holds true. In fact, developing countries suffer more by other measures too, such as the percentage of gross domestic product the disasters cost.

That Chinese dam collapse is near the upper end of the scale for "natural" catastrophes with a frequency reckoned in decades or less. The Asian tsunami of 2004, caused by a huge undersea earthquake, resulted in more than two hundred thousand deaths. The last century saw one epidemic that was on a much larger scale (the flu outbreak of 1918), calamitous famines in the wake of ill-advised land agricultural reforms in Russia (poorly documented, but with perhaps ten million deaths in 1932–33) and China (thought to have seen thirty million fatalities in 1959–61), and two world wars. But viewed globally, even these were temporary setbacks in a century which overall saw massive population increases and matching growth in economies and, for more and more people, living standards.

ARE DISASTERS ON THE INCREASE?

It sometimes seems as if there is a disaster every day. Earthquakes, tsunamis, hurricanes, floods and mudslides are all transmitted to worldwide audiences. But are disastrous events actually getting more common? And are individual disasters getting worse?

The answer to both questions is partly the result of the number of people living on this planet. Future disasters will often be worse than those in the past because more people will be caught up in them. The size of populations, and how and where they live, are big factors in casualty rates. The same causes are producing larger effects. If an epidemic had killed everyone living in Mexico City in 1900, it would have wiped out around three million people. Today, such a calamity would threaten twenty-two million. At the same time, more people now live in high-risk zones, such as flood plains or on the slopes of volcanoes, meaning that events that would once have passed unnoticed can now cause multiple casualties. Floods (see p.264) are an important case in point.

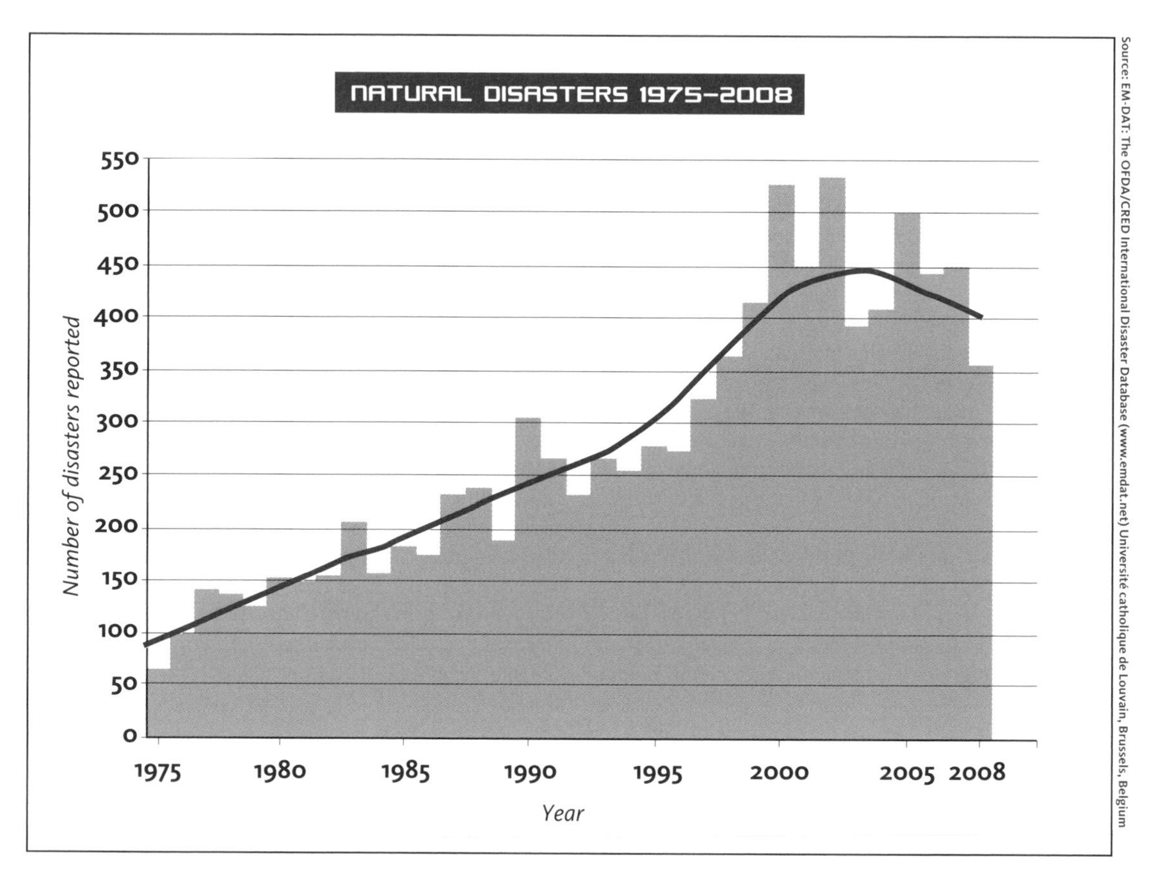

The number of recorded natural disasters is increasing – but is the world getting more dangerous or are there more people living in risky places?

The UN Centre for Research on Epidemiology of Disasters (CRED) classifies natural disasters into three groups that make up an impressive catalogue of bad things. They omit rivers of blood and innumerable frogs, but most would have been familiar to the Egyptians of Exodus. First come water and weather (or hydro-meteorological) disasters, which include floods and wave surges, storms, droughts and related disasters (extreme temperatures and forest/scrub fires), and landslides and avalanches. Then come geophysical disasters, divided into earthquakes and tsunamis, and volcanic eruptions. Finally, there are biological disasters, covering epidemics and insect infestations.

Their figures up to and including 2008 show a gradual increase in geological and biological disasters since the middle of the last century, and a steeper rise in hydro-meteorological disasters. The rate of increase is accelerating, too. The first year the data shows more than three hundred natural disasters across all categories was 1990. The first year the total broke four hundred was 1999. It exceeded five hundred the following year, and has done so again twice since then. The latest increases are largely due to more flood disasters – the number of storms is roughly constant.

Is the increase real? Yes, but it may be amplified in the figures. Communications have improved, so some hitherto unseen disasters may now register. The reinsurance industry provides much of the data, probably leading to a bias towards parts of the world where they do most business. And the improved organization of international relief may encourage reporting of disasters. Could climate change play a role, too? That, not surprisingly, is controversial. All the predictions indicate global warming will worsen floods (sea-level rise rather makes that inevitable), as well as storms and droughts. On the other hand, increased population densities, especially on flood plains, also affect the figures.

And the future? As emphasized in Chapter 7, the climate change scenarios that are likely to come about if we fail to curb carbon dioxide emissions in time make the second half of the twenty-first century disaster-prone to an extent rarely seen in planetary history. Storms, floods, droughts, and water and food shortages could threaten hundreds of millions if not billions; acidification of the oceans could devastate marine life; and a range of positive feedbacks could drive runaway warming, altering the world almost beyond recognition.

Even the less extreme estimates of the consequences of continuing to increase the amount of greenhouse gases in the atmosphere suggest that more familiar kinds of disaster will become more common. A one-metre rise in the oceans – which is often predicted for the end of the century but could come even sooner – would displace 56 million people in developing countries, according to World Bank estimates. Rich nations can probably cope with such conditions better by improving their sea defences, if they spend enough money in time. As the massive flooding of New Orleans in the wake of Hurricane Katrina proved in 2005, such measures can't be taken for granted in even the wealthiest countries.

COULD THEY GO GLOBAL?

Climate change looks like one candidate for a possible global disaster. Could there be others? Compared with disasters we are used to, calamities of considerably greater magnitude would have to appear for us to be dealing with true worldwide setbacks. Oxford University futurist and philosopher Nick Bostrom, who has a playful, intellectual approach to thinking the unthinkable, suggests that anything that caused ten million fatalities or ten trillion dollars' worth of economic loss would count as a global catastrophe, even if some region of the world escaped unscathed. He suggests that an influenza pandemic might qualify. The nearest comparison in the contemporary world is AIDS, which has caused perhaps one and a half million deaths in thirty years, with probably another three million people affected – a true catastrophe, but still not an earthshaker.

What really interests Bostrom, though, are a small subset of catastrophic risks that can be termed "existential risks". That is a grand way of saying they would threaten the species, perhaps causing our extinction. As he puts it, "An existential risk is one where humankind as a whole is imperiled. Existential disasters have major adverse consequences for the course of human civilization for all time to come."

Are any risks that bad? Yes, according to Bostrom's analysis. What's more, they are mostly new. As he outlines in a book he co-edited, *Global Catastrophic Risks* (2008), until the last century – as far as we know – there have not really been any existential risks in human history apart from those which might come from space, notably asteroid or comet strikes. Since the middle of the twentieth century we have lived with the possibility of all-out nuclear war. But he suggests this was "a mere prelude" to the existential risks that will crop up in the twenty-first century. They include, according to some experts, a variety of technological risks arising from bio, nano or computer technologies, which could generate novel dangers since they're capable of self-replicating (see box on p.93).

Even if these are left off the list, there are other existential risks which would not be, as it were, our own fault – and other low-probability/high consequence disasters that might fall short of threatening humanity as a whole, but are still pretty awesome in scope. They have a few things in common that make them tricky to deal with. First off, though we might enjoy fictional disasters in books and on film, we prefer not to think about them happening in the real world. Secondly, they're not, by definition, things we can learn about by living through them and seeing what happens. According to Bostrom, that means existing institutions, which often have to learn by trial and error, are poorly designed to deal with them. In addition, since reducing such risks is a global benefit, individual countries may well under-invest in what it might take to achieve such reductions.

A final worrying aspect of such risks, which some of Bostrom's colleagues have explored, is that estimates of the probability of these huge but unlikely events may be unusually unreliable. In order to judge such estimates, the likelihood that the arguments they're based on are sound also has to be taken into account. So if a published estimate of, say, the chances of a space monster appearing from behind the sun and eating the Earth for breakfast is one in a trillion, we should only use that figure if we also believe there's only around a one in a trillion chance that the argument it's based on doesn't conceal a big mistake. As not many arguments are that convincing, the "real" probability of the space monster gobbling us up could be higher.

The space monster example doesn't worry us because it's not that serious. But what about the chance that a giant particle accelerator like the Large Hadron Collider at CERN (aka the European Organization for Nuclear Research), near Geneva, might trigger a range of extreme physical events that would destroy the Earth? The scientific papers that have looked into these risks have come up with what

IS IT US?

At the opposite pole from writers like Nick Bostrom who want to alert governments and policy-makers to existential risks, there are those who believe that efforts to avoid future disasters miss the point. From the point of view of the rest of life on Earth, the disaster is us.

In that case, they argue, our duty is not to strive too eagerly to stay around, but rather to embrace voluntary extinction. The Voluntary Human Extinction Movement is not exactly large, as movements go. But it has a website (www.vhemt.org), helpfully translated into sixteen languages, and a motto, "May we live long and die out".

The movement is keen to emphasize that this is not an invitation to a suicide pact – no global Jonestown. They simply want people who agree with them to refrain from reproducing. They offer a series of arguments as to why this is the responsible thing to do. The general drift is that, from a Gaian perspective, we are a terribly damaging species to have around and the future of the world would be better without us. Perhaps it would be like the vision detailed at length in Alan Weisman's thought experiment *World Without Us* (2007), a book that attracted lots of plaudits from environmentalists.

The project occasionally reveals its own contradictions, however. We are told that "if any of us thinks about the situation long enough... we will arrive at virtually the same conclusion: we should voluntarily phase ourselves out for the good of humanity and planet". Finding it hard to see how the good of humanity comes into it? Watch the forthcoming feature-length film treatment of *World Without Us* while you try to work that out.

look like reassuringly low numbers. But the fact that they're based on the best interpretations of current knowledge ought to mean we are slightly less reassured. There is no way to know what is actually wrong with the arguments – that's a job for future scientists – but the possibility that they're wrong might make us demand a lower risk estimate before throwing the switch on the next new particle collider.

CAN WE PREPARE FOR THEM?

Everyday disasters are easier to cope with than the existential ones (though the latter shouldn't be completely ruled out). However, there does seem to be a rough relation between how likely particular kinds of disaster are and how bad they can get. Generally, it's easier to get better at handling the less exotic disasters, partly because their frequency makes it possible to learn from experience. That can't be done with Bostrom's extreme "zero-infinity" risks, which are cataclysmic disasters of very low probability. Here are the main everyday disasters, starting with the most common.

HURRICANES (AND FLOODS)

Tropical storms (typhoons in the southern hemisphere, hurricanes in the northern hemisphere, but the same kind of beast) are predictable to some extent. The insurance industry funds www.tropicalstormrisk.com, which estimates the chances of storms occurring over a period of several months. When they do get underway, satellite tracking allows scientists to create rough plots of their likely path, so population centres generally receive warnings hours or even days ahead of time.

Such warnings, together with what we've learned from past disasters, can greatly reduce the effects of the worst storms. How bad are they? Everyone remembers Hurricane Katrina's demolition of New Orleans in 2005, but the cyclone that hit Bangladesh in 1970 killed at least three hundred thousand people in the country with the highest population density in the world.

Both of these horrors were mainly due to flooding, rather than direct storm damage. In fact floods are the main cause of losses from natural disasters. In Bangladesh their effects have been reduced because the country has built a network of 1500 cyclone shelters (some of which double as schools), designed to provide a refuge above the level of floodwater and withstand the storm. As a result, while storms still wreak enormous damage, the death toll is now far lower. A 2007 cyclone almost as powerful as the disastrous storm of 1970 killed just 3300 people in Bangladesh. In the wake of this storm, Cyclone Sidr, the government planned to build another 2000 shelters. Other countries that have not followed suit (or cannot afford to) continue to suffer heavy casualties, with 140,000 people dying in Burma (Myanmar) when Cyclone Nargis hit in 2008.

Protective measures for hurricanes and cyclones are partially undermined by global climate change. There is no strong evidence so far that they are becoming more common as average temperatures increase, but modelling predicts that rising temperatures will make storms more intense, and this does seem to be happening. A study published in *Nature* in 2008 found that the worst tropical cyclones have been getting stronger, with higher windspeeds, for the last thirty years. It is likely, though not certain, that the main reason is the increasing surface temperature of the oceans, and that this will continue to boost storm intensity as it rises in the coming decades. Worse, the most recent modelling suggests there will be both stronger and more frequent storms. Governments investing in storm protection will be in a race against fiercer and more damaging events.

EARTHQUAKES

Earthquake tremors are caused by the grinding progress of the great tectonic plates which form the Earth's crust, setting up tensions that are suddenly released. They're predictable, but only up to a point. Despite intense efforts, scientists cannot tell you when an earthquake is going to happen. They can only give a statistical estimate for a particular period.

In some regions, that estimate is confident enough to constitute a call to action. Most people know about the likely occurrence of future quakes in southern California, for example. Less widely known is the high probability of a big earthquake in Istanbul, where the risk of a 7.6 magnitude quake before 2030 is put at 65 percent. If and when that happens, it could be one of the biggest future disasters. With less money available to invest in buildings that are more likely to remain standing, the city is well behind Los Angeles in terms of preparation. Tehran and New Delhi are in a similarly uneasy position.

It's likely that a major quake in Los Angeles would kill several tens of thousands, while an earth movement on the same scale in one of the other cities could see as many as a million deaths. All that would take in Istanbul – a city with a population of fifteen million, rising by nearly half a million every year – would be a quake coming without warning

Ruined buildings and bodybags line the streets of Beichuan, China, after a massive 8.0 magnitude earthquake struck Sichuan Province on 12 May 2008 – killing at least 68,000 people and leaving some 4.8 million homeless.

during the night and causing ten percent of the buildings to collapse. New homes for the rich are advertised as earthquake-proof; the poor are left to hope for the best.

Geologists' risk estimate for Istanbul rose after detailed investigations in the wake of a severe nearby quake in 1999. These led to a better knowledge of the fault structure underlying the whole region. The most drastic response proposed so far is a new "Istanbul 2", to be built half an hour's journey away from the current city. This plan, sketched by a civil engineer at Purdue University in the US, would relocate four million people at a cost of $50 billion or so. In the real world, nothing is likely to happen before the old city is wiped out by the quake.

The 7.0 magnitude earthquake that hit Haiti in January 2010 offered a horrifically vivid illustration of the damage such an event can cause in poorly prepared, densely populated regions. According to Haitian government estimates, more than two hundred thousand people were killed; around three hundred thousand injured; and a million were left homeless.

Cases like this have prompted more extensive efforts to refine risk estimates, especially for regions which have not been surveyed in great detail for seismic hazards. Surprisingly, serious work in this vein is extremely recent, with major insurance companies and development organizations the main supporters. The OECD's newly hatched Global Earthquake Model public/private partnership, an institution planning to compile and publish a worldwide assessment of quake risks, was just getting down to work in 2009. So in a few years, even if you cannot afford to earthquake-proof your house, you will at least be able to go online and get an objective, up-to-date map of where the best and worst places to build a new one might be.

TSUNAMIS

Everything about earthquake prediction applies to tsunamis, which are most often caused by undersea quakes. The best that can be hoped for is early warning, which is getting much more attention since the Asian tsunami disaster of 2004. This is not prediction, because the earthquakes will have already happened. But as the great waves they cause take time to travel oceanic distances, it's not too hard to set up a system to work out where they're headed, and to warn people on the threatened coastlines hours or days ahead of time to move to higher ground – assuming they have some handy. In the wake of the huge earthquake that hit Chile in February 2010, for instance, nearly 150,000 Hawaiians were instructed to head to higher altitudes, though fortunately the resulting tsunami was too mild to cause damage by the time it reached the islands.

RARER BUT SCARIER EVENTS

All of the disasters above are the kind that anyone can expect to see in their lifetime, though probably not in their neighbourhood. What about the bigger, rarer ones? These, on the whole, would be harder to deal with.

MEGATSUNAMIS

In one or two places, the local geology suggests the possibility of a megatsunami – a vast wave sweeping across the ocean with incalculable consequences when it makes landfall. The chances of this actually happening, though, are very low.

Much has been made of the precariously poised mass of the mountains of Hawaii and the volcanic Canary Island of La Palma. Such mountains sometimes experience "slope failures", in which very large amounts of rock crash down the sides, and in this case straight into the sea. If that happened in La Palma, some claim, fifty cubic kilometres of rock might be dumped in the sea at once. As a result, a giant wave would sweep across the Atlantic and devastate the East Coast of the US.

However, most geologists stress that there are no known megatsunamis of volcanic origin. Such slopes usually slip more gradually. And even if there were sudden failures the resulting waves, though large, might not travel nearly as far as the US. In short, the megatsunami makes a good TV documentary, with lots of CGI waves, but seeing is not believing. The most likely trigger for such a wave-borne disaster remains an asteroid plunging into the ocean (see below).

SUPERVOLCANOES

Like a megatsunami, this is one of the disasters we are familiar with, on a much bigger scale. Earth has certainly seen vast disruptive volcanic events in the past. A super-eruption 75,000 years ago in what is now Indonesia shot enough dust into the atmosphere to cause a 5–15°C drop in temperature over the surface of the Earth. Another such eruption would destroy world agriculture. Looking further into geological and climate records suggests that a super-eruption takes place once every fifty thousand years on average.

The most famous supervolcano location is the Yellowstone Caldera in the US, home to a massive eruption some 640,000 years ago and still a volcanic hotspot. TV programmes and books typically describe the next super-eruption as overdue, but no one really knows if or when it might happen. The Yellowstone Volcano Observatory is trying to read the signs, though. What to do if they get clearer? Aside from arranging to live on the other side of the world and hoping for the best, in this case disaster planning seems pretty pointless.

ASTEROIDS

A collision with a fast-moving chunk of the solar system is a real future possibility that has been taken seriously for the last few decades. That's partly because we now have better evidence of past impacts – most famously the now-buried Chicxulub crater in Mexico, the signature of a body whose collision occurred in suspiciously close proximity to the time when dinosaurs disappeared, 65 million years ago. Their disappearance was part of the last mass

PREDICTIONS FILE

Ben Wisner

Research fellow at Oberlin College, Ohio, and vice-chair of the Commission on Risk and Hazards of the International Geographical Union (IGU)

Highest hope: A world where the complex links between social justice, non-violent conflict resolution and the intelligent use and care of the Earth are commonplace. Urban dwellers will not be divided into the few who live protected and the many who live in slums. Cities will be centres for innovative Earth care and renewable energy. Violence of all kinds will be reduced. Nation-states will acknowledge that protection from natural hazards is a human right. The ongoing effect of accumulated climate change will still produce instability and more intense hazards, but people will adapt.

Worst fear: Earthquakes and volcanic eruptions will continue as always, but increased urban concentration will place more people at risk. Meanwhile, the interactions between violent conflict on various scales (mafia and gang activity, intra-state civil conflicts and smaller regional wars) and natural hazards will increase. It will be harder to get assistance to people suffering through drought, flood and cyclones. Climate-related hazards such as heat and cold waves, flooding, intense coastal storms and drought will be an increasing burden, particularly on low-income nations.

Best bet: I fear that the kind of complex thinking required for people to grasp these interconnected challenges and opportunities will remain blocked by greed and political expediency. So in the end I have to adopt the attitude Gramsci prescribed more than fifty years ago: pessimism of the intellect and optimism of the will.

extinction in the history of life before the human-induced one in progress now.

The solar system was a more hazardous place earlier in its history, before the giant planets cleaned up much of the debris lying around. Most of Earth's old craters have been erased by the slow but inexorable renewal of the surface through the action of plate tectonics. But celestial bodies which don't have the Earth's system of moving plates – like the moon and Venus – bear the scars of many more past impacts.

The basic physics of how asteroids travel is reassuring to some extent. A global threat would only come from an unusually large – and rare – asteroid. Anything smaller than a few tens of metres burns up in the atmosphere, though the way it burns up is still pretty spectacular. A ten-metre asteroid would produce an explosion equivalent to several times the Hiroshima bomb, depending on its composition and speed, but would have virtually no effect at ground level. This is just as well, since such projectiles reach Earth's atmosphere roughly once a decade.

An asteroid around five times bigger than that hits the planet once a millennium on average. It would have a force equivalent to ten megatonnes of TNT, so you definitely wouldn't want to be in the neighbourhood when the next one comes, but it still doesn't count as a globally significant event. That would require something perhaps a whole kilometre across, the size that comes our way once every few hundred thousand years. As well as a colossal explosion, this would cause massive tsunamis if, as is most likely, it landed in the ocean. If it hit the one-third of the planet's surface that is land, the result would be worse. The explosion would vaporize large amounts of rock, and the smallest particles would stay aloft for years and cause a drastic drop in temperature. Above a diameter of two kilometres, there would be enough dust to cause worldwide crop failure. The object that spoiled the dinosaurs' day is estimated to have been ten to fifteen kilometres in diameter.

In the 1980s, data on past impacts, combined with a good deal of campaigning by concerned scientists, drew attention to the possible dangers of "near Earth objects" (NEOs, or objects whose orbits could bring them close to the planet). Unusually, though it's common practice in discussions of future military scenarios (see p.255), science-fiction writers were heavily involved in the public discussion. The idea of cosmic collisions wasn't new (see box opposite), but science-fiction writers interested in real science began to write "realistic" depictions of the possibility. Larry Niven and Jerry Pournelle's *Lucifer's Hammer* (1977) was an early depiction of the end of civilization, brought about by the impact of huge comet fragments. Arthur C. Clarke came late to the idea with his similarly titled *The Hammer of God* (1993), but he had already introduced the idea of an early warning system to detect NEOs in *Rendezvous with Rama* (1972). In that book, his "Project Spaceguard" spots an incoming intergalactic spaceship, but in *Hammer* it swings into action to work out how to shift the incoming asteroid onto a safer path.

Somewhere between these fictional incarnations, there's a real project often referred to as Spaceguard. First proposed by NASA in 1992 and more formally known as the Near Earth Object Program, this detailed telescope survey aims to identify all the objects that might cross Earth's orbit. The current count is regularly updated at neo.jpl.nasa.gov, but has begun to level off after nearly two decades of observation. There are probably around four thousand such objects altogether, including nine hundred or so that are more than 140 metres across. Incidentally, the imaginary entity known as Planet X – which is projected by some to enter

WHEN WORLDS COLLIDE

An early classic of planet-busting fiction, Philip Wylie and Edwin Balmer's *When Worlds Collide* first appeared in 1933. Boasting lively descriptions of natural disasters of all kinds triggered by a close call with another planet, as well as the Earth bursting open under a later impact, it has a more complex plot than most simple asteroid doom narratives. The authors imagined not one but two rogue planets entering the solar system. One is on course to destroy the Earth. But the other, fortunately, looks suitable as a bolt hole for a select few. Most of the book is about the response to astronomers' early warnings, and the efforts of the League of the Last Days to build spaceships to ferry a small group of people to the second intruding world. One of the ships makes it, and the crew set out to start history over again.

The book spawned a sequel by the same authors in 1934, and inspired the classic 1951 movie of the same name. Despite having Oscar-winning special effects, the budget of the movie did not actually stretch to showing two planets crashing into each other. A CGI-age remake currently scheduled for release in 2012 will doubtless finally satisfy those who want to see what such a collision might actually look like.

Planetary collisions *do* happen. But they come about early in the life of planetary systems, when the space around a star is more crowded. The moon most likely originated in a collision between the proto-Earth and a body about the size of Mars around four billion years ago. And, in 2009, spectral analysis of a nearby (well, only one hundred light years away) star suggested evidence of debris from a long-ago collision between planet-sized bodies. Our solar system has settled down now, though the planets on course for Earth in *When Worlds Collide* came from Alpha Centauri...

our solar system in 2012 and fulfil a world-ending prophecy derived from the ancient Mayan calendar (see p.6) – is not among them. If Planet X does exist, it must be equipped with special powers that conceal it from astronomical observation.

Fortunately for our peace of mind, the survey hasn't found anything hugely hazardous so far. Asteroids are rated on the Torino scale, which measures likely impacts in the next hundred years and runs from zero (no risk) up to five (a serious chance of regional devastation) and on to ten (prepare to leave the planet). So far, all the objects tracked by the astronomers have scored between zero and two, with the twos down-rated on closer observation. The most worrying object identified at the time of writing is Asteroid 1999 RQ36, a body a little more than half a kilometre in diameter which was first spotted in 1999. The first review of its orbit indicated that there was no chance RQ36 could strike the Earth in the next hundred years. However, a longer-term calculation early in 2009 indicated that there is a 1 in 1400 chance it will collide with the planet some time between 2169 and 2199.

Aside from this, it looks as though the next couple of centuries will be free of disruption by collision with stray bits of the solar system orbiting in the asteroid belt. That still leaves the outside chance of a blast from an undiscovered comet, which would come from further out in the system and move faster. However, comet impacts are even rarer events than

Hollywood armageddon: unable to reach higher ground, people can only stand and watch as a fragment of comet is about to strike the sea off the Virginia coast in *Deep Impact* (1998).

encounters with large asteroids, and the chance of one happening in the next century is almost certainly less than one in a million.

The consequences of a major asteroid hit, even if it involves one of the smaller ones, means it may still be worth considering whether we could prepare any countermeasures. This is controversial territory since it involves putting weapons into space (another enthusiasm of some science-fiction writers, to say nothing of the Hollywood screenwriters that brought us such tales as *Deep Impact* and *Armageddon*). However, in principle there are three ways of dealing with an asteroid threat once it has been identified.

A well-targeted nuclear bomb might do the trick, but the possibility of simply breaking the thing into slightly smaller pieces which will still reach us down below is discouraging. More basically, you could just try and ram it, and hope that a small deflection happens early enough to take the asteroid off its collision course. Or, with a bit more time and engineering, perhaps a team of Hollywood-style space heroes could land on the rock and install enough rocket motors to push it off course.

Other ways of pushing and shoving asteroids have been proposed, and there are plenty of speculative design studies. The one thing all such studies have in common, though, is that the earlier the asteroid is interfered with – and the further away it is – the easier it is to alter its path. This seems like a good enough reason to keep the surveys up to date.

SPACE WEATHER

The odd-sounding term "space weather" is one way of referring to the possibility of a violent solar storm. This is emerging as a potentially more worrying threat than asteroid impacts. It has drawn less attention until recently, partly because the dramatic possibilities of the scenarios it leads to are less potentially heroic than those that involve space jockeys nuking asteroids in the nick of time.

The concern about solar storms is that they could affect the complex infrastructure which underpins

NUDGE, NUDGE

The simplest way to move a dangerous incoming asteroid off course would be to give it a whack. In the absence of a pool cue long enough to do the job, this will need a spacecraft sent in the right direction.

Although survey results indicate we will not need to rely on deflecting any really big lumps of cosmic rock this century, it might be sensible to try out the idea and see what happens. This is the plan for the European Space Agency (ESA) mission Don Quijote, due for launch sometime after 2012 and reaching a climax two years after that. The basic requirement is relatively simple, as space projects go – fly a spacecraft into an asteroid. To make the resulting smash more informative, ESA plans to have a second craft (Sancho) in orbit around the asteroid to record what happens when the first (Hidalgo) does its stuff.

The result should be detailed knowledge of what kind of rock the asteroid is made of, and how it responds to the collision. The mission is aiming at a small and fairly nearby asteroid, so it's not a full rehearsal for the real thing. Hidalgo's target will be a few hundred metres in diameter, and its reaction will help ESA decide what it would take to shift a larger, Earthbound body – in case, that is, Bruce Willis is not available.

COMPLEXITY: MAKING A DISASTER OUT OF A CRISIS?

A pandemic of, say, bird flu might not necessarily be disastrous in itself, as outlined in Chapter 12. But some argue that developed societies have become so complex and interdependent that they are also less resilient than those of the past. This could mean that a problem which is in some respects containable quickly affects other parts of the system and even creates the possibility of a more general social breakdown.

Modern societies, for example, depend on relatively small numbers of key workers to carry out essential tasks, like operating power stations or driving trucks to distribute goods. They also tend to operate with low stockpiles of many supplies, including food, and use "just in time" inventory systems to cut costs.

There is also a potential conflict between the need to keep key workers in energy and water plants, oil refineries, and transport and communications outfits at their posts and the likely advice on reducing the spread of a new virus by staying at home. Add the fact that modern media (if they keep operating) will spread news of possible shortages and promote panic buying, and the chance of vital services breaking down gets higher again. Diesel fuel, for trains as well as trucks, is a crucial link in supply chains for coal-fired power stations, for instance. And steady electricity supply underpins almost everything else, from elevators in buildings to refrigeration in food-stores, not to mention the computers on most people's desks, which operate everything from the banking system to air traffic control.

However, although there may be new fragilities in the interconnected systems we now depend on, especially in cities, we do not know how much it would take to damage them beyond repair. The evidence from some recent disruptive events is that a lot of the impact gets absorbed, in spite of media amplification (figurative and literal) of the effects. For example, SARS (late 2002 to mid-2003) had little effect on the economies of Hong Kong, China or Vietnam – all three recorded faster growth in 2003 than 2002. It has also been claimed that the three major influenza pandemics of the twentieth century (1918, 1957 and 1968) had very small economic effects overall.

advanced societies directly, rather than indirectly as in the case of an epidemic (see box above). It would happen like this: the extreme conditions near the surface of a burning star create a turbulent mass of charged particles – a plasma in physics-speak – whose flow is partly controlled by ever-changing magnetic fields within the star. Every now and again, these fields – whose behaviour is poorly understood – hurl a huge pulse of plasma into space. If it hits the Earth's own magnetic field, it causes violent fluctuations in local magnetic force, which in turn create electric currents in power transmission lines. They could easily be strong enough to melt down the transformers that are needed between high-voltage transmission lines and local electricity grids, which use lower (and safer) voltages.

We know this mainly from study of past events – and not that far in the past. The biggest documented solar magnetic superstorm took place in September 1859, when a British astronomer saw it begin as two patches of blinding light amid a

BETTER LIVING THROUGH QUANTUM MECHANICS

All our efforts to predict earthquakes and track asteroid orbits are mundane. Perhaps what we really need to avoid disasters is some exotic physics. Specifically, we need a way to use the "many worlds" interpretation of quantum mechanics. This is the reading of our most successful physical theory that leads to the idea that there are an infinite number of alternate universes that exist alongside the one we inhabit.

It says that something like an atom can exist in two overlapping quantum states, like the dead-and-alive cat in a box in Erwin Schrödinger's famous thought experiment. Observing such an atom (or cat) causes it to collapse into one, defined state. But in some other universe, the same atom collapses into a different one of its possible quantum states. This is happening zillions of times a second to zillions of atoms, and, according to adherents of this theory, all the minutely divergent universes actually exist.

Now comes the really clever part, outlined in the short paper "Changing the Past by Forgetting" (2009) by physicist Saibal Mitra. Suppose, writes Mitra, a hypothetical intelligent computer has machinery for backing up its memory, and will reset its memory automatically to the last backup when there is a system error. It might also do the same if it becomes aware of an impending disaster – he does not specify what kind of disaster.

In the many worlds view, as soon as the memory is reset, the state of the universe the computer is in becomes undetermined. In other words, the reset is a way of accessing one of the other infinite alternate universes. Some computer, somewhere, may be heading for disaster, but it need no longer be this one. In fact, says Mitra, there are almost certainly far more states of the universe in which the memory has been reset because of some minor fault than due to knowledge of imminent calamity. That means the reawakened computer consciousness, after the reset, has a high probability of being reactivated in one of the many possible disaster-free universes, and carrying on untroubled.

Our minds may be computer-like, but unfortunately there is no obvious way to use this trick with existing human brains rather than imaginary computers, so the point remains strictly theoretical. Worse, there is a sting in the tail of Mitra's logic. As he explains it, if the many worlds interpretation is true, it is inadvisable for us to test for impending disasters we can't do anything about. That is because the process of observation is affecting our path through the multiverse. Each observation we make collapses possibilities on large scales as well as small, and reduces the subset of all possible universes we could end up in.

The consequence? We actually make a disaster more likely to happen in our world, from the quantum physicist's point of view, by trying to predict it. As Mitra puts it:

> "After learning about the impending disaster all we can do is sit and wait for the disaster to happen, while we could be almost sure that we would not have to endure the coming disaster had we not tested for it in the first place. It was the detection of the impending disaster that trapped us in the wrong sector of the multiverse."

Well, they said quantum theory was weird.

projection of the sunspots he was observing. The visual sign faded in minutes, but the effects – most conspicuously worldwide auroral displays as bright as daylight – went on for days. In addition, operators of newly installed telegraph systems in Europe and the US reported sparks coming from their equipment, electric shocks, and paper tape catching fire. Disconnecting the batteries did not help – the telegraph wires were generating their own current from the magnetic storm.

The effect on new-generation power grids would be worse. A less violent solar event in March 1989 cut off power to six million people in Canada, though only for nine hours. A report from NASA published at the end of 2008 estimated that a "severe" storm could knock out half the US power system in ninety seconds, cost $1.5 trillion in the first year, and take up to ten years to recover from. The GPS and other satellites would also be fried, but the storm's greatest danger is its basic threat to electricity supply. The risks in Europe are probably somewhat smaller, since the continent's power transmission lines generally cover shorter distances.

How likely are such storms? No one really knows. But traces in Arctic ice cores (which preserve nitrogen atoms that have been affected by the energetic particles involved in such storms) suggest the 1859 event was nearly twice as big as any other storm of the past five hundred years. We also know that solar flares, which indicate that stars have hurled plasma into space, peak with the sunspot cycle, which normally lasts roughly eleven years. The next peak is expected in 2012. Solar space missions (NASA's interest in these helped spur the production of its report) are planned to help improve our understanding of our star, and perhaps predict when space weather is going to get really bad. Until then, a solar geomagnetic storm is a low-probability/high-consequence event. The question, then, is how much money to spend designing power lines and transformers to withstand the worst-case scenario. As most serious renewable energy plans involve major extension and upgrade of electric grids (to get wind, tidal and solar energy from where it is captured to where people want it), this is likely to become a serious issue. It may also strengthen the argument for local energy generation, where possible, as a more resilient long-term option.

IT CAME FROM OUTER SPACE

There are one or two more exotic possibilities on the list of existential risks. There is a small chance, for instance, that some star, or rather a pair of stars, will do us all in. Astronomers observe occasional bursts of extremely intense gamma radiation in distant galaxies. "Intense" here means a billion trillion times as powerful as the sun, for a few seconds anyway. They are poorly understood, but the top theory is that they're emitted when two collapsed stars smash into each other, resulting in a colossal release of energy.

We have never seen such a burst in our own galaxy, but if it did happen – even if it were a thousand light years away – the radiation would probably eliminate life on Earth. Oh, and it would come without warning, since it travels at the speed of light and no one knows how to detect the bodies that might cause such an explosion before it happens. Plans to build really, really deep shelters in case such a fate awaits us have yet to be drawn up.

Similarly unlikely, but not completely impossible either, is our running across a wandering black hole,

ENDGAMES

How would we react to the end of the world? That probably depends on your, ahem, world-view, according to websites targeted at Armageddon enthusiasts, like the Rapture Index (see p.18). In addition to fiction's longstanding fascination with the end of all things, various websites give us an idea of how people might react to the end of worlds. Not real worlds, of course, but ones hosted on computers. Several online gaming worlds have attracted appreciable numbers of users, but not quite enough to keep going, and have been shut down.

In February 2009, the managers of the online role-playing game *Tabula Rasa* gave their players a farewell treat, turning the decision to shut down their world into an apocalyptic final conflict, announced a few months in advance. Players went through the final day – when unstoppable alien invaders overwhelmed humanity – with a combination of excitement, enjoyment and regret, freighted with nostalgia for a familiar environment soon to be lost.

Of course, the players knew that if they missed this world, they could sign up to another, albeit different game, to get their thrills. If that option was closed to them, their reaction might have been quite different. But that isn't just a fictional possibility. Nick Bostrom argues (see p.261) that there is a real chance – assuming there is intelligent life elsewhere in the universe – that we are all living inside a simulation built by some superior intelligence for its own inscrutable purposes. If it gets bored, then the ultimate switch-off might be the result. The only consolation for this rather bleak, if fanciful, scenario, is that we would never know it happened.

or more accurately it running across us. Our galaxy may harbour ten million of these remnants of collapsed stars, and they don't stay in one place. The chance of one barging into the solar system is low, but we wouldn't know it was coming more than a few decades ahead of time. If one did travel our way, its intense gravitational attraction would mean the end of the Earth, the sun and all the planets as independent celestial bodies – leaving behind an even more massive black hole and nothing else.

Finally, physicists have speculated upon more theoretical possibilities. They go by names such as "phase transition" and indicate that there is always a chance that space-time – the pattern of particles and forces that makes up the universe as we know it at the most fundamental level – might undergo some rapid change of state, a bit like a supercooled pond icing over instantly if a small crystal forms. The result would be that matter in its present form, and hence life, would no longer exist.

Physicists seem to have no more trouble sleeping than the rest of us, though, so we can probably assume that they don't take this kind of theory too seriously. On the other hand, the physicists who built the first atomic bombs considered the possibility that they might accidentally create an all-consuming fireball that would destroy the Earth's atmosphere. They eventually decided the risk was tiny and gave the detonation the go-ahead. Still, they do seem to have an alarming habit of identifying apocalyptic risks, and then concluding that they are really, really small, so not worth worrying about.

FURTHER EXPLORATION

Florin Diacu **Megadisasters: Predicting the Next Catastrophe (2009)** An analysis of both mega- and slightly smaller disasters (including Hurricane Katrina and the 2004 Asian tsunami) from the point of view of a mathematician interested in chaos theory.

Robert Kovach and Bill McGuire **Philip's Guide to Global Hazards (2010)** Comprehensive guide to natural and human-made disasters, past and future. McGuire, the British partner in this Anglo-American duo, has a clutch of earlier books on really big bad things.

neo.jpl.nasa.gov/neo NASA's Near Earth Object program site tells you all about space rocks.

www.unisdr.org If a global bureaucracy can make the world safer, it will be through the UN-led International Strategy for Disaster Reduction.

www.wmo.int The World Meteorological Organization's site tries to make sense of weather, climate and water hazards.

15 life, society and values

life, society and values

I did not say the future could be foretold but I said that its conditions could be foretold.

H.G. Wells

So far, this guide has covered what might happen in fields related to things we can measure objectively, more or less. With energy, food or water we can recognize some of the main problems and how they might be dealt with. But what about aspects of the future that involve elusive qualities such as values, allegiances, traditions and feelings? We can be less definite about these, but they will undeniably shape what life will actually be like for future generations.

Let's suppose most of us have a roof over our heads and can keep warm, fed and clothed. The ways we might do those things tell you quite a lot about the kind of futures we could live in. But they say less about what sort of life it would be. How will we work, study or enjoy ourselves? How will our governments work? What freedoms and restraints will shape our day-to-day existence? What issues will matter most to us and those close to us?

The answers are varied and hard to predict, but we persist in trying to envision these important details of possible futures. This is territory where novelists have as good a chance of enriching our image of the future as anyone. Imagination might be as apt to anticipate the politics, culture and emotional landscape of the future as any analysis-derived conclusions. Some things will certainly endure – sex, sports and storytelling, for starters. But broad social trends, let alone the details of future lives, elude us. A quick look at past images of, say, future fashion, is more likely to provide amusement than anything else.

An added difficulty, perhaps, is the widely held notion that change is speeding up, and that the future will get weirder at a faster pace than we can easily track. It is easy to exaggerate this, or make fun of it – and it is often merely a comment on the rate of turnover of shiny consumer gadgetry. It does seem harder to keep up with new developments in some areas, especially the ones connected with information technology examined in the next chapter. But whether this is a general feature of the times is open to question (see box overleaf).

THE DAYS OF OUR LIVES IN THE NEXT GENERATION

So what can we say about everyday life that doesn't come from the pages of futuristic fiction? Let's begin with one formal futuristic exercise that looked at some crucial aspects of work, education, culture, community and government. The British educational think-tank Futurelab spent a couple of years pulling together ideas about how current trends might affect everyday life. Although its brief was to outline implications for education, it created three "worlds" as background for that discussion. To sharpen the focus on values, it held some other things constant, adopting a bunch of mid-range assumptions from other futures work. So all its worlds feature the following constants:

- There will be climate change, but not enough to cause catastrophe.
- There will be new technology, lots of it.
- There will be a lot more people in the world, but global population will be levelling off.
- Most people will live longer, and the proportion of old people will go up quite a bit.
- Resources, especially water, may be stretched, and food supplies tight.

People of tomorrow: ideas about future lives and fashions from the 1950s now seem ridiculously dated.

STOP THE WORLD...

Seeing change happen in the span of a lifetime was one of the things that altered images of the future when the industrial revolution began a few hundred years ago. But how much change can people cope with? And is the pace of change speeding up? Journalist-turned-futurist Alvin Toffler gave a big "yes" to the second question in his bestseller *Future Shock* (1970).

If his book was persuasive, it was partly because his definition of the condition he termed "future shock" was vague. And so was his definition of change. He substituted vivid writing for clarity. He first came up with the term future shock in a 1965 essay. In the midst of a turbulent decade – especially in the US – he saw something akin to culture shock afflicting his fellow citizens, defining future shock as "the shattering stress and disorientation that we induce in individuals by subjecting them to too much change in too short a time".

He spent the next five years researching the state of the world and – surprise! – found confirmation everywhere that future shock was "a real sickness" already affecting "an increasingly large number". It was a "psycho-biological condition" induced by what he described with metaphorical overkill as a "roaring current", a "firestorm", a "hurricane", an "avalanche" or a "flood of novelty about to crash down on us".

Apart from colourful language, Toffler's case for extreme change was built on three main lines of argument. There was the 1960s' widespread questioning of established values, whether those were rooted in religion, family, nation, community or profession. There was the air of transience about the culture, with people moving house more often and a high turnover of ideas, images and products. And there was the onrushing technological momentum which indicated much more change to come.

Forty years on, Toffler's breathless book is a bit of an odd read. Quite a few of the trends he emphasizes are still recognizable. But our brains haven't exploded yet (mine hasn't anyway), and he seems to have underestimated the resilience of many institutions, habits and ideas. Even new technology, on which he rested much of his case, alters things less than people often imagine. A car in 2009, for example, is laden with technology absent from its 1970s predecessor, but the experience of driving it is not all that different. Past technologies are often not superseded, but continue alongside or underlie the new ones.

www.alvintoffler.net

▶ The Internet will go on connecting everything to everything else more and more – a prospect examined in more detail in the next chapter.

Looking twenty to thirty years ahead, Futurelab's three worlds, or scenarios, differed according to the dominant values over that time. As usual, these are not predictions, and there is no expectation one of these worlds will come about. They are "ideal types" and what actually happens will be some mix of these pictures, along with other things they do not cover. But the three worlds are drawn so that they are broadly internally consistent, which makes it easier to imagine which one you might prefer to live in. They are worth summarizing in some detail, starting with the overall picture, then looking at a number of particular areas of life. The project is aimed at British education policy-makers, but the value priorities are the kind that could apply anywhere – some countries already look more like one of these pictures than others. (The whole project and lots of supporting material is laid out at www.beyondcurrenthorizons.org.uk.)

FIRST TAKE: AN INDIVIDUALISTS' WORLD OF THE FUTURE

World one goes under the heading "trust yourself". This is a world where the individual comes first. People take charge of their own lives, and rely on the state, meaning the government and all its agencies, for as little as possible. The downside is that if you do not look after yourself, no one else will. The upside is that you can create your own opportunities and solve your own problems, with little need to depend on others.

So good citizens here provide for their own needs. They take responsibility for themselves, and take little from society. The conventional wisdom is: don't be a burden – look after yourself! Life focuses on individual and family interests and needs. Society is in the background, but people are not that worried about the wider community, which they do not want to interfere with their choices. Organizations and collective schemes like a national health service or pensions are "opt in", not "opt out". As much as possible is voluntary, not compulsory. But there is not much choice about accepting this way of doing things. Taking good care of yourself, and your immediate family, is a strong obligation. If you don't, and expect the state to pick up the tab, you will be disappointed.

With such a minimal role for the state, there's not much for national government to do. It interferes in people's lives as little as possible. Basic responsibilities for law and order and defence remain with the state, and it has to regulate monopolies – where one organization controls too much of a market or activity – and try to defuse conflicts and squabbles which can't be sorted out at a lower level. In politics, there are many small parties, representing lots of different interests. Few issues are dealt with by simple majority – there is lots of bargaining before any decision is agreed. There is little support for welfare payments, seen as unearned "handouts".

All three worlds use the global Internet for wider political discussions involving debate, game-playing to develop options and even referenda. But the individualists in this world tend to vote to be left alone, if at all possible. Economically and culturally, this is a highly globalized world, but in an individualistic way. Increased mobility within and between countries means that people often move for work, increasing competition. Global competition for jobs that can be done remotely by computer is even fiercer.

The competition reinforces work's centrality to most people's lives, generating the money they need to get services. Employers provide on-the-job training, and deal with people who really all want to "be their own boss". Freelancers flourish, as do those who get what they need from the informal economy by trading labour and skills and managing with little money. That's accompanied by a strong DIY culture, with "leisure" time taken up with pastimes like growing food, making clothes, other crafts, and "digital cottage industries". Individualism carries over into solitary sports and outdoor pursuits – running, climbing and hiking are popular, team games less so.

Complementing the shrunken state is a stronger role for the family. Family is central to life, and the only group most people belong to for any length of time. It is the first source of support and care for all,

PREDICTIONS FILE

Sohail Inayatullah

Futurist at www.metafuture.org

Highest hope: A real working global governance system, overseeing a transformed economy with a triple or quadruple bottomline – prosperity, sustainability, social inclusion and spirituality. This would go with transformed identities, softening commitments to nation-states, religions or groups, and strengthening loyalty to *Homo sapiens*.

Worst fear: Endless fear, endless poverty, endless loss of the spirit... continued nationalism, crisis after crisis with the inability to see the links, deeper causes or patterns of crises.

Best bet: My highest hopes, but in terms of a plausible future, more moderated. So, global governance but still not full world person to person voting.

including financial support. But there are fewer old-style "nuclear" families, and more in new combinations. That's partly because there are lots and lots of step-parents, step-grandparents, and half-brothers and half-sisters, and partly because there are more same-sex couples and children born with technological help.

These new, more complex families are also a source of tension. Belonging can be restricting. Kinship can tie you down and conflict with freedom to follow opportunities elsewhere. But kids generally have more autonomy, experiencing "free-range" parenting, whether they like it or not.

The creative and cultural side of this world is – you guessed it – individualistic. There is lots of creative work because people are brought up to believe in self-expression. But career artists need other jobs, or patrons, because old ways of making money by selling media products, such as books or CDs, no longer work. Branded sponsorship of bands and writers is common. Most information is stored on remote computers linked to the Internet, and plucked from the "cloud" when needed, but many people pay for a more secure alternative for personal stuff – the family "data vault".

Finally, qualifications matter a lot for getting jobs. But education is also about developing the individual, and is shared by the state, employers and the family. Only the family is really interested in the learner's long-term wellbeing, though. There are lots of ways to get educated, some highly personalized for each student, and "informed choice" is the order of the day.

SECOND TAKE: A WORLD OF FLUID ALLEGIANCES AND LOYALTIES

The second world sketched in this project goes under the heading "loyalty points". People work a lot with others, but are often torn between what they want and the demands of the groups they belong to. Relationships between people, employers and the state are carefully worked out, and more and more aspects of life are regulated by contracts. Your personal reputation and what everyone thinks of you is vital and must be carefully managed.

One of the attractions of these restrictions is that you can know where you stand in most situations, with guidelines covering what you have to do and what others will do for you. On the other hand, to get on in this world you have to get really good at negotiation, and keep track of all those agreements.

Citizens get involved in many groups and organizations, usually in ways ruled by one of those contracts that spells out what they need to do. Nationality is less fixed than it is now, as more people move around between countries in pursuit of a better deal. For those staying in one place, the way organized groups and associations loom large in everyone's life makes reputation both an asset and, potentially, a burden. Groups of all kinds, not just employers, want to know whether you can be trusted, and what you might contribute. You, on the other hand, are always looking at the options and thinking whether you really want to join.

On the whole, people have strongly interconnected social lives, knit together through their roles in lots of different kinds of organizations. These include groups linked to their religion, work, local area, interests, health or education. Local communities are generally strong, except in areas dominated by the very rich (who can manage without getting much involved with others who are not) and the very poor (who may have less to build community with).

In this world, as in the individualistic one, the role of the state has also changed. Much of what used to be organized nationally is now done locally under contracts (again) whose general terms are set by the state. It also provides basic security, but even policing is now often local. The many new associations are regulated and registered by the state, which also keeps an eye on its systems for getting and sharing data about its members, and keeps key records itself.

Old-style politics has a restricted range in this world, since so much of people's lives is covered by specially negotiated contracts. Laws set the framework in which the contracts operate, but do not say what has to be in them in detail. Global issues need intense negotiation, which is closely monitored by individuals and groups interested in the outcome, but actually done by professionals.

Globalization is still a force in this world, but tempered by other trends. Personal mobility is important, but then so is belonging to groups. Some of the groups people join (or are allowed to join) actually meet, some are virtual. Either kind can easily get involved in larger networks online.

Work is more likely to be on contract, often to more than one organization at a time. Groups form and re-form around particular projects, and are often managed online. It can be pretty insecure if you are not that good at what you do, or people do not like you much. Local work centres offer a haven for mobile workers and their laptops, enabling people to both work online and continue to benefit from real-world contact with others. Domestic work and care work are valued, but still not high status.

Families remain important, but who is in the family can vary a lot. As well as the new kinds of arrangements in the individualistic world, some families are close-knit, while others are more extended through relationships based on friendship as well as kinship. Some households comprise families who relate less to each other than to all the people they know online. Most families place a premium on caring and keeping members healthy, as this brings benefits such as discounts on health insurance.

Children's leisure is highly organized, and some react against this when they get older by taking risks with extreme sports or drugs. Adults tend to drop these as they know their health care is not covered if they hurt themselves. They go shopping instead. Leisure products which boost health and fitness and make you look good are popular because they enhance your reputation.

When it comes to the ever-growing production of information of all kinds, people keep a close eye on their personal data because managing reputation is so important. There are strong rules about what information organizations can share about you. But sharing knowledge also helps promote reputation for people and communities, leading to lots of entertaining and original communication. In cultural production, content is mostly free, the reward for creators coming as kudos. As before, this means there are problems making a living from creating new stuff.

Because it takes so long to make the network of connections necessary to get through life in this negotiation-heavy world, the age of leaving compulsory education has come down, while full adulthood comes later, creating a long "extended playground" in adolescence. Once people are looking for work, their online reputation and identity matters more than old-style paper qualifications, and they find their skills need constant updating. The general goal of education is to help people "find their niche" in a complex web of organizations and groups, and equip them to contribute to the ones they want to belong to.

THIRD TAKE: A WORLD OF COMMUNITY AND CONNECTION

The third world in this exercise goes under the heading of "only connect", and is one where people see themselves as members of a wider community first and individuals second. At the opposite pole from those dwelling in the individualistic world, they believe that risks facing one group are faced by everybody, and that people depend on each other to tackle them. There is also a renewed sense that the talk about serious issues that happens in public – whether in the media, in pubs and clubs, or in the playground and the staff room – can really matter. The Internet makes this kind of discussion far richer, and more interconnected.

An attraction of this world is that everyone can get more involved in the things that really matter to them. On the downside, as with the world in which defined relationships are paramount (only more so), reputation is vital, and continually revised. Everyone is judging everyone else's actions all the time, and there can be relentless pressure to "do your bit". Good citizens in this world earn their reputation by helping maintain the supply of common goods. These are things that people cannot own for themselves, like a nice environment, decent government, or services that allow everyone to use important things, like roads, the national power grid or the Internet.

Citizens keep an eye on their local and national representatives, get involved in democratic discussion (not just voting) and are ready to take action, even if only in small ways. The ruling idea is that many small contributions will get the job done.

The main role of the state is to maintain civil society – all of the bits which fit between the family and the national government. As much as possible, from childcare and rubbish collection to community policing, gets managed at the local level, often by local groups. Taxes have risen to support benefits, which are paid to retired people and those not in paid employment who still work to get these things done.

People live and breathe politics, but it has a different flavour from our own. They are directly involved in discussion of the things they really care about, and – as far as possible – in decisions, too. Arguments about important issues are no less heated, but it is more taken for granted that people will discuss them in detail. This is a never-ending opportunity for some, a tedious chore for others.

Work is still important, but it is not the only thing, and not just about the money. Success is defined less in terms of how much you earn and spend. People

HOMES OF THE FUTURE

The house of the future has been a staple of exhibitions for half a century, usually featuring heaps of automation, labour-saving features and new materials. But what will homes actually look like in coming decades?

There will certainly be a complicated mix of dwellings. Ranging from hovel to mansion, the most common twenty-first-century household accommodation will be in a densely populated city, and probably in an apartment rather than a house. Newly built homes will come in lots of shapes and sizes, and old ones (the majority, in most places) will get a range of updates. Within that mix, here's a likely menu of features. Which ones get taken up will depend on money, personal preference, policy and cultural trends.

- **Flexible homes** Any one development will have spaces of different sizes, and the rooms will be easy to reconfigure. One room might be a nursery when children are small, an office for wired home-working later on, then a safely furnished space for an elderly dependent relative. Living spaces could be open-plan for a couple, subdivided as they acquire children or granny comes to stay.

- **High-tech homes** New technology increases flexibility, for work and leisure. Computer and communication connections will be as essential as electric power. Wired, and wireless, connection will be available at home – though its ubiquity elsewhere could also make home less distinct from other places. Constant upgrades will be available for new-builds and Victorian piles that survive to the end of the twenty-first century.

- **Eco-homes** Some of that technology will organize domestic life so it has a smaller environmental impact. Old and new materials already exist to insulate homes and control their temperature, but millions of old homes don't yet use them. Rising energy costs and concerns about global warming will encourage upgrading. And sensors for every fixture and fitting – from the shower to the wall-screen TV and computer display – will help reduce energy use. Further add-on technologies will ensure electricity comes from the cheapest source on the grid at any moment, help with recycling and probably generate some of the energy used. Sensors will also help the home-dwellers organize their lives and keep track of what everyone's doing, to children's annoyance and parents' relief.

- **Radical homes** Most of the housing people will occupy in twenty or thirty years' time has already been built – in developed countries at any rate. But there is a growing catalogue of possibilities for new homes which are anything but traditional. Big developers who start from scratch, like the backers of the new city at Songdo in South Korea, can put in energy-efficient glazing and ventilation, linked under intelligent control, on day one. They can add underground pneumatic pipes to deal with solid waste, as well as rooftop water collection and grey water recycling for irrigation, along with spaces for urban farms.

- **More individualistic projects** These might pick up on ideas such as architect Mitchell Joachim's vision of a living tree house growing around a frame woven from carefully trained elm or oak, though you would have to wait a good few years to move in. More plausibly, you might choose a modular home made from processed, recycled plastic. At the moment, buildings made from Thermo Poly Rock (as the Welsh producer calls it) only last eighty years. But by the time they need replacing there may be something equally environmentally appealing, yet more durable. Hundreds of architectural models for successful homes of the future already exist – the question is which of them will get reproduced in quantity.

The UK's largest "eco-village", the Beddington Zero Energy Development (BedZed) in London, was designed by architect Bill Dunster to support a more sustainable, low-carbon lifestyle.

care more about "quality of life", and employers support this in lots of ways. They may offer flexible jobs, and sign up for good causes that their employees support. They accept that it is in everyone's interest to help workers contribute to society as well as to the organization that pays their salary.

There is also a less clear-cut distinction between work and leisure. This goes beyond the blurring of boundaries introduced by having access to the Internet everywhere all the time, which is common to everyone. Employers may take an interest in employees' pastimes outside work. Some parts of what would once have been leisure time pursuits may be introduced into work, in the interests of having well-rounded employees. It can be a bit contrived and forced, but most people feel there is a better balance between work and life.

Family, like everything else, is as much negotiated as given. The "traditional" model of a couple caring for their own children still exists, but is only one option among many made possible by changes in law and technology. The new flexibility is easier to deal with because some of what used to be family responsibilities – like caring for those who cannot look after themselves – are now shared more widely.

The cooperative cast of this world makes people feel more comfortable with the new possibilities for culture created by new technologies. Old divisions between producers and consumers have pretty much disappeared, and artists expect their work to be changed, remixed, sampled and mashed up with other creations. Only the rarest and finest stuff is preserved in its original form. Once again, organizing who pays for what is complex and, for the artists, so is making a living. It is easy to contribute, but getting an audience – whether consumers or co-producers – has become harder in some ways. Education is mainly aimed at helping people discover how to get on with others, in every sense. That includes learning about different ways of seeing things, how people depend on each other, and the value of working together.

WHAT WILL SHAPE VALUES?

Those three sketches give some idea how different dominant values might shape life in the next couple of decades. But what will shape the values? Can we really choose between these three kinds of worlds? It seems as if there is less choice about some things than others. Let's consider some of the main likely influences on future life in this light.

GLOBALIZATION

The economic and cultural connectedness of all parts of the globe seems to be a feature of all of these worlds, though maybe tempered by moves back to local activity and community because resources are tight or people want that emphasis in their lives.

But at the moment, globalization is the stronger trend. The volume of world trade has risen fast, from eight percent of world output in 1950 to twenty-five percent in 2000. More people travel further, and more are connected through electronic and media networks. And some problems – climate change being the most pressing example – are inherently global because actions in any location contribute to effects everywhere else, and so call for global responses.

Globalization as a shorthand term also covers other things. In *Going Global* (2008), Michael Moynagh and Richard Worsley suggest considering the phrase in three respects: physical globalization (all that trade), cultural globalization (in which more people share values and aspects of their identity), and perhaps virtual globalization, an emerging third phase which takes place entirely online, and includes parts of both the other two.

All three look likely to continue, assuming there is no great collapse of the economy, but there are some countervailing trends. For example, local and national cultures, and subcultures, can be reinforced by the Internet connecting like-minded people all over the world (and in the same town). And corporations with a global reach can recognize regional variations. McDonald's, for example, will sell you salmon in Norway, lobster in some Canadian cities, and rice and beans in Costa Rica. But the logo, the staff's greeting and the taste of the fries will still tell you that you're being served by McDonald's wherever you happen to be.

Meanwhile, the rising pitch of world trade, cross-investment and economic involvement with global markets, and abolition of distance via electronic technologies suggests that "opting out" of globalization is not in fact an option. Responding to competition by declining to compete is a recipe for losing. Until a fully developed alternative comes along, global competition looks set to remain the economic order of the day, and perhaps the context for cultural production as well.

From a long-term perspective, and in spite of the rise of China and India as proto-superpowers, the globalizing trend looks like the continuation of a trajectory which began around five hundred years ago. Culturally, it looks in some ways like the continued dominance of the West – Europe was the first region that became "modern" in the generally accepted sense of the term, thus it also got to define its own terms. Capitalism, even in its modified form in China, and mass culture are in the ascendant in more and more countries. They go along with urban life, accompanied by styles of clothing, film and TV, music, and eating and drinking which take on local cultural colouring, but are increasingly patterned on models first popularized in Western cities. The latter, of course, also get more cultural products from other parts of the world, authentic and ersatz, and their own culture is influenced by new imports.

The optimistic view is that this is not just two-way traffic but something more complex and interesting, and destined to become more so. The aspects of culture that people value, and identify with, are chosen from a broader menu. Old order and tradition may decline in importance, but were generally the preserve of a minority. New, broader-based cultures, driven by mass markets, will often generate product which looks dumbed down (and is). But the scope for creating more satisfying work will be broader and richer as well, and the size of the audience educated to appreciate it will continue to grow.

THE IMPLODING, OR AT LEAST SHRINKING, STATE

Globalization helps reduce the role of the state in each of these futures. While there's a desire for more local autonomy in some respects, in other ways we face being ruled by global forces over which individual states have less and less influence. How drastic a change would that be? Political scientists emphasize that the nation-state is a modern invention (usually dated from the treaties of Westphalia in 1648), and one unsuited to a globalized world in many respects. Economics, free trade and problems that take no heed of borders (climate change again) all reduce national governments' influence over their own affairs. Having said that, although the financial system is global, the banking crisis of 2008 was mainly tackled at national level.

People also still seem to want to live in a nation-state defined in geographic, ethnic and cultural terms which make sense to them, rather than in some country whose boundaries were set by others, or to identify with some superstate like the EU. But the loosening of loyalties to nationality or place which Futurelab's three worlds envision also sounds right, at least for some people in the more affluent parts of the globe. For others, economic necessity driving migration in search of work may achieve the same effect.

Within states, whatever form they take, there is good reason to hope that more of the globe will be democratically governed. The basics of democratic government – free elections, secret ballots, and an independent judiciary – seem pretty well established. And they are spreading. There are now one hundred and twenty or so electoral democracies, nearly twice the number there were twenty years ago.

This has made less difference to the percentage of people who live in democratic states than it might as one of the very largest, India, has been a democracy since independence in 1947, and another, China, remains a one-party state. The future of government in China is much the largest uncertainty in estimating democracy's reach in the near future. Recent history indicates that capitalism, of a kind, and rapid economic development can go ahead without any move to more democratic government. How long might that last? Francis Fukuyama, the man who claimed that capitalism and democracy go hand in hand in *The End of History and the Last Man* (1992), now says that democracy in China will happen eventually, but that "the authoritarian system will keep going and get stronger" for the next few decades. That might just be good for the rest of us, as it may soon be possible to judge whether China is better placed to deal with carbon emissions, for example, by taking measures which would be harder to implement in a democracy.

Whichever way China goes, representative democracy has increasingly become the standard. In most places it still operates with lots of imperfections, as well as the continual need to try to ensure free access to the media and prevent money buying votes or parties (eternal vigilance, as they

PREDICTIONS FILE

David Brin

Author and futurist

Highest hope: We finally come up with a useful, eclectic definition of sanity that can help humans – at long last – to live happy, diverse, creative, tolerant and joyfully cooperative/competitive lives.

Worst fear: A return to the "classic" form that dominated nearly all other human civilizations – some type of feudalism, dominated by a conniving oligarchy of kings, priests, nobles, commissars... Our Enlightenment, only a couple of centuries old, has propelled us in new directions filled with hope and possibility. But the old ways tug at us and keep trying hard to return.

Best bet: Almost at the last moment, we'll remember how committed we are to a Great Experiment in science, compassion, democracy, more science and a marvellously ironic interplay of competition and cooperation, neither of which has ever done this well, all by itself.

say). But might the day-to-day business of governing get more democratic than giving citizens the right to choose a new lot of governors every few years? It clearly could, if people either demand or are offered more regular ways of engaging with legislators, and their opinions can be taken into account without the system seizing up. Some of the new ways of wiring everyone up will contribute to this – a topic for the next chapter.

FUTURE IDENTITIES: RELIGION

Globalized cultures dissolve some boundaries, and could make it less compelling to identify yourself as a citizen of a particular country or be proud of a nationality – a position which seems increasingly reserved for the tired, recycled rhetoric of the extreme right. That does not mean there will not be powerful contributors to personal identity, and sources of values and convictions which reflect and amplify it. One which clearly has a strong future is religion.

Discussion of the future of religion is especially prone to wishful thinking. We would all like to think that others will come to share our most deeply held beliefs. But it is safe to predict that people will want answers to the same "big questions". Often, they will come from the same sources they do now, the hopes of secularists (like the author of this guide) notwithstanding.

Some modernizing societies have seen religion take a smaller role, at least in Europe. Others, such as the US, see little or no weakening of religious allegiance. And where science challenges particular points of religious doctrine directly, or even appears to, it still risks coming off second best. Moreover, challenges to established religion can induce a move to more fundamentalist variants of existing religions as a defence against erosion of religious values.

The future course of these trends is murky, though. In the UK, for example, the futurist-oriented social science review the Tomorrow Project (www.tomorrowproject.net) suggests that "the future of religious identity hangs in the balance". The country that has seen the most extensive secularization in old Europe could still go one of three ways, they reckon. Religion might continue to decline in importance, the decline might halt or even reverse, or – perhaps more likely – there could be an increase in "disorganized" religion, in which personal brands of faith are more important to more people. The project's authors suggest that one good bet is religion becoming more diverse:

> "Identities based on organized faith ... [will] polarize between 'fundamentalist' and 'open' expressions of belief. In the case of the churches, this polarization will loom larger in believers' lives than differences between the denominations."

That comment may be relevant globally, and has a political correlate. The modern, metropolitan middle classes favour more restrained sects and denominations. Among the increasing numbers of the urban poor in many parts of the world, though, the fastest-growing religion is Pentecostalism, a brand of evangelical Protestantism which is an altogether livelier and more involving affair with an emphasis on vocal worship and local self-help.

The growth in new-style Pentecostalism is not likely to be enough to prevent Christianity in all its forms from falling to second spot in global popularity. Accurate figures for adherents are hard to come by, but mid-range estimates suggest that there are just over 2 billion Christians worldwide, with 1.5 billion followers of Islam, a little more than a billion non-religious folk, and 900 million Hindus. Demographic projections suggest that the numbers in mainly Islamic and Hindu countries will increase, and Islam may overtake Christianity in absolute numbers by the middle of the century.

However, this is a pretty uncertain extrapolation, depending on being right about both future birth rates in different regions and the continuation of patterns of belief down the generations in rapidly changing circumstances. There does seem to be a tendency for fundamentalists to have more children than adherents of less strict doctrinal interpretation. But suggesting that strong beliefs will gain ground depends on an unlikely assumption that fundamentalists find it easy to raise their children to reproduce their beliefs, generation after generation. A gradual reversion to less exacting creeds seems more likely.

The role of religion in the longer term is even less certain, but there are still some efforts to sketch the terrain. A report on the future of religion from a 2006 conference of Boston University's Pardee Center for the Study of the Longer-Range Future began with an overview from Peter Berger, a leading sociologist of religion. He drew two conclusions. First, earlier views that modernization leads to a decline in religion were off the mark – instead, it leads to an increase in religious pluralism, with people in most countries facing a choice between more competing faiths and creeds. Second, as he put it, "religion will continue to be a centrally important factor on the world's scene".

The increase in pluralism could include completely new religions. Science-fiction novelists make up religions fairly often and at least one – Scientology – has real adherents, and annoying proselytizers. There are plenty more, exemplified by the folksy aphorisms of Bokononism in Kurt Vonnegut's *Cat's Cradle* (1963), but most are parodies, makeovers or satires of existing religions. More inventive possibilities include the creed known as "The Truth" in Iain Banks' novel *The Algebraist* (2004), which holds that everything we see is part of a computer simulation.

Moving out of science fiction proper, the idea of a forthcoming technological singularity (a greatly increased speed of technological change) has distinct religious overtones, as noted in Chapter 17. Robert Geraci's *Apocalyptic AI* (2010) lays out the similarities between the visions of some popularizers of robotics and intelligent computers – who foresee the advent of super-intelligent machines that end up furnishing new homes for human consciousness – and longer-established, religiously rooted depictions of transcendence. Perhaps a new religion that promoted the construction, or even worship, of such machines could one day be entrenched in worldly reality in ways which the old religions cannot claim.

There are other science-related possibilities, such as celebrating the Big Bang as a new creation myth, in a story which begins at the origin of the universe and culminates in the evolution of life and consciousness. And there is the prospect of even more direct involvement of science in religious discussion as scientific theories of evolutionary origins or the neuroscience of religion offer new accounts of where beliefs come from. But it will not take much theological sophistication to reconcile this kind of explanation with continued religious belief.

HOW FAR DO VALUES GO?

Whether values are rooted in the family, the nation, the global community or an old or new religion – or more likely some combination of all these – they do not lead directly to economies or societies turning out in a particular way. This is partly because the values people claim to uphold often conflict with

PREDICTIONS FILE

Ziauddin Sardar

Editor of *Futures* and cultural commentator

Highest hope: Deep and wide-ranging reforms will take place within Islam, leading to the re-emergence of a dynamic, knowledge-based and forward-looking Muslim civilization.

Worst fear: Muslim thought will become even more ossified and obscurantist and, as a consequence, Muslim societies will fall further into despondency and strife.

Best bet: Both my highest hopes and worst fears will be realized simultaneously. There will be reforms in certain areas and retreat in others. Chaos will reign but new ideas will also emerge from the edge of chaos. However, in aggregate there will be positive progress, we will see some genuine reforms within Islam and attempt to rebuild the Muslim civilization step by step.

their actual behaviour. On many issues, the values espoused as a source of hope for the future point to some noble goal that commands widespread assent, but which may or may not get any closer to reality if self-interest continues to hold sway.

A MORE EQUAL FUTURE?

Gross global inequality is on most lists of things we'd like to reduce in future. It is prominent among the targets defined by the United Nations as Millennium Development Goals. In true UN fashion, these were formulated in bureaucratic, but satisfying detail: eight goals, broken down into twenty-one quantifiable targets, assessed via sixty indicators (see www.undp.org/mdg). They include goals aimed at reducing economic inequality, such as improving gender equality and raising educational standards. But the basic measure of personal income is perhaps the most crucial. The declared goal for 2015 is to reduce by half the proportion of people living on less than a dollar a day.

Based on current trends, it looks as though more people, and a larger proportion of the total population, will be better off in absolute terms. The very poor are being lifted out of poverty by economic expansion in China, India and parts of Africa. But the same economic expansion that carries this trend has been accompanied by increasing inequality in incomes, between and within countries.

This is tricky territory for researchers, but a World Bank study found that in 1988, the top 5 percent of the world by annual household income were 78 times as rich as the poorest 5 percent. Just five years later, they were 114 times richer. Within countries disparities are even greater. Before the economic slump of 2008, chief executives in the US were being paid more than five hundred times average earnings – up from a mere two hundred times in 2003. Although some salary and bonus packages at the very top have since been curbed, the overall trend is still that economic expansion, which is the main factor in reducing poverty, also allows the wealthy to grow still wealthier.

Factors influencing this pattern in the future include the increasing visibility of extreme wealth and how this affects tolerance for such disparities, and the course of campaigns to rein in overconsumption in favour of more fulfilling pastimes. There is also, perhaps, political mileage in increasingly solid evidence that people are happier, on average, in less unequal societies. But it is less clear whether these include the people who actually have the major wealth.

A CULTURALLY DIVERSE FUTURE?

We would all like to preserve the riches of the world's existing languages and cultures, and see them feed the cultures of the future. But globalization and other forces are reducing some kinds of diversity. Languages are disappearing in a manner reminiscent of endangered species, and a small number of tongues are set to dominate future communication. Of these, one is in a lead it is unlikely to lose.

It is not the world language yet, but English is emerging as the only plausible candidate for a tongue that is understood by at least some people all around the globe. The highest estimates are that about a quarter of the global population now can speak English, whether fluently or a bit, and its use is spreading fast. Those who speak English as their first language tend to overestimate the importance of this, as they are less prone to speak more than one language than people in other countries. Nevertheless, English as the default second language of business, diplomacy and science will become more entrenched, reinforced by the embedding of the language in much of the world's stored electronic information and by the global reach of the Internet.

Again, China contributes the most serious uncertainty to this picture. It is itself home to many languages, but retains Mandarin as its official language. That currently makes Mandarin an indispensable second language for people who want to do business in China. On the other hand, some twenty million Chinese study English every year, according to the British Council, so it's possible that situation may change.

As some languages grow, others die. Linguist David Crystal estimates that if current trends continue, half the current six to seven thousand languages in the world will disappear by the end of the century. The ones which have never taken written form (around a third of that total) are lost forever if the last speaker dies. A written record can allow a language that has fallen out of use to be revived, so efforts to document threatened languages are crucial. Here, the Internet is a help, as it means a scattered language community can communicate more easily.

A LESS MATERIALISTIC FUTURE?

In the countries on the wealthy side of current inequalities, pleas to recognize the downside of the pursuit of material gain are common. Appeals abound to shift priorities, seek other kinds of satisfaction, and lead a more fulfilling life than the one which seems to follow from accumulating more stuff. This chimes with the official prescriptions of all religions, of course, and with numerous calls for more environmentally sustainable ways of life.

What are the chances of it happening for most people? Modernity, capitalism and (maybe) democracy go together, but one of them shows alarming signs of being self-limiting. After the former Soviet Union and the Eastern European nations fell into line with the idea that the market is the best motive power for their economies in 1989, the notion of alternatives to capitalism seemed largely theoretical for a couple of decades. Then the banking crisis of 2008 prompted new questions about the future of the world economy. But there were already more serious questions about whether capitalism's continual drive for growth would breach natural limits (see the "limits to growth" debate in Chapter 4).

Opportunistic environmentalists have drawn up plenty of plans for addressing these concerns through investment in ecologically beneficial technologies to boost employment – a "green new deal". But even if such plans were implemented, there remains a question about whether the entire world system might grind to a halt if growth falters. There seems little sign of any future path that resolves this doubt.

A 2009 report from the UK's Sustainable Development Commission (a body that advises the government but does not really expect its advice to be heeded) reviewed the prospects for "Prosperity Without Growth". It makes a familiar call for the more thoughtful definition of "prosperity" that would recognize the externalities, as economists term them, of current production – climate change being the most troubling example. It would also take into account the drawbacks of high-earning economies, which tie wage-earners to a consumerist treadmill by creating new needs to maintain profits, and may even be responsible for the low birth rates that will make it hard to balance budgets in the ageing countries of Europe and Asia.

But it also asks whether any stability could be found in an economic system somehow adjusted to manage without growth, in the already rich countries anyway – an adjustment which is needed because "responses to the crisis which aim to restore the status quo are deeply misguided and doomed to failure", both economically and environmentally. These are good questions. However, the report's main author, economist Tim Jackson, said when it was published that "the reason why nobody asks the difficult questions that we are asking here is because nobody really has any answers to them".

In that case, the immediate future is likely to see capitalism trying to respond creatively to these different pressures. How? Markets and trade will continue, and perhaps extend into new areas, regulated in new ways – carbon markets to help tackle climate change being the most obvious example. And the system will continue to adapt to new demands. When economist Joseph Schumpeter described capitalist economies as subject to continual "creative destruction" in the 1930s, he was talking mainly about companies and products, but the same phrase may apply to institutions. The contemporary political theorist Geoff Mulgan suggests that forecasting how the response to the triple pressures of ecology, globalization and demographics will evolve in detail is pointless. But he also argues that new technologies have the potential to remake capitalism "more clearly as a servant rather than a master, whether in the world of money, work, everyday life or the state".

What about the older ideas of capitalism collapsing under its own contradictions? The predictions of Karl Marx now seem to have few adherents, but there is a more recent world system theory, much influenced by Marx, from the American analyst Immanuel Wallerstein, which maintains future changes will be much more drastic than more conventional commentators envisage. He is endearingly vague what these changes will be, but surer, as Marx was, that the current system is destined for collapse. And he cites the same Marxian reason – that "surplus value" (what employers extract from employees) declines so fast that it gets too hard for anyone to make a profit.

The outcome, according to Wallerstein, will be decades of turbulence, during which there will be struggles over whether the replacement non-capitalist world system is hierarchical and exploitative, like the one we have now, or more democratic and egalitarian. And then? Ah, well, "it's intrinsically impossible to predict what the outcome will be; the only thing we can be sure of is that the present system won't survive and that some outcome will occur". Um, thanks for that.

Those are some of the questions that set the framework for what life will be like. They still do not quite get down to the nitty-gritty – the sex, sports and story-telling. Those, like everything else, will be affected by the new, wired culture in which more and more of our lives will be embedded. That does not mean we will be having sex with robots (though programmer and author David Levy believes that it's on the way, and something many will welcome). But it does mean that many aspects of our relationships, working lives and leisure will be changing as new technological possibilities emerge.

FURTHER EXPLORATION

Thomas Friedman **The World is Flat: A Brief History of the Twenty-first Century (2007)** A vision of globalization, from a US standpoint, taking in economic, social and technological influences.

Alvin Toffler **Future Shock (1970)** The breathless diagnosis of a future that would feature stressful social and technological change – from forty years ago. It hasn't been reprinted recently, but there are plenty of copies around.

Fareed Zakaria **The Post-American World (2008)** A useful exploration of the shape of geopolitics and economics in a world where the US is rivalled by India and China, by an Indian-educated, New York-based foreign affairs writer.

www.beyondcurrenthorizons.org.uk Website of documents from the UK Futurelab project on educational and social scenarios for 2020.

www.tomorrowproject.net Summaries of the results of futures projections on a whole range of social trends. Focuses on the UK but considers issues of wider relevance, such as globalization.

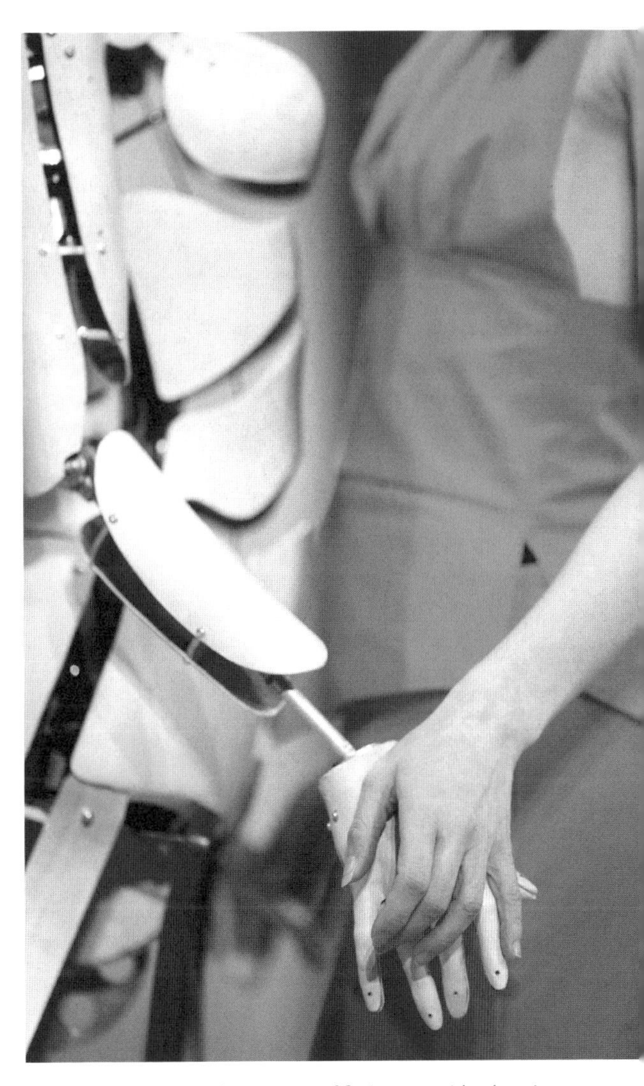

Although it's currently the preserve of fiction, sex with robots is likely to be a feature of our future lives.

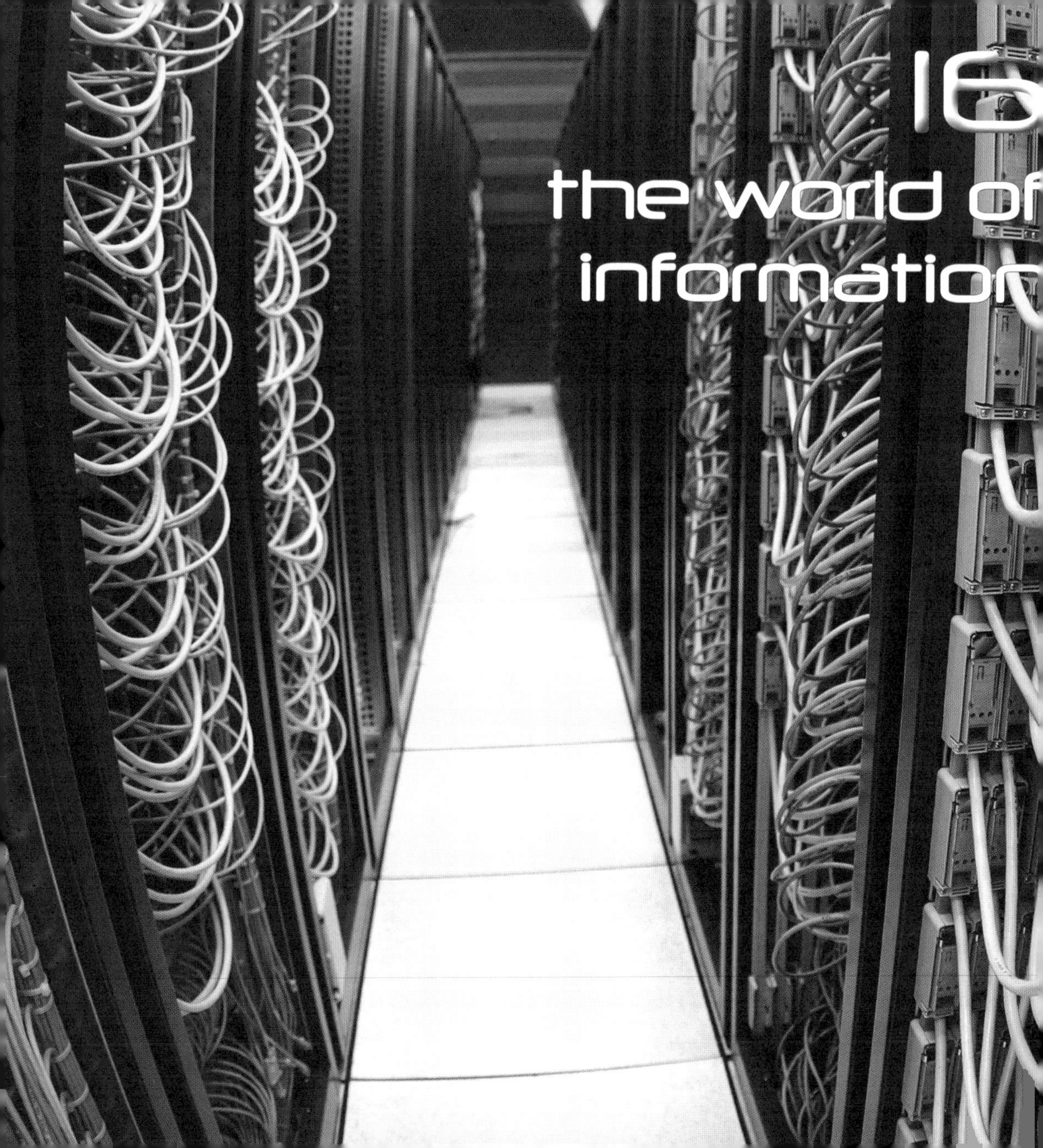
16
the world of
information

the world of information

Computers in the future may weigh no more than 1.5 tons.

Popular Mechanics, 1949

Thinking about life in the future means thinking about culture, and the way we experience culture is in the middle of some big changes. The way people connect with each other, and the way we use information, are being enriched – and often altered – by fast-changing technologies. In the new world of technologically mediated connectivity, is predicting the outcome of changes you can already see in progress any easier than guessing the fate of ones not yet underway? Probably not. But we be more certain about which may be the most important potential developments.

CULTURE: TALKING THE TALK

Culture is a slippery concept. You might say it means all those things which involve human beings talking to one another. Only talk? Well, there are images and music too, and they are often compelling in ways mere talk cannot match. But they also give us more things to talk about. And aside from the arts and sciences, education, politics, law, and even economics are the preserve of talk and its developed offshoots, reading and writing. Culture also includes things which are built, made, grown or fabricated, and these depend on ideas. So commerce, fashion, science, engineering and technology are also the province of talk.

Culture is also about information – an idea incorporated into our current talk about information technology, or even the information society. Confusion can arise from paying too much attention to superficial changes in the way information's shared and stored, while the actual content often remains pretty much the same. Just think of the transitions from vinyl records to tapes, CDs, MiniDiscs, MP3 players and then simply computer files streamed over

PREDICTIONS FILE

Richard MacManus

Founder and CEO, ReadWriteWeb.com

Highest hope: Our entire living environment – the houses we live in, the objects inside them, the services we use such as health care and transport, the clothing we wear, perhaps even nano devices implanted in our bodies – is connected in a human-centred but data-driven network, the "Internet of Things", that makes the way we live and work much smarter and more efficient.

Worst fear: People's private data being misused by unscrupulous corporations or paranoid governments. Privacy will be a very important issue when so much data about our daily lives, preferences and actions is flowing through the Internet.

Best bet: I think there will be teething problems and controversies regarding privacy over the next fifty years, as more and more personal data about our lives gets networked. For example, police may be able to trace your every move in a fully networked smart city. But these issues will be overcome within fifty years, and privacy policies will be in place that prevent abuse by people, corporations and government agencies.

the Internet. All are different ways of doing the same thing: recording sounds – most often music – which can then be heard again in a time and place separate from the live performance. The original information is captured, and can be used again and again.

New ways of making this happen can be extremely powerful, and we are in the midst of big changes in the ways information is handled. The fact that we can see this particular change going on around us is prompting innumerable attempts to sketch information futures. It helps to ignore the technological details, however high they score on the geek-thrillometer, and think more about what kind of information will be around, who can get it and how it might be used.

One logical endpoint of current trends is that everything will be available to anyone, all the time. This is the explicit mission of outfits like Google, which is aiming "to organize the world's information and make it universally accessible and useful". If you have broadband Internet access now, you can dimly glimpse what this will be like. There is a large "shadow Internet" which search engines do not reach, but you can already locate the information that people want to make accessible with relative ease. And you can find it again quickly, as you'll know if you've ever used a search engine to locate something rather than trying to remember where you wrote it down or last saved it. The ultimate vision involves a supercharged information depository-cum-search engine that's always on hand.

Yale university software guru David Gelernter suggests thinking of an "information beam", which you will be able to tap into at will. You might need a computer to access it, but computers will become so small, powerful and ubiquitous that they will often hardly be noticed. Waxing lyrical, Gelernter states that instead of a desktop computer, you might have "a scooped-out hole in the beach where information from the cybersphere wells up like seawater".

What will the beam deliver? All the information that ever there was, but tuned to select the stuff you actually use every day, or that you need now for some specific task. A personal key will tell whichever device you happen to be using how you like your individual infoverse configured. New and old information of all kinds will be accessible on demand, from the Library of Congress and the latest satellite views of the Earth to stock prices, flight times and every piece of music or video ever recorded.

TOTAL, PERSONAL INFORMATION

So far, so obvious – this is merely the sum of what we can already do on the Web. But if constant, complete availability of information is one endpoint (information from "out there", if you like), there's another, more personal one. Call it information from "in here". This requires a bit more extrapolation, and the assumption that Moore's law (see p.82) holds true for a few more decades. If the extrapolation remains correct, much of the information will be personal in a way which is a radical extension of any previous documentation of individual lives.

We can now capture and store information so easily that it will soon be feasible to archive your life. As Lancaster University computer scientist Alan Dix first worked out in 2002, a good quality video and audio stream can be compressed to around one hundred kilobits per second. (A bit, or one binary digit, is the standard information currency.) Assuming a biblical life span of three score and ten, our lives last two billion seconds. That means a complete recording of one life, a kind of super home movie, needs around two hundred terabits, or two hundred trillion bits.

Now compare that with the cost of computer memory. You can buy a one-gigabyte flash memory stick now for a few Euros or dollars (we've shifted here to bytes – one byte is 8 bits – because that is what we are used to buying). In ten years' time that will buy you one hundred gigabytes (a gigabyte equalling a billion bytes). Another ten years and it might be ten terabytes.

So our hypothetical memory stick can now store more than a year's worth of video, say, or several different streams over a single year, as well as any other personal information you care to record – a heart monitor, perhaps, or a GPS trace, not to mention all the emails you send and receive. It's no longer a home movie collection: it's a lifelog.

Extend this trend a little longer, and a whole lifetime's experiences might be kept on an even smaller, cheaper device. In fact, if such lifelogs come into use, memory sticks will be obsolete. The Moore's law assumption means that storage will be so cheap we can leave copies lying around to be accessed from the information beam at our convenience. In the ultimate info-vision, the entire population ought to be able to do this without overloading the system. You live your life, and you have a back-up copy of the life as it was lived.

At that point, everyone's relationship with history changes. What will you want to look at again? How will this external archive relate to your "memories", as we used to call them? What use might it have when your own memory begins to fail? And, perhaps less personally but ultimately important, will actual historians be able to find uses for this material, or just be overwhelmed by the sheer volume of stuff? Bear in mind that it may not just be people who become able to record everything that happens to them.

A DRIVE-IN INFO-WORLD

As outlined in Chapter 5, the information capabilities of objects will be boosted hugely in the next few decades, and they will talk to each other too. Visions of what it might be like in a future in which objects respond to you, learn about you, and record the state of their worlds tend to portray homes packed with state-of-the-art technology: fridges that can read sell-by dates and so forth. But homes – brick-built ones anyway – turn over slowly, and technology may or may not infiltrate every nook and cranny. A more likely scenario might be based on a designed object that is mobile, gets replaced more often, is supplied through a market in which competing manufacturers are constantly upgrading "standard" equipment, and which is subject to lots of rules and regulations aimed at keeping the system it is part of running more or less smoothly. That is, a car.

As with the transition from the phonograph to the MP3 player, the technology changes but the purpose remains the same, if more fundamental. But changes in the way personal transport happens alter the experience in more ways. Think, first of all, of the horse and buggy. As long as you owned one, or could hire, borrow or steal one, all you had to do apart from that was feed and shoe the horse and ride the buggy. OK, a horse is a hulking animal and can demand a lot of care and attention, but it was your own business.

An automobile in the second half of the twentieth century was a different matter. It acquired more and more documentation to satisfy the authorities – a licence for each driver, a logbook with a record of owners, insurance documents, perhaps a road test certificate, and a mileage and service record.

An electric car for the near future. Its most radical feature may be the fact that it's being developed and designed by Dutch-based group c,mm,n via an open source community.

But that collection of paperwork is nothing compared with the near-future car, electronically equipped to send, receive and compile information, and to interact with its users. It will be easy for a vehicle to record its own journeys and the identity and number of passengers; to monitor how it is being driven and how this relates to fuel economy or carbon emissions; and record readouts from sensors of oil pressure, engine temperature, gear changes and tyre pressures. It may be able to monitor driver behaviour and play back near-misses; respond automatically to speed limits and alerts about adverse weather conditions; track reports of accidents or congestion; and plot more efficient routes.

On a more personal level, it could register authorized drivers of the vehicle, monitoring them for alcohol or drug use; check when the driver last stopped for refreshment; remember a driver or passenger's taste in music or radio programming; and record personal memos and reminders related to the current route. It will obviously register toll payments and prepayments, parking fees and traffic fines. Perhaps it will know what you bought the last time you visited a drive-through fast-food counter.

None of this is more than a modest extension of what a modern car already does, with its GPS unit, engine management system and automated toll card. Commercial airliners and Formula 1 cars, which positively drip with telemetry, are already much more comprehensively equipped, informationally speaking. The augmented version of the regular family car is unimaginative in the sense that it assumes the continued dominance of personal over public transport, though perhaps it runs on electricity from a hydrogen fuel cell. But it also makes the leap to the next step, extracting and recording all personally relevant information and detaching it from the car itself. If it is simply deposited in the information beam, and can be retrieved on demand by your personal digital assistant, then any car – borrowed, rented or booked from a city car club – can become "your" car for the duration of each journey. The information becomes, in a sense, more important than the actual vehicle.

That will not mean that people no longer prefer to drive a Ferrari sometimes instead of a vanilla saloon. But they might just care less about ownership of either. Or perhaps, as is often predicted, the car will drive itself, on a wired highway (less fun for Ferrari owners), though a vast infrastructure shift would be necessary. Someone has to bury the guide cables in all those wired highways, so this would be a long-term project.

That is just one example of the kind of things which it is easy to suggest are likely, since they are more or less linear developments from what is already happening. There will be lots of other possibilities and consequences generated by the changing worlds of information, and the details of most will doubtless elude us until they appear. Some of the key questions about the costs and benefits of all this, however, are already apparent. Before going into some of those, though, let's mention a couple of items from a list offered by David Gelernter of "natural laws" which inform the right way to think about what to expect on the technological front.

First, in all these areas of computing, communication and information processing, software is the key to new departures and new capabilities. In other words, our future lives are likely to be more obviously affected by what our computers can do, rather than how quickly they can do it. Yes, there may be much more powerful, or at least smaller, cheaper and faster computers, but if they work in much the same way as the older, slower ones, without enabling us to do new things or offering some kind of innovation, we probably won't notice the difference. What would make a real difference? We may not know it until we see it, but something like really convincing voice recognition and speech synthesis, for example, would greatly alter our relationship with computers.

Also, we replace things when we find something better, not something newer. A good example is the printed book, a technology established in its essentials around five hundred years ago. Beating the book in terms of readability, convenience, ease of access to the bits you want, and instant indicators of how much you have read, is still a little way off for computerized devices, and has been a little way off for around forty years. They are just so damn convenient. The latest e-readers are finally gaining significant market share, but still don't quite match ink on paper in bright sunlight, for example.

Those two generalizations seem more or less true. Other questions about the information future have less certain answers. Here are a few of them.

INFORMATION OVERLOAD OR COGNITIVE SURPLUS?

The cry, echoing Alvin Toffler, that the world is moving too fast is still heard regularly. Now it tends to be linked to complaints about information overload. There are too many books, films, bands, TV channels and websites. Everyone gets more email, tweets, Facebook updates, RSS feeds, voice messages and texts than they can cope with, and tries to cram more and more into the day. There are too many choices, too many decisions and too little time.

However, just as the dangers of "future shock" were overblown, the problem of information overload may not be as bad as it is made out to be. True, there is a knowledge explosion – stuff that might actually be useful for someone to know exists in ever larger amounts, and researchers struggle to keep up with their own fields, let alone anyone else's. In addition, in many scientific fields there are data mountains piling up as more and more powerful instruments capture information which is worth analysing. But researchers also have more powerful tools for reviewing and retrieving what is out there.

For the rest of us, the proliferation of messages and media can seem like a mixed blessing. But the access to the riches of all the world's cultures which comes along with the unlimited amount of ephemera and dross is some compensation. Just as search engines have kept up with the growth of the Internet, so technologies for filtering out what each of us finds interesting from the information torrent will keep getting better. There are educational challenges in developing our ability to distinguish the worthwhile material from all the rest, but they aren't completely new either.

There are also new and evolving possibilities which will help sift, organize and improve the surfeit of culture we enjoy. Internet guru and NYU professor Clay Shirky speaks of a new "cognitive surplus" arising in affluent societies since the middle of the last century. Harnessing that surplus, he believes, will be a powerful cultural force which can be amplified by the new tools for interaction and participation coming into use on the Web.

For the first time, contends Shirky, millions of people – rather than just a privileged few – have large amounts of free time. In the second half of the last century increasing affluence, and longer, healthier lives gave us far more unstructured free time than any previous society enjoyed. At the same time, TV has grown into a mass medium and soaked up this free time. People in developed countries, who often claim to be time-poor, watch around twenty hours a week. That is serious free time, amounting to half of a standard working week.

So far, so passive. But now TV viewing figures are falling. People with free time, especially younger people, are using other media, which allow for interaction and participation. Some of that interaction will be as trivial as most TV was and still is, but if just a little is devoted to collaboration on more serious stuff, the potential is impressive. The entire contents of Wikipedia needed perhaps one hundred million hours of thinking time to compile. The Internet-connected population of the whole world still watches roughly a trillion hours of TV every year.

If just one percent of that TV time was diverted to doing something more active, it could support one hundred Wikipedia-sized efforts annually. Of course, once the TV habit loosens, the time can be used for lots of different things. The point is that there is a large untapped resource – the cognitive surplus – which can now be used creatively. That might be enough to process all the mountains of new information in ways which are of some use.

Shirky's ideas, elaborated in *Here Comes Everybody* (2008) and *Cognitive Surplus* (2010), are supported by works by other authors, including Jeff Howe's *Crowdsourcing* (2008), Don Tapscott and Anthony Williams' *Wikinomics* (2008) and James Surowieki's *The Wisdom of Crowds* (2004). They all basically argue that new forms of collaboration, often self-organized, enable more people to make worthwhile contributions to new projects, whether for business or general social benefit.

In 2009, technology analyst Kevin Kelly went further, suggesting that the best way to sum up the kinds of organization which are supported by the new connectivity is to call them a new brand of socialism. In a long article in *Wired* magazine, Kelly argued that evolution of new social media points toward emergence of a global collectivist society. This begins with sharing – of photos, videos, bookmarks – which steps up through cooperation to organized online collaborations, such as those which have supported open-source software projects. The final stage is collectivism, "where self-directed peers take responsibility for critical processes and where difficult decisions, such as sorting out priorities, are decided by all participants".

What is special about the new technological collectivism, Kelly reckons, is that it resolves the contradiction between the power of the individual and the

PREDICTIONS FILE

Jonathan Zittrain

Professor of Law, Harvard, and author of *The Future of the Internet – and How to Stop It* (2008)

Highest hope: There will be a comfortable standard of living for every human in the world.

Worst fear: There will be permanent divisions among and within countries between the privileged and the ignored.

Best bet: I'm for hope (if not optimism), so am counting on the humanizing dimensions of information technology to help people see that they're not so unlike others who otherwise would seem unfamiliar or even alien.

power of the state which crippled earlier experiments with socialism. And so another utopian vision is born. The evidence that it can be realized in practice remains scattered, and a bit thin. But the possibility that digital platforms offer new ways of meeting needs which go beyond the standard alternative to old-style collectivism, the market, sounds enticing.

WILL WIRED SOCIETIES BE MORE OR LESS DEMOCRATIC?

Imagine this device, described by the computer scientist John Holland in 2002 and already sounding like something you will be able to buy in the shops in a year or two. It combines in a wristwatch-sized package a "world communicator", video camera, computer, global positioner and notepad. It also has a 3-D projection display allowing easy control through a video game-like interface (that's the bit which isn't quite there yet). Add advanced simulation software, of the kind already seen in games like *SimCity* or *Spore*, configured to allow modelling of complex systems. The whole package would be a "planner", allowing anyone to amplify their experience and intuition about the options they face.

According to Holland, the post-video game generation will be able to "examine the consequences of familiar actions, using realistic and controllable 3-D interfaces that have adapted to the capabilities of the user". They will be able to use their planners as simulators for testing the possible consequences of social and political decisions. A democracy in which a substantial portion of the electorate had such devices, or were simply able to discuss the results of others' outputs, might be very different from one where policy is made in the more haphazard, suck-it-and-see way we have relied on up to now.

But how different? There might be a constant churn of criticism in the blogosphere, attacking this or that simulation for being inaccurate, using objectionable or unjustified assumptions, or just denying that it makes sense at all. It is hard to imagine the effects of this real-life politics without actual examples of what might be written into the simulation. Are we talking about the sort of things you can play around with in *SimCity* – tax rates, zoning laws and the like? Or will they extend to, say, regimes for regulating carbon emissions or trying to influence drug use or dietary habits? That would depend on some pretty far-reaching advances in social scientific understanding, as well as fancier information technology.

PLAYING THE FUTURE

Such simulations could emerge gradually from existing computer gaming conventions. Multiplayer games might offer one way to harness all that cognitive surplus. Those who fear the effects of action-oriented computer games suggest that they could turn everyone who plays them into violence addicts with poor impulse control and a low attention span. This is like assuming that all novels are trashy thrillers. Could computer games play a more constructive role? Writers for the Millennium Project, the US-based clearing house for futures ideas, certainly think so.

The Millennium Project developed ideas about how custom-designed games might help us explore future possibilities with its scenario for Great Cyber Games, which included links to databases of "global problems, opportunities, challenges, strategies and tactics". These sorts of games are played to find better answers than those already available, and those competitors who come up with good answers get to move on and play in the real world. One scenario written for the Millenium Project describes how this might work:

> "The Great Cyber Games were attractive to policy or other kind of decision makers because it filtered out all the noise of computer conferences, journal articles, and got right to the person with the ideas. The players liked it because they had the potential to see their ideas realized and earn a living at meaningful work."

Suggesting this is not entirely fanciful, in 2008 the Palo Alto-based Institute for the Future ran a multiplayer online game called *Superstruct*, which put players into a carefully contrived world of 2019. Those who signed up were given details of five threats to the stability of their society – disease, climate refugees, food shortages, information terrorists and the struggle to secure energy supply. Their job was to find new ways of cooperating to try to deal with or even overcome these threats. Behind the game was an attempt to demonstrate that collective intelligence might come up with possibilities that hadn't been thought of before, and which might even be usable in the real future. The results are archived at www.superstructgame.org. According to the organizers: "We told more than one thousand stories about our lives in 2019 … focused on the future of energy, food, health, security and society."

WILL WE LOSE TOUCH WITH THE REAL WORLD?

The lovingly crafted environments of computer games, whether online or offline, are an aspect of another trend which worries some – the rise of virtual realities. Will they seduce us away from participating in the real world?

If everything is, in some sense, information, then with enough information you can reproduce everything. In other words, a computer can contain a virtual world which is some kind of match for the real world. This includes computer modelling for research and simulation, but extends to worlds which a computer user can "enter" by interacting with the machine. One ultimate vision of the IT future is access to virtual worlds which are as rich as, or richer than, the everyday world. Is this a promise or a threat?

VIRTUAL HQ

We are very proud of our AVW Virtual Headquarters with its meeting rooms, auditorium, and social and networking areas. Please come in, explore, enjoy. All AVW members have free use of this exciting facility.

To enter **LOBBY** click here or paste this link into your browser
http://www.3dxplorer.com/view.php?k=MzEyMQ

To enter **AUDITORIUM** click here or paste this link into your browser
http://www.3dxplorer.com/view.php?k=MzEyMA

The Association of Virtual Worlds website showcases its virtual headquarters, including networking areas, meeting rooms and an auditorium: all the accoutrements of a stylish, if rather corporate, professional club.

This is a tricky subject because the discussion, like virtual worlds themselves, relies heavily on our imagination. Some writers have moved on from the notion of cyberspace – the ensemble of connected computers and databases that characters in novels such as William Gibson's *Neuromancer* (1984) can navigate when they "jack in" to the system. They talk of the "metaverse", another term popularized by a science-fiction writer – this time Neal Stephenson, who coined the word for *Snow Crash* (1992). In that tale, humans enter a virtual world as avatars, and can interact there with other characters, who may be avatars of fellow humans or pure software creations.

But a quick read of Gibson reveals that the "real world" version of cyberspace does not bear much resemblance to the one in his books. Nor do the various attempts to create "real" metaverses get close to the immersive version imagined by Stephenson. Spend a little while in *Second Life*, the widely publicized design-your-own-avatar online world launched in 2003, and – while it is amazing that a "three-dimensional" world which you access through your computer screen can work so well – it quickly seems pretty creaky.

But perhaps that doesn't matter. The same imagination which leaps to the full "metaverse" whenever a simplified online environment is unveiled is also very good at filling in the missing details of such environments. Go back to an older technology and think of what it is like to read a good novel. The information content, formally speaking, is low. But it is easy to lose yourself in this much older virtual reality space for hours or even days. Novels, though, are not (yet) interactive.

The new computerized worlds can be classified into three types, including fully realized virtual worlds. The most basic kind are "mirror worlds", the title of a 1992 book by David Gelernter. These are ultra-detailed models of the actual world built from vast data stores, Google Earth being the best-known example so far. They're more than just souped-up maps, as the user can

PREDICTIONS FILE

John Moravec

Futurist, University of Minnesota, and editor of www.educationfutures.com

Highest hope: We will develop into a learning and innovation society, where new knowledge production and its creative, beneficial application take place everywhere, all the time. We will use technologies purposively and adaptively to help us learn, create and communicate more effectively. Our first step will be to transform our schools and universities from institutions of learning into institutions of discovery, creativity and innovation.

Worst fear: We will continue to operate our educational systems as nineteenth-century factories, producing minds that are fit for industrial work, but not capable of thinking and designing their own futures. Future technologies that we will design will outperform us in all of these areas.

Best bet: We will fail to act. Our education systems will collapse as we build transhumans, post-humans and AI that are not only smarter than us, but are also more creative and innovative. They will outperform us. At this point, natural selection will determine humanity's future, not human development. While the new configurations of intelligent life on the planet move on, we will be forever trapped at the event horizon of the Technological Singularity.

move around "inside" and manipulate them in various ways. Such worlds lend themselves to a kind of virtual tourism: Google Earth includes Mars, for instance.

Somewhere between mirror worlds and immersive virtual environments where you can send your computer-controlled avatar are arrangements for "augmented reality". Here you view the world through some device that provides an overlay of useful data on top of what your unaided eye would normally allow you to see. This kind of thing started out as the "heads-up" displays developed for fighter pilots. For civilians, it matters less whether you take your eyes off what is in front of you. So scenarios in which information comes through your glasses, or perhaps a wired contact lens, blend into simpler ones in which information you might need is displayed on interactive surfaces on the buildings or vehicles you are passing in a city street, or just on your phone. As the "Internet of Things" (see Chapter 5) emerges, ubiquitous embedded devices will communicate with us, or with things we wear, as well as with each other.

As processors get smaller and smarter, bandwidth increases and displays get fancier, it's likely these slightly different types of computer-mediated reality will blend into one another. Some of the time, at least, people in countries where access is well developed may experience similar things whether they are exploring a mirror world, walking down the street, or immersed in a full virtual reality environment. But how much time will they spend deep in the metaverse?

INTERNET FUTURES: WHAT THE EXPERTS SAY

Some stabs at answers to questions like this can be found in a 2008 report from the US-based Pew Research Center. Updated every couple of years since 2004, its survey suggests possible futures and asks Internet experts to comment on them. For example, we can imagine that by 2020 virtual worlds, mirror worlds and augmented reality will be popular network formats, thanks to the rapid evolution of natural, intuitive technology interfaces and personalized information overlays. Most well-equipped Internet users will spend some part of their waking hours – at work and at play – at least partially linked to alternate worlds or augmentations of the real world. This future lifestyle involves seamless transitions between artificial reality, virtual reality, and the status formerly known as "real life".

This is basically the computer-enhanced world envisaged by the collective authors of the Metaverse Roadmap (metaverseroadmap.org), an enthusiasts' effort which appeared in 2007 to chart the course of these technologies until 2025. Most Pew respondents thought this sounded quite likely. At one extreme, a respondent was confident that "'Real life' as we know it is over. Soon when anyone mentions reality, the first question we will ask is, 'Which reality are you referring to?'" But a significant minority disagreed.

Believers saw the shift into increasingly virtual lives being driven by the freedom of a new frontier (there are no bureaucratic requirements to satisfy if you want to organize a rock concert in cyberspace), and the more mundane energy-saving appeal of cutting travel to meetings. The doubters reckoned the limitations of the virtual experience will dampen its appeal to regular folk who still value the texture of the everyday world and the richness of full human interaction. This reflects glib assertions like one in an April 2009 report from US consultancy Research 2.0 that "more powerful computers in the years ahead will allow software designers to replicate the real world in-silico".

The cheat word there is "replicate". All these efforts are replicating aspects of the real world. The question is whether the ones they choose are the important ones, and whether we are prepared to do without the others or fill them in for ourselves using the software already in our heads, as we do with a novel. Why, in the end, would you want to have most of your conversations through an electronic hand puppet?

The appeal is certainly mysterious to some. Gert Lovink, an Amsterdam-based academic, suggested that "most people are not interested in avatars. They have trouble enough managing their first life." On the other hand, the ability to distance yourself from the first life you are having trouble managing may strengthen the appeal of virtual worlds – for all of the people some of the time, and for some of the people all of the time.

The middle-of-the-road forecast is perhaps that within a decade or so, virtual spaces will have developed far enough to be destinations of choice for some kinds of entertainment. But augmented reality will be more significant. As Anthony Townsend of the Institute for the Future put it, "The real world is a fascinating place – overlaying information and cues from digital spaces will make it even more compelling – for socializing, traveling, playing games, and working. It will still be real life [although enhanced] in the sense that people who wear eyeglasses still see real life, just a refocused version of it."

That does not rule out the growing importance of more complete virtual realities in the decades after 2020, especially considering how rapidly this technology is changing. But how far it will get, and how fast, is easy to exaggerate.

SPECULATIVE INTERNET-ENABLED TRENDS

A few of the other possibilities posed by the Pew researchers divided opinion still more sharply. Take the idea that social tolerance may increase because we know more about how everyone else thinks and feels thanks to the Internet. The reasoning is that this would help reduce violence, especially sectarian violence and hate crime.

Around a third of the experts believed (or hoped) this rosy scenario would come true, but 56 percent mostly disagreed. One main reason for scepticism was the scarce evidence that knowing more about people you don't care for increases warm feelings towards them. Another was that the Internet can encourage inward-looking communities of the like-minded – including fanatics and crazies – as readily as it spreads truth, enlightenment and good fellowship. A bit like all previous communication technologies, in fact.

There was also disagreement as to whether intellectual property can be secured through a combination of clever copy-protection systems and strengthened legal controls. About sixty percent thought this unlikely. This highlights one of the great continuing puzzles about the new information world, and the root of many present and future conflicts. How to persuade people to pay for stuff they can get for free? And if they won't, how to secure some kind of income for the people who create it? There are plenty of ideas, but no one really knows. The general view, though, seems to be that this will be negotiated through experiment and social evolution. Technical fixes will be attempted, but will merely pose a challenge to hackers, and lead to endless bouts of geek arm wrestling.

What about privacy in the new, transparent world? Will people be comfortable with lots of personal information going into the infosphere, and the general access that goes along with greater tolerance of others' behaviour? Perhaps transparency equals forgiveness. This was one of the most divisive questions of all, with opinion evenly split. That is not surprising as the potential loss of privacy – and the level of information of new kinds other people will be able to obtain – is one of the biggest challenges in store. How it plays out will depend on complex interactions between people and technology, and on changes in attitude as people grow up experiencing life online. Transparency sounds good, unless you interpret it as exchanging the anonymity of a big city for a new version of the visibility of village life, where everybody knows everyone else's business. Loss of privacy for other people (politicians, celebrities) pleases us. Loss of our own is less attractive.

Truman Burbank (Jim Carrey), sleeping on the big screen in the prescient movie *The Truman Show* (1998), has yet to awake to the value of his absent privacy. Is this becoming true for society in general?

Some speculate that if there is equal loss of privacy for everyone, we will mind less. Or, faced with the prospect of mutually assured humiliation, we will take more care about what we say, and when. It could also lead to most of us becoming less interested in knowing personal stuff about others. Consider the scenario of the spooky 1998 film *The Truman Show*: if one person's life was available to watch 24/7, it might command an audience, as in the film script. If everyone's life was available, who would watch?

These possibilities will obviously play out in complicated ways, as respect, esteem and reputation are such crucial parts of people's lives. Many speculate that there will be so much information about everyone, in so many databases, that there will be specialists in "reputation maintenance and repair" who will scour the electronic universe for stuff about you and clean it up. The richer you are, of course, the better job you will be able to afford. The more sophisticated approaches to the new problems of reputation and privacy will include multiple digital identities and perhaps, for the really well-off, disappearing from public digital space altogether. Criminals, of course, will try to do the same, and the line between privacy willingly surrendered and surveillance will continue to be hotly debated.

Whatever the course of this debate, it seems plain that privacy, and its loss or protection, will be key future issues. There may well be variation between different generations, groups and even professions, in what they are willing to tolerate in terms of openness, what they are required to divulge, and how hard they work to keep some things to themselves. And the amount of material which will be available about everyone who does not opt out of the online world altogether, and the ease of general access to it, will mean that reputation management becomes a serious preoccupation – both as something you are taught how to do yourself, and something you can pay specialists to manage on your behalf, perhaps. Power

PREDICTIONS FILE

Richard Eskow

Consultant, writer and CEO of Health Knowledge Systems, Los Angeles

Highest hope: We will eliminate most major diseases, extend the human lifespan and improve our physical and cognitive abilities. More importantly, new medical technologies will turn us into a "human network" which allows us to experience and communicate a shared worldwide reality. New technology will enable us to communicate using thoughts alone, and share information and sensory experiences "telempathically" with anyone anywhere in the world. The same technology will personalize the informational and entertainment experiences now on the Internet, create new art forms and give us mental control over mechanical and knowledge-management devices.

Worst fear: We will fragment into two or more widely divergent social groups, resulting from different groups' abilities to purchase added physical and mental ability. We will divide into a new "First World" and "Third World" – a division based on our accumulated physical and mental resources, rather than traditional economic and geopolitical divisions (although these divisions will likely be amplifications of those we know today). Conflict and even genocide could result from these divisions.

Best bet: We will eliminate many major diseases, and access to mental and physical enhancement techniques may even be considered a human right. But we will start to suffer the environmental consequences of our neglect, which will hurt our global health and force us to fight rear-guard battles to protect our medical well-being. Economic and social divisions will be exacerbated by improvements in medical technology, but the disparities will be eased by the rapidly falling costs of some new tools (drugs, external enhancement devices, etc).

will still matter (of course) and will now include control over what reaches the enlarged public domain. But while it is easy to invent a term like reputation management, that doesn't mean it will be easy to do. For the first time in history, nothing written or recorded will ever disappear – and it can always be held against you.

Privacy is also likely to play out differently in different parts of the globe. Conceptions of identity, and the nature of and reasons for shame, vary from place to place – and future Internet growth will not necessarily lead to the same social consequences in China or Indonesia as it may in the US or Europe. And in some countries, the stakes in protecting privacy and anonymity are higher. The Internet may make it easier to spread expressions of dissent in a state ruled by a dictatorship, but the rulers will be trying to use it to track their opponents.

The trade-offs between privacy and security will also be bones of contention in more democratic states. Total transparency may enhance security, but do we want it to be a requirement? Will people who try and retain a measure of privacy be, by definition, acting suspiciously? This is an extension of the argument which has generally been used to justify the installation of CCTV cameras in countries where they have proliferated wildly – such as the UK: "if you have nothing to hide, you have nothing to fear". On the other hand, in their personal lives, everyone has something to hide, some time – don't you?

ABOLISHING BOUNDARIES

There's also contention around the consequences of weakening the boundaries between work and leisure, and whether this will help us manage our time. The Pew experts mostly (56 percent) agreed that this would happen – not surprisingly, as it is a pretty straightforward extension of what we see now for the well-connected, BlackBerry-toting classes. They were much less clear on whether it would be a good thing.

On the plus side, the nine-to-five day is a recent invention, a carryover from the new industrial work discipline of the nineteenth century into the modern office. There can be a new freedom and flexibility in being able to do anything, anytime, anywhere; the prospect of cutting out commuting hell by working from home or in the local café; and catching up whenever it suits you. But would cybernomads find that work interferes with other stuff they want to do? If so, then permanent access becomes more of a burden than a benefit.

People will need to defend themselves by setting up new boundaries, and resist corporate control of their lives – which could be more easily achieved if 24-hour online access enables continual surveillance of work habits and productivity. As one sceptic put it, "It will not be a net-positive for anybody but Type As and geeks – people who didn't have a social life in the first place." The most important feature of a new communication gadget may be how good its "do not disturb" function is.

WE DON'T NEED NO EDUKASHUN...

Educating the next generation (and keeping the old generation up to speed) is pretty fundamental to culture, so it is no surprise to find that the educational possibilities of the new info-world draw some of the most striking hopes and fears. More deeply, the cognitive future is also a deeply divisive topic. In principle, we will have access to an astonishing world of knowledge. Everything anyone ever thought or said will be available – but will we be smart enough to make any sense of it? Or will we become slack-jawed, dumbed-down creatures with poor attention spans, incapable of sustained thought, and happily distracted by porn, comic books and infantile tosh on YouTube?

This pessimistic view gets a lot of press, partly because the press tends to favour contributions from those neuroscientist Susan Greenfield calls "people of the book". She used the term in *Tomorrow's People* (2003), to label a group who are about to lose their cultural ascendancy. That would be people who are addicted to extended narrative or argument, linear prose, and solitary, silent reading. Perhaps those have always been minority sports, but writers have tended to be drawn from the ranks of such readers, and now often lament the apparent threats to the book.

As the writer, and readers, of this book, we may accept that this technology, whose modern form appeared more or less as it is now in the fifteenth century, embodies much of what we value about our culture. But is it, or what it represents, under threat? Greenfield certainly thinks the threat is real.

Greenfield is a member of the British House of Lords, a platform she used to air her fears that "social networking sites might tap into the basic brain systems for delivering pleasurable experience". She went on, "As a consequence, the mid-twenty-first-century mind might almost be infantilized, characterized by short attention spans, sensationalism, inability to empathize and a shaky sense of identity." As she seems unable to cite actual evidence for any of this, the suggestion seems itself to indicate a tendency to dumb down public debate. The thesis

Just as twentieth-century commentators fear the Internet is dumbing down thought and culture, Plato (depicted, left, with Aristotle in Raphael's *School of Athens*) suggested, 2,500 years ago, that books would inhibit true learning.

has supporters, though, including US commentator Gary Small, in his book *iBrain: Surviving the Technological Alteration of the Modern Mind* (2008).

A more measured, and in some ways more persuasive, point of view was offered by Nicholas Carr's article "Is Google Making Us Stupid?", in the US magazine *The Atlantic* in mid-2008. He reported that he, and many other writers and literary types of his acquaintance, were aware that their reading habits were changing as they spent more and more time browsing the Internet. It seemed to be getting harder to make time to digest anything long and complex, and to concentrate on it properly if you did make time. Add the recent emphasis in neuroscience that the brain is highly plastic – that is, its actual wiring changes depending on how you use it – and one is bound to wonder where all this will lead.

On the other hand, as Carr conceded, this kind of speculation can easily fall prey to the fact that "just as there's a tendency to glorify technological progress, there's a countertendency to expect the worst of every new tool or machine". Writing, for example, got a bad report in Plato's *Phaedrus* 2,500 years ago – it would avoid the effort of memorization, Plato had Socrates argue, and substitute mere book-learning for hard-won wisdom.

In addition, while technological and cultural environment obviously does have effects on cognition, we have little clue in advance what these may be. As an illustration, there was no advance warning of the startling 1980s finding by psychologist James Robert Flynn that IQ scores had been rising around the world ever since the tests began, in the early twentieth century. Why have they been rising? As Flynn argues in *What is Intelligence?* (2009), it's not because we have all become smarter. Instead, we became better at dealing with the kind of questions that appear in IQ tests. That's because a larger portion of the population have learned the kind of abstract, analytical problem-solving skills that used to be confined to a minority, and which are encouraged by literacy and book-learning. The implication is that, even if the people of the book are set to assume a less dominant place in the culture, they do so from an extraordinarily strong position.

People who spend more time (though not all their time) doing other things, such as playing computer games, will come to do better on other cognitive skills. There is already evidence, for instance, that video games improve visual skills like tracking independently moving objects. Is that useful any other time? That depends what you want to do. But it underlines how as the technologies people interact with change, there will be gains as well as losses, and it is a mistake to fixate on one or the other.

Most commentators seem to agree that old-style fact-based education will need to change because the facts, well, they are just there for the taking. Books and other recorded text are a way of remembering stuff outside an individual human mind. To that familiar extension of culture, a technology has now been added that can bring you whatever is stored in the collective external memory, whenever you need it, wherever you are.

This leads to some extreme scenarios. According to computer scientist and learning theorist Roger Schank, there will be no teachers, classrooms or textbooks. Instead, education will consist of entering a virtual world of your choice and learning to do things there. In his view, within fifty years "our homes will be dominated by virtual experiences; our schools will have been replaced by them". That seems to neglect all the other things schools do by way of socialization, child care and even social control. Some of those things, perhaps, can also be arranged via the new, enhanced Internet, but we do not yet really know how.

WHO DOES KNOW?

Experts may have direct involvement with the development of the technologies and applications which are the new vehicles for culture. But the speculation about these issues still mostly comes from people who grew up with an older culture of books, pens, telephones joined to the wall by wires and even (like the writer of this book) manual typewriters.

The people who influence which way all these questions get answered in practice will be the ones who learned to take the possibilities of the new technology for granted much earlier in life. And there is some evidence about how they will be inclined to use it.

In *Grown Up Digital* (2008), Canadian author Don Tapscott reports on a study of almost eight thousand people in twelve countries who were born between 1978 and 1994. He paints an optimistic picture of this new generation, who in general are socially involved, adept, adaptable and highly skilled at using new media, not dumbed down. He reckons their expectations will differentiate them from the generation which grew up with passive, one-way media like TV. They put a strong emphasis on freedom to choose, and on customizing stuff. They are questioning, and want integrity and openness, from politicians, employers, manufacturers and service providers. They are habitual collaborators, and expect entertainment in work and education as well as outside of them. They are also impatient, expecting stuff to happen fast, and attuned to constant innovation.

History suggests that none of these attitudes and attributes will stop them, as they get older, from becoming adults who bemoan the condition of future youth. But for now, there is at least some evidence that the direction of future culture is positive, at least for the digital natives.

FURTHER EXPLORATION

Nicholas Carr The Shallows: How the Internet is Changing the Way We Think, Read and Remember (2010) According to Carr almost everything about the Internet makes things worse, for our attention span, our ability to develop long trains of thought, to analyse information critically and to reflect.

Clay Shirky Cognitive Surplus: Creativity and Generosity in a Connected Age (2010) US internet analyst Shirky details his optimistic view of the power of the Web to unleash collaboration and overcome the passivity of old media consumption.

www.eff.org The Electronic Frontier Foundation is a US non-profit trying to preserve freedoms in cyberspace in the face of corporate and government assaults.

www.elon.edu/predictions The Elon University project "Imagining the Internet" has a vast quantity of discussion and opinion on the evolution of the whole shebang, including the successive Pew surveys of experts.

networkcultures.org/wpmu/portal Interesting research from the Netherlands-based Institute of Network Cultures, mostly published in English. Favourite title so far? *Technobohemians or the New Cybertariat?* (2007).

www.roughtype.com Nicholas Carr's blog charts the evolving thinking of one of the most interesting commentators on the effects of the Internet. Bad things are to come, he thinks.

17

moving on, moving up

moving on, moving up

The future is going to be a fast, wild ride into strangeness.

Damien Broderick, *The Spike*

Some say the old Enlightenment vision of progress is fading. But there is one strain of futures thinking which still takes it very seriously, and envisions not so much the perfectibility of humankind, but its replacement by the next stage in the evolution of intelligence. This might turn out to be through genetic enhancement, augmenting our brains with computers or replacing "meat machines" entirely with much more powerful artificial creations. Any of these could herald a new era: the post-human. And the staunchest advocates of its coming believe it will arrive this century.

CHANGING LIVES, OR CHANGING LIFE?

It can be easier to imagine a vastly different future if we can perceive real change in the past, but how much difference have past innovations really made? A person (like me) who lives in a solid Victorian house in a mid-sized city can access wonders that would astonish the original owner – health care that largely works, high-speed travel, all manner of gizmos and gadgets, information and entertainment in vast quantities, and so on. But is the pattern of my life all that different? My working, eating and sleeping habits, my powers of thought (limited), my physical capacity (even more so): most of my day-to-day interactions would seem unremarkable to whoever enjoyed the view from my window in 1870 or thereabouts.

In other words, technology, or at least the bits of it we like, has mostly been applied to making life more comfortable. But as our ability to apply new knowledge has increased, the possibility that technology might change the conditions of life more radically has gripped the imagination. As noted in Chapter 2, the advent of evolutionary theory invited the question: what comes after *Homo sapiens*? Writers like H.G. Wells thought such change would come about naturally, courtesy of evolution through natural selection. Later authors such as Olaf Stapledon or J.D. Bernal speculated that we would take evolution into our own hands, through technology.

More recently, some commentators – they tend to call themselves "transhumanists" – have gleefully forecast that the next stage of life on Earth will involve moving beyond the limitations of big-brained bipeds that live for eight or nine decades. They want to go beyond the lives we lead now into a new realm of possibility. If they can deliver, then the twenty-first century – though it would be equally amazing in the twenty-second – offers an entirely new possibility. Not an apocalypse, like the asteroid that may have wiped out the dinosaurs, but transcendence.

Those alternatives sound almost biblical, and there's certainly a near-religious zeal in some of the transhumanists' manifestos. But their point is that the answer is now outside the religious realm, and can be anticipated as a result of real, material progress here on Earth. So why do they believe this? Could they be right? And if they are, when will it all happen?

NEW TECHNOLOGIES, NEW POSSIBILITIES...

The important thing to notice about technological change, say the theorists of the coming transition, is that it gets faster. But it isn't just getting faster, the rate at which it's getting faster is increasing too. In other words, technology feeds on itself. Not only are we creating more new technologies, but each of those discoveries also allows us to learn something new and, therefore, to create even more new technologies, and so on. This ever-quickening rate of change means that the amount of progress we have seen in the past one hundred years could repeat itself in just a few decades.

PREDICTIONS FILE

Danny Belkin

Special projects manager at Abbott Labs, scientist, futurist and blogger at dannybelkin.com

Highest hope: The achievement of an effective link between, and the ultimate successful integration of, humans and AI. By effective, I mean useful, permanent, high-bandwidth communication between the two. The creation of "runaway" (self-improving) AIs without the direct integration of humans would mean humanity will miss the opportunity to make an evolutionary leap – up to a higher-level organism made of a multitude of human individuals linked via machine communication, possessing a single overall consciousness.

Worst fear: Runaway AI concluding that humans are worthless or irrelevant and relegating us, from an evolutionary standpoint to a bacteria-like status.

Best bet: Unfortunately, the advance rate in the biological and medical sciences lags behind that of AI and related fields. Ultimately the link will be made, but the question is whether it will still be relevant by the time it happens. However, technological breakthroughs do happen, and if this field receives the attention and funding it deserves, the possibility of humanity linking organically with AI will increase profoundly.

The notion of "accelerating change" has been addressed by many futurists over the years, including Alvin Toffler, who claimed in his 1970 bestseller *Future Shock* that people become disoriented by rapid technological change (see box on p.280). But the theory's most important proponent is inventor and futurist Ray Kurzweil, who developed the argument that technological progress increases exponentially in his book *The Singularity is Near* (2005). He explains that you can look at quite a few areas of technology and invention and plot rising curves of performance. Energy use might be one, transport speed another, but the key fact is that each new development is a little more fleeting than the one before. Walking, horse-riding, boats, trains, cars, planes, rockets… each invention leads onto another, faster method of travel, and the intervals between them keep getting shorter.

This example may not be entirely convincing: there aren't many points you can plot on a graph for transport because it hasn't changed that many times throughout history. But Kurzweil argues that this kind of trend amounts to a new law of technological development: he calls it the "Law of Accelerating Returns". (His 2001 essay of the same name is available at www.kurzweilai.net.) By proclaiming it a "law", Kurzweil gives his theory a sense of inevitability. And although this is debatable, it is true to say that there have been exponential improvements in two key areas: information technology and DNA analysis.

Kurzweil and a few like-minded others argue that these impressive advances mean that technology could soon begin to change us into a new kind of creature. This builds on the possibilities for human enhancement (discussed in Chapter 12) that might develop from current medical treatments, but they see it going much, much further. Transhumanists believe this might lead to qualitative changes in human life, even ushering in a kind of post-human existence.

HUMANS 2.0

The main possibilities for this would arise from the convergence of three game-changing areas of technology: genetics and biotechnology, brain science, and computer science and IT (see Chapter 5). Developments in biotech are here. As well as leading to possible enhancements in human lifespan, this technology could be combined with better knowledge of the brain to allow us to improve mental function. This could come from altering genes that affect the production of neurotransmitters (the chemicals which help convey messages between brain cells), for example.

The most often cited precedent for this is Princeton neurobiologist Joe Tsien engineering a strain of "smart" mice in 1999. These "Doogie" mice (named after a precocious genius from an American TV series) were given extra copies of a gene that allows production of a receptor for a neurotransmitter essential for learning and memory. The mice didn't start scratching equations in the sawdust in their cages, but they did outperform their peers on simple tests of recall, meaning they were "smarter", according to Tsien.

Two years later (in less benign experiments), mice modified in the same way were shown to learn faster, remember better and solve puzzles more efficiently than their less well-endowed cousins – they were also more sensitive to pain. This research remains an important signpost of the possibilities for mankind, not just rodents, but it's clear that the brain and the nervous system are wondrously complex, and evolved along a path with a lot of trade-offs. That means there is no reason to believe that either of them have yet reached the best they can be.

Another possibility for improved mental function might come from more research into the role of genes in the formation of the cerebral cortex – the wrinkled outer layer of the brain that is much larger in humans than in less agile-minded mammals. This difference in size seems to result from the action of just one or two genes during early brain development. If we could keep those genes switched on for a little longer, so that they cause one more doubling of the neurons (or nerve cells) in the cortex, for example, perhaps we would have the making of even smarter brains?

Again, the story is bound to be more complex. Even if we could cram all those extra neurons into the skull, they still have to get connected up in useful ways, and survive the colossal pruning that also goes on as the brain develops. Besides, if we really knew how to boost brains in this way, we'd be trying to do it by now. What's important it that there are already ideas about how to improve mankind which aren't completely implausible, and there will soon be lots more.

In the meantime, there are plenty of enhancement avenues to explore aside from getting smarter.

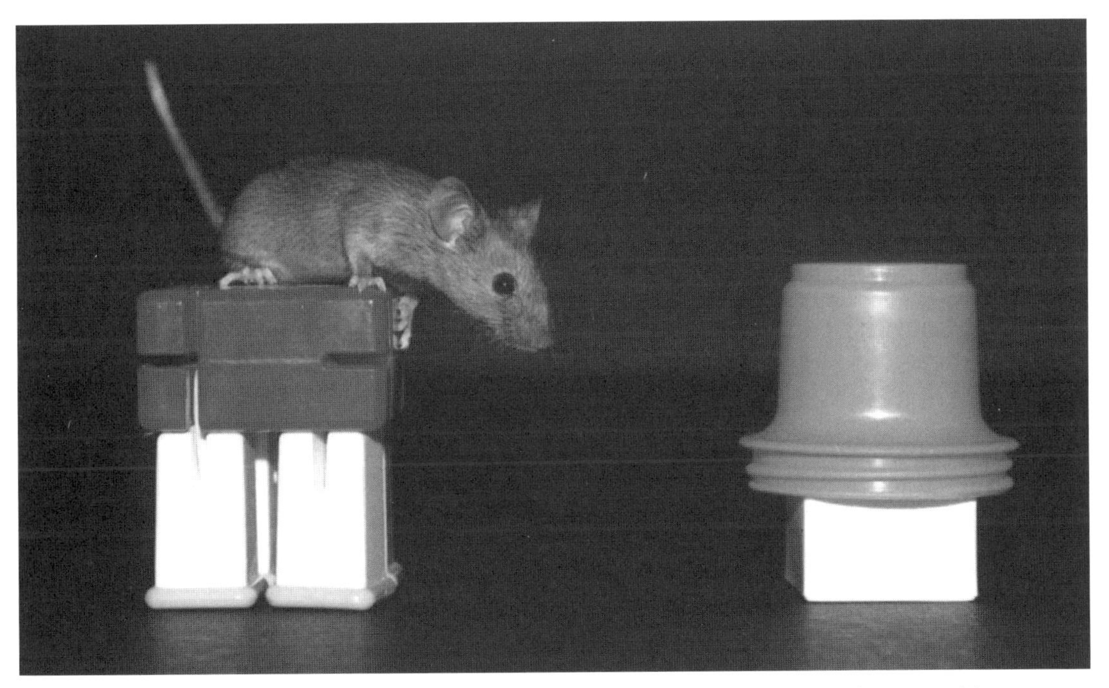

Resourceful rodent: a genetically modified "Doogie" mouse proves its advanced intelligence during an object-recognition test.

How about new senses? Mice (again) have already been given genetic tweaks that allow them to see colours they can't normally perceive, by modifying them to carry a human gene for a photosensitive pigment they cannot make. Amazingly, the cells in the back of the mouse eye which made the new pigment, now existing in three varieties instead of the mouse's normal ration of two, were incorporated into the signal processing of the mouse brain without trouble. Scientists have effectively replayed a tiny part of recent evolution and given the mice a genetic upgrade. But the most general interpretation of the experiment is that enhancing sensory inputs can produce immediate changes in sensation. That invites speculation about upgraded humans being able to experience new tastes or smells, or perhaps even extending our visual range into the ultraviolet or infrared spectrums.

These developments in biotechnology and neuroscience could also be combined with the expected increase in computing power, and perhaps even artificial intelligence. This union might come about either by artificially emulating neural networks, or by creating a different kind of intelligence in computer circuitry. In this scenario, a post-human world might be one in which genuinely intelligent machines match – and then surpass – humanity's mental capacity, perhaps rendering humans obsolete. Or perhaps, if we discover how to hook up our brains to artificial technology, it would be a world in which augmented human intelligence has unimaginable powers of thought, far beyond that of a normal human brain.

Along with all the potential developments in bio, neuro and computer sciences (most of which we haven't even thought of yet), there is the prospect of nanotechnology helping them converge into a new form of science and technology – one of intelligent systems that can develop (and self-replicate) in completely new directions.

Convergence... and the coming of the cyborg

There is more than one kind of technological convergence, but the one that preoccupies the prophets of the post-human is the convergence of people with technology to create that favourite science-fiction hybrid: the cyborg.

The term was coined in 1960 in a technical article about keeping humans alive in outer space. Since then, a swarm of Cybermen, Terminators and Cylons have invaded popular fiction, and a wealth of commentary has used the term so loosely that having false teeth, wearing spectacles or being given a hip replacement might qualify you as an early cyborg. There are, of course, more impressive mechanical and biochemical enhancements becoming possible – usually, as with internal pacemakers to stabilize heartbeat, to restore normal function. As implanted devices get better, they may open the way to enhancements more like the "bionic", or superhuman, often portrayed in fiction. But a true merging of man and machine would involve the nervous system. Can that ever really be achieved? And if so, will we still be able to see the join?

The original idea was that a cyborg (or cybernetic organism) was a human–machine hybrid, and our increasingly intimate relationship with information technology does seem to be bringing the cyborg horizon closer. British academic Kevin Warwick got lots of publicity in 1998 when he had an RFID (radio-frequency identification) chip implanted in his arm that could be read by the doors, lights and other equipment around his university department. It was a rather trivial stunt, no different to wearing a chip on a bracelet or tie clip, but when he got his surgeon to repeat the exercise a few years later with a mini-electrode array connected to the nerves in his arm the result was more impressive. Surely this was a step on the road to direct links between our computers and our brains? And that would be a new departure in our relationship with technology, wouldn't it?

It depends on how you read the history of technology. As with most histories, you can see it as a tale of continuity or discontinuity. The case for continuity is put by philosopher Andy Clark in *Natural-Born Cyborgs* (2003). He argues that it is in our nature to try to enhance ourselves. In his view, our technological future is simply an extension of a long history that began with the use of speech, passed through written text and printing, and is now moving towards digital technologies.

He explains that these successive "mindware upgrades" are "cognitive upheavals in which the architecture of the human mind is altered and transformed". But that transformation is external as well as internal, as the brain is influenced in subtle ways by outside aids. In other words, even something as simple as a notebook and pencil, used to jot something down so you can look it up later, is a cognitive enhancement, in this case an external memory aid. In fact, properly appreciated, the notebook is part of your mind.

This rather startling conclusion is philosophically controversial. Clark could well be wrong; his verdict does seem to involve changing the definition of the mind too radically to be a helpful way of understanding what is going on with the newer technologies. But you do not have to buy that part of his argument to see sense in his more general conclusion:

> "Our sense of self, of what we know and who we are, is surprisingly plastic and reflects not some rigid preset biological boundary so much as our ongoing experience of thinking, reasoning and acting within whatever potent web of technology and cognitive scaffolding we happen currently to inhabit."
>
> *Natural-Born Cyborgs*

On the one hand, this suggests that humanity will change as we interact with machines in new ways, reinforcing the fears of those who don't want to lose the old ways of thinking (usually the ones supported by silent reading). But on the other hand, it suggests that this kind of change has always gone on, so talk of a technological leap to a "post-human" existence is completely untrue. We were never human, in any stable, enduring sense of the word in the first place, but simply configured to take advantage of whatever memory, communication and information processing aids were around at the time.

It is possible, however, that the current round of technology-induced changes are happening faster than people have previously experienced. Clark does concede that they have "gathered momentum" recently, and that this has implications no one really understands. But his basic point still stands. Direct connection with information technologies – as in the cyborg vision – is not the most important consideration. Our brains are well equipped to merge their information-processing capabilities with the technologies they invent through normal sensory channels, and although direct wiring is an intriguing possibility, it isn't really necessary.

Those normal sensory channels do have their limitations, though. We might be able to outsource our memories, for instance, but there are limits to how fast we can process the information that a device provides when we retrieve it. It's those kinds of limits that are on the minds of the forecasters who predict a closer convergence between people and technology. Direct hook-ups between the human brain and nervous system and electronic sensors and processors are an essential part of the cyborg vision. We do not know if that can be achieved in ways that allow complex meaning to be communicated (because we have little idea what language the brain uses internally to encode such things), but if it can, what kind of external intelligence would the brain be tapping into?

THE SINGULARITY: THE BRAVEST NEW WORLD

Beyond the transhumanists' basic ideas about human enhancement – and beyond most of our imaginations – the future enthusiasts foresee a whole new historical and technological era for humanity. Their big idea is summed up by a word borrowed from the physics of black holes: the "singularity". Black holes are mysterious objects created by the collapse of matter under its own gravity whose internal workings lie hidden behind an "event horizon" from which no information can escape. For physicists, this is a singularity, in which equations show a hypothetical point at which some crucial quantity suddenly leaps to infinity. The technological singularity is a point at which the acceleration of technology passes into a new, self-fuelling phase with essentially no limits on what it can achieve.

The notion that technological advance, specifically in computers, might eventually pass a crucial threshold was first argued in detail by the statistician I.J. Good in his paper "Speculations Concerning the First Ultraintelligent Machine" (1965). Good, who worked as a code-breaker for British intelligence during World War II, declared that an ultra-intelligent machine would be better than a human being at any intellectual task. And as intellectual tasks include designing intelligent machines, it would be able to design an even better one, as would the next machine, and so on, leading to an "intelligence explosion". His bottom line was: "the first ultraintelligent machine is the last invention that man need ever make".

I.J. Good thought it probable that such a machine would be built in the twentieth century, and that it would transform society "in an unimaginable way". The timescale has since been stretched, though only a little, but this is still the essence of the claims made by enthusiasts of the idea now known as the "technological singularity" (or more dismissively as the "rapture of the nerds"). Interestingly, the singularity is both a prediction and a kind of anti-prediction. It is coming, it is coming soon, and it will change everything, say its adherents. But what comes afterwards will be so different, they can say literally nothing about it. As Australian futurist Damien Broderick put it in his 2001 book *The Spike* (his term for the singularity):

> "At that point, all the standard rules and projections go into the waste-paper basket… The coming world of the Spike is, strictly, unimaginable."

The assumption that the outlook is unforeseeable isn't often emphasized by more conventional futurists, because it makes their jobs obsolete. Most forecasting is geared towards steady, linear change, but the singularity prediction isn't like that: it imagines exponential increase roaring ahead on various technological fronts. This immense acceleration of the pace of change means that looking even a short distance ahead becomes impossible, or at least the equivalent of looking hundreds or thousands of years ahead at other times in history.

Furthermore, the kind of beings who will inhabit the post-singularity world will not just think faster than us (or more than us), they'll think differently. They'll have their own ideas about where to go next and what the future will bring, and will pay no heed to the quaint notions of their inferior ancestors. Turning to Broderick again:

> "By the end of the twenty-first century, there might well be no humans (as we recognize ourselves) left on the planet – but, paradoxically, nobody alive then will complain about that, any more than we bewail the loss of Neanderthals."

That seems a bit hard on the Neanderthals, a subspecies whose loss some people at least do bewail, but still, you get the point. In time, if we are in some way recognized – perhaps on a good day, even venerated – as worthy ancestors, our lives might seem as curious to those to come as the lives of, say, meerkats seem to us. Cute, amusing, worth a well-made documentary every now and then, but not, you know, especially relevant to our own culture.

TIMETABLES FOR TRANSCENDENCE?

So when is all this going to happen? Those who believe in the singularity, like campaigners for research on life extension, tend to want it to happen in their own lifetime. It would be annoying to know you were one of the "last mortal generation", to borrow a phrase from Damien Broderick. Ray Kurzweil, the most gung-ho of the prophets of the singularity, certainly wants to make it personal. He sees the crucial breakthroughs as coming in artificial intelligence, and maintains they will be with us in forty years or less. He is doing his utmost to hang around long enough to get the benefit (he is 62 at the time of writing), courtesy of an impressively complex regime of pills and dietary supplements that he hopes will preserve his ageing biological apparatus until it can be replaced by an electronic successor.

Bold technical predictions often acquire a timeless quality, receding into the future at a roughly constant rate. Like fusion power or genetic engineering that actually works, they are always twenty to twenty-five

PREDICTIONS FILE

Giulio Prisco

Physicist and computer scientist, and former senior manager in the European space administration. He blogs at giulioprisco.blogspot.com

Highest hope: That mind-uploading technology – such as the copying of human memories, personality and consciousness to non-biological processors – and AI that is sentient and smarter than humans will be operational realities in fifty years. This will trigger a positive singularity.

Worst fear: A global, benevolent and politically over-correct nanny-state dictatorship may paralyse all relevant development.

Best bet: A combination of these two trends, with a mild singularity takeoff if we manage to minimize the effects of the second possibility above.

BYE-BYE, BODY?

Most of the scientists who reckon lives can be made longer in the near future are talking about using biotechnology. But if that doesn't work, some have a plan B. They believe we can experience endless life by "uploading" our brains into computers – otherwise known as "whole brain emulation". These supermachines will be so powerful that they won't just run fully working duplicates of our minds, they'll also provide all the inputs our brains now enjoy, and new ones we can hardly imagine. So this new, virtual life will be infinitely better as well as longer. Lots of space tourism and really good virtual sex usually figure at this point too, with a bit of instant language-learning or playing Beethoven sonatas for the intellectuals.

Uploading brains isn't meant to be read as a fantasy, though, but a serious prediction. According to the best-known advocate of machine life, computer-scientist Ray Kurzweil, it could happen in just a few decades. He argues that the numbers support the idea. First, the number of neurons inside the human skull is fixed – there are roughly a hundred billion of them. If you also know the average number of connections each one makes, and their rate of firing, the maximum processing power of a standard-issue *Homo sapiens*' brain is easy to work out.

Computer processing power, on the other hand, is growing exponentially and will soon get ahead of the biological computer in our heads. But there are additional assumptions that others are more dubious about. One is the idea that the brain works like a digital computer. It's possible, but we're not really sure. The brain is usually modelled against the fanciest machine around at the time – a fast-weaving loom, a telephone exchange, a calculator, a personal computer... When we invent something cleverer still, you can bet the brain will start to look like that too.

Another assumption is that really speedy computers also mean really smart computers (or at least as smart as humans). But that depends on how they're programmed, not just how blindingly fast they are. And besides, the field of artificial intelligence seems to have come up against many hard problems, and few breakthroughs, since a slew of optimistic predictions in the 1960s.

Finally, there is the crucial question of what actually gets "uploaded", or how the artificial version of the human brain gets inside the computer. Kurzweil's view on this has evolved, but his latest scenario is based on "scanning". Originally this idea envisaged a super-detailed MRI scan, which could locate individual neurons and their connections. Later, he incorporated the new idea of nanobots sitting inside the brain and reading out what is going on, to ensure that we'll be able to measure the brain completely and copy it inside a computer. You could then create a new mind identical to the one scanned.

The idea intrigues, but how would the new mind feel about discarding the old one? (How would the old one feel about being discarded?) And would the new, virtual life carry meaning for the mind experiencing it? These kinds of "brain in a vat" questions have long entertained philosophers interested in consciousness and identity. Kurzweil's technical imagination poses them in a new way, but that doesn't mean his technology is realizable in the manner – or on the timescale – he envisages.

The old-fashioned way of transferring the mind, aided by Marty Feldman in *Young Frankenstein* (1974). Today's vision sees brains scanned and uploaded.

years further off, no matter when you ask the question. So full marks go to the other noted singularity guru, Vernor Vinge, the man who first popularized the idea. His original best guess for when the great convergence would occur (offered up in 1993) was 2030. In 2008, in a useful recap of the debate in the engineering professionals' magazine *IEEE Spectrum*, he stuck by his prediction that 2030 is the key date.

PREDICTIONS FILE

Robin Hanson

Economist and futurist at George Mason University, Virginia

Highest hope: The big important change coming is whole brain emulations (WBEs). More likely than not it won't be here in fifty years, but if it does arrive my highest hope is that we will foresee and pre-adapt to this change, rather than being blindsided by it. To adapt well, individuals should invest in forms of capital expected to remain valuable, in anticipation of their reduced future ability to earn wages.

Worst fear: I fear racism and xenophobia toward WBEs, wherein some places enslave them, while others drive them underground into second-class citizenhood, and still other places destroy themselves by successfully banning them. WBEs might well respond to such repression violently.

Best bet: Large local disruptions, but a rapid global transition to a world dominated by places that embrace WBEs.

Vinge bolstered his position with a rundown of the different ways the singularity could come about, an interesting sideline on the idea of the grand . He said that the exponentially growing technologies in play do not need to converge on a single point, and proposed a variety of ways they might evolve or combine to create newly intelligent entities for us to share the planet with. Superhuman artificial intelligence tops the list, but he also highlights:

- Intelligence amplification (IA, not AI), which would involve increasing human abilities by linking brains directly to computers.
- Biomedical enhancement, or increasing intelligence by neurologically tweaking our brains.
- The super network, in which humans, their computers, databases and networks all merge to become one superhuman being.
- A slightly different computers-all-get-together scenario he calls "Digital Gaia", in which the microprocessors which will, by then, be embedded in absolutely everything, coordinate themselves to create an emergent superbeing.

This list is a counter to the many scientists and technologists who argue that this or that idea favoured by the singularitarians is unlikely, or even completely impossible. It's not hard to credit the argument, for example, that molecular nanobots will get gummed up swimming through the morass of molecules inside the human body, or that "downloading" a mind into a supercomputer is largely a fantasy (see box on p.324). OK, says Vinge, but even if those arguments turn out to be true, no individual achievement is a prerequisite for the singularity. There are other routes, and one is likely to prove feasible.

As with most opinions about the future, the conclusions people draw from where we are now are as much about temperament and aspiration as logic. Vinge says as much himself when he considers the views of science writer John Horgan, who has made much of the limits of science in his books *The End of Science* (1996) and *The Undiscovered Mind* (1999). When they look at the current state of neuroscience, brain–computer interfaces and neural prostheses (artificial substitutes for parts of the nervous system), says Vinge, he and Horgan view the same facts, but draw different conclusions from them. The more conservative Horgan cites them to illustrate how distant we are from anything resembling the singularity, while Vinge sees them as signs of progress being made.

TRANSHUMANIST POLITICS

At the moment, the reason the proponents of the singularity tend to get attention is because their predicted timescales are so short. The idea that technology may take us to some higher stage of evolution isn't that hard to believe. If it's going to happen further in the future, even if that only means in the next century, most people are likely to remain fairly relaxed about it. But the suggestion that it might occur in our own lifetimes, or our children's lifetimes, is more startling. What might happen if technological trends move far enough to ensure more people took that idea seriously?

The advent of some kind of artificial intelligence that exceeds our own would be the dawn of a new era. Maybe it would lead to the technological singularity envisaged by Vinge, Kurzweil and Broderick. Or maybe it will lead to a war to the death between the machines, their acolytes and the remnants of unenhanced humanity. It sounds like the script for yet more movie sequels to *The Terminator*, but this scenario has actually been argued by an author with serious credentials – Hugo de Garis, whose day job is developing artificial brains.

His book *The Artilect Wars* (2005) predicts that there will be a world-shattering conflict before this century is out between "Cosmists", who are in favour of building artilects (artificial intellects) and "Terrans", who are dead against it. The artilects would not just be intelligent, they would operate, in singularitarian style, "trillions and trillions of times above the human level". The vision of inevitable war aside, de Garis's suggestion that the question "Should humanity build godlike, massively intelligent artilects?" will come to dominate global politics does seem worth considering – whether or not his timescale is realistic.

The development of artilects will be driven by a range of influences, de Garis believes. The industrial and military benefits of supercomputers will sustain fierce competition. And their supporters will invoke the human instinct to explore, the possibility of expanding into the universe, and the urge to reach a new stage of evolution – call them all cosmic destiny for short. Their opponents will prioritize preserving the human species as it is, which they fear will become redundant in a world of superintelligent machines. They will argue that whatever safeguards are built into the first generation, the artilects' behaviour will be beyond our ken, and hence inherently unpredictable. De Garis predicts that the argument will escalate until the Terrans are driven to make a preemptive first strike against the Cosmists, before they succeed in their designs. Billions will die.

This rather simple way of dramatizing the arguments seems unconvincing. However, de Garis is not the only one who thinks this century's crucial political struggles will grow up around human enhancement. James Hughes, former director of the World Transhumanist Association, argues that what he calls biopolitics – though he means the term to include computerized enhancement as well as genetic tweaks

– defines a new dimension of political debate that goes in a different direction from the two we are currently used to: economics and cultural politics. The result, as he lays out in detail in *Citizen Cyborg* (2004), is the development of a more complex map of political positions.

Hughes takes the side of what he calls social-democratic transhumanism – you might call it sensible transhumanism. A fair sample of what that looks like is the "Transhumanist Declaration" written by Humanity Plus (see humanityplus.org). They want to see enhancement technologies developed under democratic control, which maximizes personal choice and minimizes risks, especially those annoying existential risks.

More, shall we say, unrestrained versions of transhumanist politics come from quasi-religious outfits such as the rather wonderfully named Order of Cosmic Engineers. Appropriately, perhaps, their main efforts to date have taken place in the online world of *Second Life*. Beyond their vision, if there is a beyond, lies the Mormon Transhumanist Association, who consider the ideas of the singularity enthusiasts as part of Mormon prophecy.

Whether or not Hughes is right about transhumanism becoming a key element of the political landscape in the twenty-first century, he does provide a useful guide to the technologies that will undoubtedly fuel future passions. He suggests that, among all the things technology can do for us, there are four fundamental desires that drive innovation. They are the urge to control the body, live longer, get smarter and be happier.

Cutting across those desires, and often fuelling opposition to new developments, are the technologies that blur the established cultural boundaries. We prefer to take some limits, or essential laws, for granted, and consider it a transgression when technology seeks to challenge them. That is why the cyborg is such a potent image – it violates the boundary between human and machine. According to Hughes, other ways in which things will generally get mixed up – or messed up – include shifting the boundaries between:

- Animals and humans
- The living and the dead
- Real and artificial
- Young and old

There could be others – man and woman, parent and child, species and hybrid, gay and straight – but that seems a fair enough list to start with. It's easier to get beyond the particulars of a new technological trick if we can see why it might get a strong reaction, for or against. In other words, if you thought gay marriage or IVF for lesbian mothers were tricky issues to get your head around, you ain't seen nothing yet. Hughes's conclusion is that democratic transhumanism needs to expand the idea of what a person is to include all kinds of strange modifications, hybrids, upgrades and add-ons. Rights for robots and quality schooling for superchimps? In Hughes's future these are important political goals, not just ideas for philosophers to ponder.

However, the variety of positions held should not disguise the fact that the number of true believers in full-blooded transhumanism is still small, although some of them do get a lot of media attention. The explicitly religious affiliations of some transhumanist adherents support critics' suggestions that the idea has some of the characteristics of a cult. Activist and technology enthusiast Dale Carrico offers a rational

SINGULAR FICTIONS

If the technological singularity is a real possibility, we haven't a clue what happens on the other side. Not surprisingly, this means that much of the writing on the subject is science fiction. This is even less surprising when you learn that Vernor Vinge, the man who first used the term "singularity" to describe a possible technological future, is an award-winning science-fiction novelist, as well as a former professor of maths and computer science.

Even so, it's pretty hard to write about any period after the singularity and still make sense – but that is kind of the point. Stories that find a way around this include Vinge's *Marooned in Realtime* (1986), which treats the singularity a bit like the apocalypse, in that the story is about the few people who are left afterwards. Then there are the post-singularity tales which address the fairly plausible idea that ultra-powerful intelligences come to exist in computers, but immediately lose interest in humans and retreat into their own virtual spaces, leaving us to get on with our lives much as before.

More involved with the actual singularity is Cory Doctorow's *Down and Out in the Magic Kingdom* (2003) – which features a choice of paths to immortality – cloning for all who want it or universal mind-backups on computers. The characters dwell in a post-scarcity economy in which money has no place and the economy is based on a currency of reputation. The plot features a conflict over how to run Disneyworld in a society where the rides can be hooked up through brain interfaces.

Charles Stross is not necessarily a believer in the singularity (he compared the magical technologies envisaged by some writers to "pixie dust"), but his novel *Accelerando* (2005) takes on the challenge of writing about the time beyond the singularity directly. It does that by focusing on three generations of the same family, who are all around at the end but mostly in post-human form. The fact that the first sections of the book, which outline the build-up to the singularity, work a lot better than the end, perhaps confirms how hard it is to say anything that makes much sense about life after humans. If you can't get on with the book, there is a web glossary of all the technical terms that Stross, a one-time computer scientist, throws in for the cognoscenti (en.wikibooks.org/wiki/Accelerando_Technical_Companion).

Other notable efforts to imagine the unimaginable include Rudy Rucker's *Postsingular* (2007), which sets up two contending singularities and their human agents, one featuring nanobots, the other a kind of distributed quantum computer, and Greg Egan's *Diaspora* (1997). The latter explores a post-singularity world (not imminent – it's set in the thirtieth century), in which the solar system is a unified, computer-linked network, and "people" are mostly either software entities or computer intelligences inhabiting robot bodies. A few "fleshers" remain, though, to give us something to identify with.

Finally, there's a good collection of stories, old and new, about future humanity and post-humans, in uber-anthologist Gardner Dozois's collection *Supermen: Tales of the Posthuman Future* (2002).

middle-ground position. He accepts that big technological transformations are on the way, but regards the transhumanists as a bunch of dreamers, heavily invested in wish-fulfilling ideas which are, in essence, old fantasies in new clothes. For him, glorifying the advent of immortality and robot gods is a diversion from tackling existing real-world problems, and from managing the development of realizable technologies that can do some good – if they are used correctly.

CRYONICS: A ONE-WAY TICKET TO THE FUTURE

If you are really desperate to see the singularity, but not sure if it will arrive soon enough, there is one way of achieving a kind of time travel. Lots of old stories hinge on allowing someone to see the future by falling asleep for decades, Rip van Winkle style, and nowadays a small number of people are hoping to see the things to come by signing up to be frozen after they die. They hope that if they can be preserved long enough, technology will have advanced sufficiently to be able to revive and repair them. Of course if that works, we will have to redefine what we mean by "dead".

The scheme to achieve immortality through cryonics was first outlined in 1962 by Robert Ettinger in *The Prospect of Immortality* (see box opposite). Since then, discussion of the possibilities of nanotechnology has encouraged supporters of the procedure to believe that damage to essential organs, especially the brain, will be repairable using future technology. As cryo-enthusiast Ralph Merkle, a computer scientist and nanotechnology guru, explains:

> "Cryonic suspension is a method of stabilizing the condition of someone who is terminally ill so that they can be transported to the medical care facilities that will be available in the late twenty-first or twenty-second century."

At the moment, being deep frozen down to -130°C under controlled conditions, and funding

Trans Time, Inc., a cryonics company based in California, freezes humans and animals in liquid nitrogen. Company president John Rodriguez inside one of the empty cryon tanks used to contain the bodies.

the maintenance of the equipment to keep you that way, costs around $250,000. It's not just a matter of turning into ice, the process now in use for would-be cryonauts involves replacing the water in your body with other liquids. The outfit that does most of the freezing, Alcor, calls it "vitrification". A death certificate has to be issued before the process can begin, to keep on the right side of present-day law. But the upside of that is that if you plan ahead, you can use your life insurance to pay for the transfer to a flask instead of a coffin. Some of those currently in store have their heads and bodies in separate flasks, a procedure Alcor calls "neuroseparation".

A steady trickle of customers come up with the money, though the number of people preserved at Alcor and elsewhere remains in the low hundreds. (Despite persistent rumours, that well-known futurist Walt Disney is not among them.) Some hundreds more are signed up for the procedure when they die. According to the *Wall Street Journal*, some of the wealthier ones have made arrangements for their investments to be managed on their behalf while they're in suspended animation. Then, they hope, if the technology pays off, they not only get to see the future, but wake up in a hundred years or so far richer than they were before. Presumably they'd then have to

DEATH: DON'T JUST ACCEPT IT – DO SOMETHING ABOUT IT

In his 1962 manifesto *The Prospect of Immortality*, physics teacher Robert Ettinger made much of the then-recent work on freezing and reviving organisms, or parts of organisms (eggs and sperm are the cells frozen most often, and they seem to survive fairly well).

The book envisages the frozen dead being rescued from their "dormantories". They awake in a world of super-advanced technology, and discover innovations which sound like the most magical of nanoassemblers (see p.86): "intelligent, self-replicating machines" that can "scoop up earth, air or water and spew forth whatever is desired in any required amounts."

In a later work, *Man into Superman* (1972), Ettinger developed one of the early visions of transhumanism. As he put it in the preface:

> "Most people think the future will be just like the recent past, with maybe a little more chrome and somewhat higher prices. They need to pay closer attention to what is happening. What is happening is a discontinuity in history, with mortality and humanity on one side – on the other immortality and transhumanity. With a little encouragement, many of us can make the transition."

The point of the book was that the life awaiting the newly thawed would not just be more of the same – a common objection to the immortalists' programme. It could be far richer and more rewarding than the life the preserved had known before.

Despite this persuasive rhetoric, few have so far signed up for what Ettinger presented as a wager with no downside. To his bemusement, people seemed happier to accept death than to arrange for their maintenance in a bath of liquid nitrogen. As he said, many are cold, but few are frozen.

At the time of writing, Ettinger himself was still alive, and in his early nineties, but had presided over the cryopreservation of two wives and his mother. The website of the Cryonics Institute, which he founded, refers to them as "patients". Ray Kurzweil is also prepared to be frozen if the singularity fails to arrive before he dies.

PREDICTIONS FILE

Natasha Vita-More

Transhumanist theorist and designer

Highest hope: The realization of a transhumanist world-view, which is based on the desire to defeat death and reverse the damages caused by somatic and cognitive ageing. This world-view includes an ardent hope that human behaviour develops beyond physiological rigidity of discrimination, intolerance, hate, and other attitudes and actions that stand in the way of advancing human rights and diversity.

Worst fear: That humanity never evolves – that we, as we are now, are the final cut.

Best bet: It could be that within fifty years we begin seeing a kinder humanity – if we can first resolve the outstanding issues of religious and political hegemony, poverty, disease and pollution. If we can do this, the rewards will be manifold because we will be inspired and proud of actually accomplishing what ought to have been accomplished millennia ago.

repay the life insurance, though. If you're at all interested in following in their footsteps, the legal vehicle you need is a "personal revival trust" – which gives a whole new angle on retirement planning.

FURTHER EXPLORATION

Andy Clark **Natural-Born Cyborgs: Minds, Technologies and the Future of Human Intelligence (2003)** A philosopher's take on minds and technology, arguing that we have always used our own creations to enhance our abilities, and will go on doing so.

Joel Garreau **Radical Evolution: The Promise and Peril of Enhancing Our Minds, Our Bodies – and What it Means to Be Human (2005)** Roundup of current and near-future technological possibilities by a *Washington Post* journalist, who is optimistic about their potential for transforming ourselves.

Ray Kurzweil The Singularity is Near (2005) This lengthy tome is the most detailed argument for the singularity. Two previous, similar, books, allow some judgement of the author's powers of prediction – *The Age of Intelligent Machines* (1990) and *The Age of Spiritual Machines* (1998).

www.alcor.org The largest cryonics organization, Alcor performed its first human cryopreservation in 1976, and also offers neuropreservation for people who don't have neuroses about becoming a "brain in a vat".

www.cryonics.org Get in touch with the Cryonics Institute in Michigan, US, if you're interested in being preserved alongside Robert Ettinger's family. (Number of human patients in cryostasis at the time of writing: 95; number of pets: 67.)

humanityplus.org Humanity Plus make an effort to consider the downsides of technologies for souped-up humanity, but they're basically advocating getting on with it. Their website and online magazine *H+* are good ways to keep up to date with the discussion.

singinst.org The Singularity Institute (or the Singularity Institute for Artificial Intelligence, to give it its full name) express the opinion that although "smarter minds are harder to discuss than faster brains or bigger brains [that] does not show that smarter minds are harder to build."

18

infinite in all directions

infinite in all directions

Looking to the future, we give up immediately any pretence of being scientifically respectable.

Freeman Dyson

If you were around for a decent portion of the twentieth century, you may remember that, until near its end, the year 2000 – or perhaps 2001 for emphasis – seemed an excitingly futuristic date. Similarly, now that we're in the new millennium, serious discussion of the future tends to fall silent on anything beyond the end of the twenty-first century. Most of this book has focused on that discussion, looking at the prospects for the next fifty to one hundred years. But can we go further? And, if we can, how far can we go?

Limiting talk about most human affairs to the next five or ten decades is prudent in the face of vast, proliferating uncertainties. But on a larger scale, some things are still well established as reasonably probable. As Martin Rees, Britain's Astronomer Royal, is fond of pointing out, stars and galaxies are simple systems compared with organisms and societies. As long as our current science is more or less right, we can tell what they will do over very long timescales. And, since the Victorians discovered deep time (see p.8), there has been a tradition of speculation – sometimes pretty wild speculation – about other aspects of the far future. It's fun to try and go there in our imaginations, with whatever ideas seem helpful at the time, and see where they might take us.

It also, perhaps, yields a new perspective on the problems of the present. If there is a possibility that life will continue for millions of years, there is more to lose by messing up the planet now. On the other hand, if everything we know points to Earth remaining habitable for a finite amount of time, however large, some will want to plan for a move elsewhere.

THE LIE OF THE LAND

For life on Earth as a whole, there is a lot of time, though not unlimited. Some things change very slowly. Continents, or rather the rafts of rock they sit on, move a few centimetres a year. The basic pattern of land and sea will look the same in a thousand years, though sea levels will probably be higher. Even after a million years, the main land masses will only have shifted a hundred kilometres at most. Appreciable continental movements are reckoned in tens or even hundreds of millions of years. Geologists believe that all the present-day continents began as a single land mass 250 million years ago, and will be reunited, in a reincarnation of the ancient mega-continent known as Pangaea, around 250 million years in the future.

Big predictions rest on big assumptions. In this case, the minor assumption is that present movements – Africa moving north into Europe, Australia heading for Southeast Asia – will continue. The major one is that the current broadening of the Atlantic Ocean will reverse, through the formation of a new zone where the sea floor is forced beneath the Earth's crust, and the Americas will begin to move slowly back towards Europe. But will there be life aboard those drifting continents, and in the air and seas around them, over all those years?

Probably, but not necessarily human life. Mammalian species in the fossil record last a million years or so. We think we're different of course, and we may have gone transhuman or post-human long before then (see Chapter 17), but if not, we will be vulnerable to other changes that come over the Earth faster than continental movements. The magnetic field changes direction every two to three hundred thousand years, on average, and the next flip is overdue. This probably poses no great hazard, though the field declines to nil during the changeover, which could destabilize the ozone layer. A much bigger challenge will be the one that was more prominent in the media before it became apparent that global warming was in progress: a new ice age.

THE NEXT BIG FREEZE

The climate we short-lived humans think of as normal is really just an interlude in a shift to a much colder planet. In the medium term, geologically speaking, the ice which has dominated Earth in the recent past will reassert its grip. The reasons why it comes and goes – more than a dozen times in the last 2.5 million years – are not completely understood. The immediate trigger is probably a cyclic variation in the amount of solar radiation that reaches the planet, amplified by an increase of ice in the north (where most landmass is found) which bounces more heat back into space. But the longer-term cause is climate scientists' current preoccupation: atmospheric carbon dioxide (see Chapter 7).

On average, CO_2 levels have gradually been falling for the last four hundred million years, reducing greenhouse warming. The fall is due to a combination of increased absorption by plants and the chemical weathering of rocks on landmasses that are slowly being increased in size by the action of plate tectonics. Although solar radiation has been increasing over the ages, and will dominate the picture later in the planet's history, it has been counteracted by the loss of carbon dioxide. Yes, CO_2 levels have increased since industrial times, but this is a temporary halt to a long decline. There is now only a twentieth as much of the gas in the atmosphere as there was at its peak level.

The geological record suggests the next ice age is due in a few thousand years at most. Cold spells have predominated for over two million years, and can last for a hundred thousand years. The milder periods in between, the interglacials, are typically shorter, around twelve thousand years on average. And the last ice age ended fourteen thousand years ago. On the other hand, there have been longer interglacials on occasion, so we cannot be sure when this one will end. But end it will, and the result will be a very different geography for the planet.

How different is clear from the traces of previous freezes. If the next ice age follows the same pattern, most of the landmass of the north will be covered by an ice sheet several kilometres thick. The water locked up in the ice will reduce sea level by 400 feet (120 metres). Add in prolonged droughts and constant winds howling along the margins of the ice sheets and much of the Earth becomes inhospitable. When the last glaciation kicked in, there were probably only a few million humans trying to survive. If the present billions were faced with similar changes, they would have nowhere to go.

Assuming future humans do manage to negotiate this little obstacle, and make it into the next interglacial, they will face repeats of the same process for some millions of years thereafter. However, it is possible that human action has already broken the cycle. In some models, ocean acidification caused by the current fossil fuel-induced increases in atmospheric CO_2 will lead to the additional release of the gas as the shells of marine organisms dissolve. This could be enough to delay another ice age for, say, half a million years.

This is, in effect, unintentional geoengineering, and is small comfort in the short term as we contemplate the other effects of global warming. However, if we become convinced of the need for intentional geoengineering (see p.148) because of the looming prospect of climate chaos in this century and the next, perhaps our descendants will take it for granted that regulating global temperature, or trying to, is part of good government.

HOTHOUSE EARTH

Ice ages may return, but the even longer-term prospect is for a resumption of global heating. As palaeontologist Peter Ward and astronomer Donald Brownlee outline in *The Life and Death of Planet Earth* (2003), science points to a planet that gradually grows hotter, and becomes less and less capable of sustaining life. The two crucial factors are temperature and carbon dioxide levels – the first gradually increasing, the second going down over time. The weathering of silicate rocks will eventually be the main influence on the latter. Some time between 500 and 1000 million years from now, on their estimates, it will reduce CO_2 so far that all plant life will end.

The main surviving life would then be bacteria, living in the ocean or deep in the rocks. Soil and atmospheric oxygen would largely disappear a few million years after the plants faded out. Any remaining animals, presumably somehow adapted to grazing on algae and bacteria, would slowly suffocate.

In any case, temperatures will continue to rise steadily, driven by the more intense radiation from the Sun predicted from our understanding of solar physics. In a billion years' time, the Sun will get hot enough to boil off the oceans. The steam that rises from them will decompose in the upper atmosphere, and the hydrogen liberated will drift off into space. The end result is a baking hot, bone dry, lifeless planet.

Some recent research has suggested possibilities for extending the timescale for the different stages in this sequence – life's use of nitrogen could reduce atmospheric pressure and slow the temperature rise,

EARTH ENGULFED?

When a star like the Sun nears the end of its life, it expands and turns into a so-called red giant. It will then be so much larger that it was assumed, until recently, that it would engulf all the nearby planets: Mercury, Venus and Earth. Then it was pointed out that the ageing star will also be losing mass. That means its gravitational pull on the planets will weaken, and their orbits will gradually move outwards.

For a while, astronomers suggested that this would allow the Earth to stay just out of reach of the Sun. But the latest refinements of the calculations tell a different story. A 2008 paper by astronomers Klaus-Peter Schroder and Robert C. Smith suggested that our planet is just too close to escape. Its orbit will not expand fast enough to elude the swelling star, which will overtake it around half a million years before it stops growing.

Once that happens, Earth will carry on ploughing through the tenuous outer layers of the star for a while, before the drag starts to slow it down and it spirals inward instead of moving further out. Finis, the end, in seven to eight billion years' time.

Does this matter? Not a lot, except to students of mechanics. There won't be anyone on the Earth to find out if the final prediction comes true. It will be baked to a cinder three billion years or so before this endgame is played out.

for example. This might allow some sort of biosphere to exist for an extra billion years or so. But the final result remains the same. The planet will be barren, until its inevitable end when the Sun finally nears the limits of its nuclear fuel (see box above).

That is the most likely course of events, from a geophysical point of view, but there are other scientifically plausible ways for the Earth, or life, to end. Asteroid or comet impacts (discussed in Chapter 13), or sudden gamma ray bursts from deep space, are more likely to enter the story as time passes. There are also computer simulations of the future of the solar system which suggest there is a one in a hundred chance that the orbits of the inner planets are irregular in ways that would allow Mars or Venus to collide with Earth in three or four billion years' time. Oh, and our galaxy is on course to collide with one of its neighbours, M31, in three billion years' time. But that will be less of a collision, and more of a gradual merger, so it's unlikely to directly affect the fate of the Earth.

CHANGING THE PLAN: MOON BASES AND SPACE COLONIES

So much for the current state of the science of future Earth. That story would apply to any planetary body orbiting a medium-sized star, but ours has an extra ingredient: intelligence. How might future humans, or their descendants, alter the story? Will we find our way off the planet, locate other Earths, perhaps build a new one somewhere else? Here the stories involve ideas from the realm of science fiction, but that doesn't mean they could never happen.

Human life so far has been tied to the Earth, but what if we left it? We already have space travel, we

Lunar living for twelve: NASA's 1989 design for an inflatable moon habitat includes living space for a dozen astronauts, a fully equipped life sciences lab, a gym, wardroom and hydroponics gardens.

just haven't got very far yet. What comes next? There is any amount of speculation about serious efforts to explore the rest of the solar system, harvest solar energy from the moon, mine the asteroids and settle colonists on Mars or perhaps one of Jupiter's handy-sized moons. More far-reaching visions extend to "terraforming" Mars, and gradually making it more like Earth.

Alternatively, we might send some people to live in artificial space colonies. The best-known idea here is the O'Neill cylinder, a design worked out by Princeton University physicist Gerard K. O'Neill in the early 1970s. This would be a cylinder four miles in diameter and up to twenty miles long. It would continually rotate to create enough artificial gravity for people to be able to live on the inner surface, and be coupled to another cylinder rotating in the opposite direction to make the whole shebang easier to manoeuvre. The centre would have zero gravity, and so provide fun not available on old-fashioned planets.

The notion of space exploration has been around for more than a century, and some of these putative projects have been worked out in great detail. O'Neill and his students designed the space colonies with sun traps, radiation protection, farms and all, and reckoned everything needed could be lifted up there by a fleet of spaceships. Space enthusiasts love nothing more than to do the math, working out how much energy would be needed to move payloads, where the energy might come from, and how fast the job could (theoretically) be done. And these ideas have been extended to clever devices such as the "space elevator", a cable-driven lifting device tethered to the Earth, which would get around the need for rockets (see box on p.340).

That's all fine, and some of it might even happen one day, but few of the visionaries that draw up such schemes have really thought about how people would fare in space. That has been the province of science-fiction writers. Perhaps the most lovingly detailed vision of how life might work out on another planet is in Kim Stanley Robinson's trilogy of novels (1992–96) about Mars (*Red Mars*, *Green Mars* and *Blue Mars*, signifying the progress of a well-described terraforming). His Mars has colonists, politics, a space elevator (which collapses spectacularly – watch out for that cable!) and a richly realized sense of what it might be like for humans to live on another world.

For all its realism in the literary sense, it's far from realistic from a scientific standpoint. The timescale for terraforming, for example, is set by the needs of the narrative, not the constraints of science. And although such an imaginative tale can fuel enthusiasm for real Mars missions, the novelist's interest lies in creating a setting for clashing ideas about social organization, colonial development and ecological stewardship to play themselves out. In other words, it's not a prospectus for a trip to Mars, but a reflection on the current and near-future state of the Earth – as is the most interesting science fiction.

The non-fiction writers who have looked into humans living in space tend to be unconvincing. Science writer Adrian Berry, for example, in his grandly titled *The Next 500 Years* (1995), was gung ho for orbital hotels, moon bases, terraforming Mars and mining asteroids. He predicted a private sector drive to move into space in pursuit of profit; space tourism could be more awe-inspiring than any earthbound journey. Imagine it: the sublime beauty of our home planet against the dark backdrop of deep space.

On the other hand, the price might be worth paying for, well, a low-gravity bonk. As Adrian Berry put it, slightly more decorously, "one of the great joys of the twenty-first century will be taking holidays in lunar hotels for the specific purpose of making love". The moon's gravity would be good for swimming and ball games too, apparently. Berry's enthusiasm for low gravity knows no bounds. He also suggests that "the moon, with its surface gravity one sixth of the Earth's, will be much favoured as a site for old people's homes". Funny, none of the science-fiction writers seem to have thought of that.

GOING UP? SPACE ELEVATORS

Space travel might be cheaper without the tiresome need for rockets to blast you clear of the "gravity well" of our planet. The idea of a permanent structure reaching into space goes back more than a hundred years. Build that, the theory goes, and you could climb up somehow, and get beyond the final frontier the easy way.

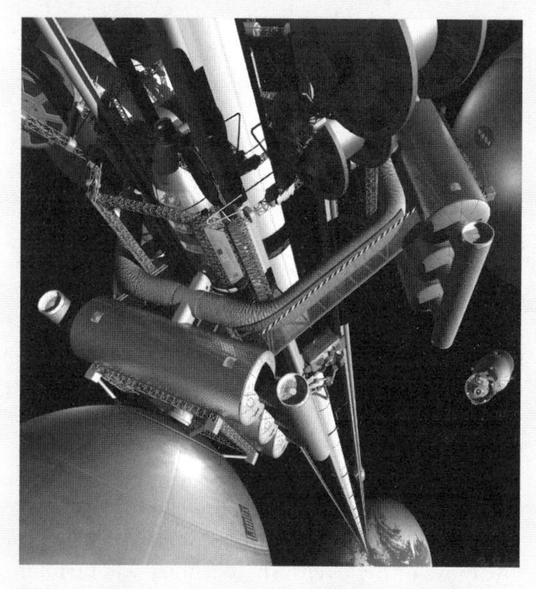

NASA artwork of a hypothetical space elevator, viewed from geostationary orbit 35,786 km above the Earth.

The current incarnation of this notion is the space elevator. The idea is that a satellite launched into geostationary orbit might gradually pay out a cable that reached down to the ground. The cable would need a giant counterweight, and the whole assembly would have to be thousands of miles long – geometry means that the geosynchronous orbit, at which a space vehicle stays above the same spot on the Earth's surface, is 22,000 miles up. Early design studies suggested that there was no known material strong enough for the elevator cable, but recent experiments with carbon nanotubes suggest they might fit the bill – if they could be made stronger still, manufactured in quantity and lifted into space.

Achieve all that, and technologists can probably manage to make a reliable machine for climbing the cable. The Spaceward Foundation is running competitions for both cable strength and climbing machines and their energy supply. They have already awarded a prize to a team who achieved their first target: building a rig that climbed for a whole kilometre at two metres per second. In the next round, they hope to see success at five metres per second. The experiments are fun, but an Earth-tethered space elevator is clearly some way off. The idea might work better on Mars, though, as that planet's weaker gravity would make the demands on materials and the lift energy required smaller.

Not all predictions of the future in space are quite so fanciful, and after centuries or millennia of development, who knows how far out into the solar system humans might get? But as we are still considering missions to return to the moon, and tentative plans to make our first ever visit to Mars in a decade or two, anything beyond that is purely speculative. Space missions with people on board are still mostly exercises on paper. Enthusiasts often blame unenterprising governments for this, and suggest that the private sector should take the lead on human expansion into space. A few examples of this may come to fruition in the next decade or so, as space tourism begins to take flight, but it will probably be on quite a small scale for the foreseeable future (see box below). So let us skip over what cannot be foreseen in detail and consider the next step. In the really long term, as we have already learned, humankind must leave Earth to survive. But simply getting off this planet is no way to evade the death of the Sun. For a more open future, we must leave our solar system. So what about travel to other stars?

BOOK NOW: A ROOM WITH A VIEW

Billionaire jaunts into space are one way to get private sector space exploration efforts moving, and the market may soon be getting larger. Two rival companies are already planning to transport customers to space hotels – and they claim the first reservations will be honoured in a just few years' time.

Four and a half million dollars will buy you three nights at the Galactic Suite Space Resort, although you'll need two months spare for training beforehand and an extra couple of weeks for post-flight recovery. The "hotel" is going to be on the small side: it's a pod that looks like a sleeker version of the old Russian Mir space-station. It will have room for two paying passengers and one pilot and will be in orbit 451 km above the Earth. Not that far into the galaxy, then.

The company behind the project is based in Barcelona, but is planning to launch its first passengers from a Caribbean island in 2012. If it flies, phase two will involve a five-module design for four passengers and two crew, which will even have a "toilet and spa" in one of the modules.

The other company vying for your space dollars is Bigelow Aerospace, which also has plans to be in orbit by 2012. They're developing technology originally envisaged (but then dropped) by NASA for making expandable modules, which can be launched and then "inflated" to create more room for passengers and crew. Founder Robert Bigelow made his fortune in the hotel business, and says he has $500 million to spend on bringing the company's plans to fruition.

They reckon they can offer four weeks in orbit for $15 million. The ticket price is high, but less than the going rate for a visit to the International Space Station. If you want to go into space right now, a week on the space station (courtesy of Space Adventures) starts at around $20 million, including a flight on board a Russian Soyuz launcher. Half a dozen people have taken the trip since 2001, and Microsoft co-founder Charles Simonyi liked it so much he went back for a second stay in 2009.

The first off-world flights with Richard Branson's modestly named Virgin Galactic will be much shorter than these cosmic holidays, but also far cheaper. They're selling seats on suborbital flights for a mere $200,000. The company will use an improved version of SpaceShipOne, a reusable craft which won the Ansari X Prize for a privately built spaceship in 2004. It will be launched from specially built aircraft that take off from New Mexico, and carry up to six passengers. The flights will last just two hours and reach heights of 110 km from the Earth.

www.galacticsuite.com

www.bigelowaerospace.com

www.virgingalactic.com

HOW FAR CAN WE GO?

A generation raised on *Star Trek* and *Star Wars* is very familiar with visions of a future in which interstellar travel is an everyday accomplishment, and most probably assume it's just a matter of time. After Neil Armstrong walked on the surface of the moon in 1969, the knowledge that people had been into space, and crossed that "final frontier", made anything seem possible. The trick was getting into orbit. Going to the planets, and then the stars would just be more of the same, wouldn't it? Maybe not. Looking at space travel realistically means confronting just how far away everywhere else actually is. The moon is, by definition, tied to the Earth. That is why it's only three days' journey away (or approximately 380,000km). No distance at all, really. But a trip to our nearest neighbouring planet, Mars, would take around nine months or so using current space travel technology.

What kind of distances are we talking about to visit other destinations? We may have learned that

IS ANYBODY (ELSE) THERE?

Any talk of the possibility of alien life soon stumbles over the question famously put by physicist Enrico Fermi in 1950 – where are they? This was his response to arguments that, amid all the hundreds of billions of stars, and probably billions of planetary systems, there must be some that could nurture intelligent beings.

Answers to Fermi's question include arguments that intelligent life elsewhere in the galaxy:

- has always destroyed itself
- is too far away to communicate with
- prefers to keep very quiet.

Or perhaps there's simply never been any. If we believed that, how would it affect our attitude to the future? If we really are alone, perhaps it's more important to look after our planet as best we can. On the other hand, if we might not be, perhaps that gives us more hope that intelligent life has a place in this vast and puzzling cosmos.

Either way, we should soon have a better sense of the possibilities. This will come from two directions. One is finding out how to brew up new life in the lab. If we get a better idea of how a living system could evolve, that makes it easier to figure out the chances of it happening elsewhere. The other is a welter of new astronomical observations. In recent years, it has become routine to log disturbances in the light reaching us from distant stars, which suggests the presence of planets orbiting round them – there are now a few hundred on the books. Most are very large bodies, but a few appear Earth-like.

US and European space missions due in the next decade will refine and improve these observations. The European Space Agency's Darwin mission, for example, planned for the mid-2010s, should be able to decode spectroscopic data (showing the absorption of different wavelengths of light by atoms of different elements) from planets outside the solar system to give some idea of their atmospheric composition. If one of those planets has an atmosphere that is far from chemical equilibrium – i.e. one that contains reactive stuff like the oxygen found on Earth – it will probably be due to the action of life.

So, in a few years' time, we may not be any nearer to contacting alien life, but we ought to have a better idea if any actually exists.

we're not at the centre of the universe, but we still measure it using a local yardstick. One astronomical unit (AU) is the average distance from the Earth to the Sun – around 150 million kilometres (or 90 million miles). The proximity of the moon is brought home by the fact that it is only one six-hundredth of an AU away.

The Earth is one of the innermost planets, but the solar system extends over thirty astronomical units – the distance from the Sun to Neptune, the outermost planet since Pluto was demoted to dwarf planet in 2006. Pluto is now classified with other small, cold objects in the Kuiper belt, a region under the Sun's influence which goes on for another thirty AU or so. Then there are solar orbiting comets, which are more or less tied to the Sun, in the Oort cloud (very thin for a cloud), which spans an impressive fifty thousand AU.

Now we're getting into space proper. But not that far. In fact, if you were to travel this path, nothing much would happen for the next couple of hundred thousand AU. The nearest star, Proxima Centauri, is over 250,000 AU distant. And there is no sign of any Earth-like planets around stars any nearer than a million AU away.

So the minimum distance for a colony outside our solar system is 600 million times the span from the Earth to the moon. That effectively rules out transporting human beings there using any known technology. Some kind of completely reliable suspended animation might get round the prohibitive energy costs of life support, not to mention the problems of staying both alive and sane over a journey of decades, or even centuries. And the standard science-fiction accessory, a faster-than-light drive, would avoid those decades and centuries in transit. But until those things exist, any colonization of the galaxy will have to be done by very patient robots. By the same token, we're unlikely to be greeting any visiting aliens, unless they're super-intelligent beings with some kind of advanced alien physics that enabled them to get here.

ROBOTS WILL BOLDLY GO...

Robot colonists would be the best bet for extra-solar excursions (those outside the solar system), and probably for visits to nearby planets as well. They would need less radiation shielding, be able to withstand extreme environments at the destination and, like robot soldiers, can be exposed to higher risks with a clear conscience.

But what kind of robots will they be? Here, speculation about space travel has cheerfully imported the ideas outlined in Chapter 17. Nanotechnology and superfast computing open up new possibilities for moving out into space. The simplest way to see where this might lead is to describe the von Neumann probe, named after the computer pioneer John von Neumann.

Suppose you could build high-function automatons (self-operating robots), perhaps even intelligent ones. Given the right materials, they could build copies of themselves. Now imagine a self-replicating automaton with human-level intelligence – or even a bit higher if the ethics committees allow it – reaching a planet in another star system. It would land, set about gathering raw materials and energy, and start manufacturing copies of itself. Give it a bit more information, and it could make other things too. Ask it nicely, and it might even set up a colony equipped for human occupation.

After making the long haul to one star system, it might as well build a new probe, and launch it on to

Roving Mars: Could NASA's robotic geologist, which landed on the surface of the red planet in 2004, be the distant ancestor of super-intelligent von Neumann probes that will colonize the whole galaxy in a few hundred million years?

the next one. So self-replication would allow a slow, expanding wave of robot exploration outward from Earth – or any other point of origin. This brand of future history does not call for any faster-than-light trickery. The probes could lie inert while travelling, and be automatically restarted when they fall planetwards. Allow a reasonable time for each hop, and for building the apparatus to support the next launch, and you could expect coverage of the whole galaxy in a few hundred million years, tops.

Some of the galaxy colonization scenarios that use these ideas preserve a role for living, breathing humans. Perhaps the probes could be equipped with information about the human genome, so they could grow a new generation of people at their destination – having, again, first built suitable living quarters. But this portrayal of a galactic future for *Homo sapiens* seems a bit illogical. A future world in which such advanced von Neumann probes could be built is clearly one in which computer- and nanotechnology have advanced far enough that the singularity (see Chapter 17) is only a matter of time.

It doesn't matter how much time that might be. If intelligence moves to new electronic hosts then colonization by automatic probes will be the same as colonization by people – or their descendants. It will import consciousness into new regions of the universe. The most ambitious proponents of the singularity have already projected the idea into that realm. According to Ray Kurzweil in *The Singularity is Near* (2005), non-biological intelligence will eventually come close to using all the matter and energy in our cosmic vicinity. There is no limit to this process, until "ultimately, the entire universe will become saturated with our intelligence".

Universe saturation aside, this notion of the future of intelligence also has implications for the future of Earth and the rest of the solar system. All that matters in this abstract conception of future intelligence is devices with enormous computing power, the stuff to build them with and the energy to keep them working. The "stuff" is what some see as the logical result of cultural evolution taking over from biological evolution – a general-purpose material that can support computational activity, and therefore intelligence. To emphasize that this might well have a different composition to the silicon chips we see around us at the moment, writers pursuing this line of thought like to call it "computronium".

As medical technologist Steven Harris emphasizes in an essay in Damien Broderick's provocative collection *Year Million* (2008), as long as it can run software, it can reproduce any desired property of life: "the distinction between living and dead matter will lose its meaning". If that's the case, then perhaps there will be no limit to the conversion of dead into living matter. In this new, unimaginably complex world, two simple requirements remain: matter and energy. The implication of all this is that one possible future for the Earth sees it get dismantled.

In that case, the planet would never be engulfed by the expanding Sun as it would long since have been dispersed. If the inert mass of the planet is converted into computronium, it will also end up spread out because this is more efficient for collecting solar energy. It will not be completely scattered because the computerized intelligences that now control things will want to communicate with one another. The result might be a kind of cloud of solar collectors and computer processors. In time, this cloud would incorporate all the other planets too, with Jupiter contributing by far the largest lump of matter for whatever manufacturing process is involved.

This kind of evolution leads to a "Dyson sphere", or less neatly, a Dyson swarm or shell. The idea, first formally proposed by physicist Freeman Dyson half a century ago, comes from the observation that the energy needs of human civilization appear to continually increase (see below). Eventually, he argued, that would drive future civilizations to try to collect as much of the Sun's output as possible. (The exponential growth law again: if power needs grow at only around one percent a year, then stellar quantities would be required within a few thousand years.) The way to do that is to surround the star with collectors, instead of allowing most of its radiation to pass out of the solar system into space.

In the more recent visions of the Dyson shell, the energy-users that make up this super-civilization

ENERGY LEVELS AND CIVILIZATION

Our energy orgy of the last 150 years is impressive. But on a larger scale some think it's only the beginning. As civilizations advance they seem to continue finding new ways of harnessing energy. That means that older, more developed, and hence more powerful, cultures could exist elsewhere in the universe. How much more powerful? In 1964, Russian astronomer Nikolai Kardashev suggested a threefold classification of advanced technological civilizations.

- **Type I civilizations** control the energy of an entire planet. Originally this was taken to be roughly where humans were in 1964, but in fact the Earth's energy budget is around 10^{17} (getting on for a million million million) Watts, whereas in 1960 humans were only using around four million million Watts. So perhaps we don't qualify for Type I yet, but we are on the way.
- **Type II civilizations** have taken a great leap, and control all the energy of their local star, perhaps by constructing a Dyson shell. For a star like ours, this would give them a thousand million times as much energy to play with as the Type I civilizations.
- **Type III civilizations** would be able to look down on Type II (assuming such advanced beings were so small-minded). They would command the energy of an entire galaxy. This would be another big gap, with an "average" galaxy containing perhaps a hundred billion blazing stars.

But what would these fantastic cultures do with all that energy? Type I would be like us, only more so. They might have weather control, or be able to prevent earthquakes. Type II will be the staples of much science fiction, with space exploration and colonies on other planets. Type III are your basic lords of the universe: masters of space and time, planet sculptors, star shifters, players of black-hole billiards, that kind of thing. They would be invulnerable to any calamity save the end of the cosmos, and might even have an idea or two about that.

Any such scheme is a standing invitation to go further, and suggestions for Type IV go from civilizations that span many galaxies to migrating to hyperspace and other suitably mind-boggling options.

Lower down the scale, much has been made of the fact we have seen no evidence of either Type II or III civilizations in our neighbourhood (we could perhaps miss a Type II, but a Type III in our own galaxy would be impossible to overlook). Perhaps the path to Type I is so fraught with hazard no one ever gets any further? Or perhaps, one day, we'll be the first lifeform to reach this level in this galaxy.

are computerized. In fact, there could be more than one layer of collectors and processors. In theory (if anything so far into the realms of speculation can be called theory), the first layer of computer intelligence would be the primary user of solar energy. The second layer would use waste heat from the first, the third collect waste from the second, and so on. Computer scientist Robert Bradbury took the name for this even grander vision from the Russian dolls it slightly resembles: a "matrioshka brain".

Would these megabrains devote a small portion of their super-thoughts and vast resources to building and despatching von Neumann probes to other star systems? The obvious answer is that they will if they want to. But imagining what it might mean for such an entity to want something, and what its wants might be, is an essentially pointless exercise – think about explaining human needs and desires to a bacterium. However, perhaps what matters is that it's possible in principle. Furthermore, if the matrioshka brains cared about their own long-term survival, travel beyond the solar system would eventually become necessary – either because all the planets had been used up or, further ahead, because the Sun was nearing the end of its life.

These and many other possibilities are discussed in other essays in Broderick's collection – which unfolds speculations extending far beyond the million years of its title. The details are fun to explore and, often, to disagree with – though it's sometimes as hard to find reasons to disagree with them as to believe them. They are so far outside our present capability that they read like fictions even when they're presented as logical outcomes of clearly stated assumptions.

However, one basic point does emerge from works like this about far-distant futures (in time as well as space). The fate of an uninhabited planet around a reasonably well-understood star is predictable in the long term with some precision. But if there is life on that planet, and it manages to survive, that makes the range of possible outcomes almost infinitely more interesting.

PREDICTING THE FURTHEST FUTURE

Just suppose we emerge from the perils of the twenty-first century with a planetary civilization in sustainable shape, that technology develops, that we eventually find reasons to move into space and colonize the solar system, or even the galaxy. That final step would mean our ultra-distant descendants would not face extinction with the demise of the solar system. But how long might they, or anybody else, then go on? Forever, some say.

Looking further ahead in time entails dealing with a widening cone of uncertainty, an expanding array of possible futures. That's why most of what this book says about the years after 2050 comes with more caveats than the forecasts for the first half of the century. But that doesn't necessarily mean it's pointless to speculate about the furthest future. We cannot know what intelligent beings – supposing there are any – might be thinking about as the universe reaches its old age, billions of years from now. But we can ask whether the conditions for them to have thoughts might still exist. If so, then there might even be some thread of continuity between their thoughts and ours.

One reason our own thoughts can range into the far future is that we explore these regions mathematically. If you have a mathematical framework set up correctly, you need nothing more except the ability to do the calculations. This is the style of investigation that Galileo and Newton convinced everyone was the best approach in the sixteenth and seventeenth centuries. The "book of nature", they argued, was written in the language of mathematics. And the predictive power of their theories, especially Newton's mechanics, seemed convincing. If an equation – or, more likely, a set of equations – contains the time variable "T", it does not matter what value T takes.

This assumes that the laws of physics stay the same forever. That assumption breaks down into two: that the same laws operate, and that our understanding of them stays the same too. The second one is questioned more often than the first, though neither is bulletproof. After all, Newton's laws turned out to be wrong. But they weren't completely wrong – Einstein's theory of gravity, which superseded Newton's, made only modest corrections to the earlier predictions most of the time.

For now, Einstein's relativistic equations are the main tools for predicting the far future of the universe as a whole. They have proven so accurate in so many contexts that it is hard to imagine them being displaced by anything radically different. There may well be deeper physics to be discovered, especially physics that will unite the large-scale theories of Einstein with the microworlds of quantum mechanics. But it will surely be a unification that incorporates these theories rather than consigning them to history.

WORLD WITHOUT END

Religion is one source of ideas about the ultimate fate of the universe. The non-religious answers come from the modern science of cosmology – which is largely a creation of the second half of the twentieth century. As sciences go, it is still a fairly ramshackle affair, but its fixation with some of the big questions of existence means that there are hundreds of pop-science accounts of cosmology – every professional cosmologist seems to write one sooner or later. Nowadays, they nearly all agree that the universe we live in had a beginning – the famous Big Bang – around fourteen billion years ago. Putting big-bang theory at the heart of modern cosmology – instead of the rival "steady-state" theory that sees it going on forever – immediately raises the crucial big question: will the universe end?

A key fact here is that the visible universe is expanding, as first pointed out by Edwin Hubble from his survey of galaxies in the 1920s. For a long while afterward, the odds were even as to whether the expansion would slow and then fall into reverse – gravity eventually dragging all the mass in the universe back together into a big crunch – or go on indefinitely. The result appeared to depend on the total amount of stuff, a hard quantity to measure, and most discussions left it that the universe might collapse in, say, ten to twenty billion years, or might not. Whether the big crunch, presumably into an all-consuming black hole, would be an end point or the start of something else (a big bounce?) was also uncertain, because theory breaks down under the unimaginably extreme conditions of the beginning and possible end of the universe.

The question is not settled beyond dispute, but opinion has shifted in the last decades because of evidence that the rate of expansion is not slowing, as we would expect, but accelerating. How do we know? Painstaking measurements of very distant – and therefore very old – supernovae showed they were receding from us faster than predicted by previous theory. The universe appears to harbour a hitherto

DARK MATERIALS

The advent of dark matter (invoked to explain why the dynamics of most of the agglomerations of stuff – galaxies and so on – in the universe makes them seem heavier than they appear) and dark energy has made the cosmology of the future more complicated. Between them they account for more than 95 percent of what the universe is made of, so it turns out we know much less than we thought we did about our cosmic habitat. The possible endgames also depend on key assumptions about dark energy, dark matter and their precise properties.

The additional possibilities, aside from the big crunch and the big chill, were usefully catalogued in a 2006 paper by Rüdiger Vaas of the University of Geissen, Germany. They include a big rip, in which dark energy causes expansion to accelerate and destroys space (and matter), and a big splat. The latter arises out of ideas about so-called superstring theory, which envisages that the ultimate particles of which matter and energy are composed exist in extra dimensions beyond the three we are normally aware of. This raises the possibility of collision with another universe, which exists in four dimensions inside a higher-dimensional space. (Don't try and imagine it: you can't.) The result, Vaas explains helpfully, would seem to us like the big crunch, followed by a new Big Bang.

There are other ideas on the wilder shores of cosmology, too. They range from the universe suddenly ending, due to instabilities in the structure of matter and energy, to taking the notion of infinity really seriously. The latter implies that all possible configurations of matter exist and, furthermore, that they all exist an unlimited number of times. In other words, the universe extends to uncountable other versions of you, reading this book, right now and in the future. These and many other such notions are summarized with brio in Marcus Chown's *The Never Ending Days of Being Dead* (2007). The trend he seems to confirm is that our ideas about the cosmos are getting stranger and stranger.

unknown repulsive force which is countering gravity's effort to halt the expansion of space that began with the Big Bang.

All this depends on something called "dark energy" which, if it exists, must account for around three quarters of the energy density of the universe. As we do not know what dark energy is, a cosmology that relies on it may not be a good guide to the future. Still, it seems at the moment to tip the balance in favour of a universe with no end in sight. This is the big chill, or perhaps the big whimper. It is not the only possibility, because our ignorance about the nature of dark energy – and the equally mysterious dark matter – leaves the field open for some other, new speculations (see box above). But it appears the most likely.

DARK, COLD AND VERY OLD

The big whimper is a far from exciting future prospect. In fact, it resembles the gradual running-down of the cosmic clockwork first theorized by nineteenth-century physicists. The heat death (that is death *of* heat, not *by* heat) of the universe implied by the laws of thermodynamics – which impressed the Victorians just as they were grappling with

PREDICTIONS FILE

Paul Davies

Physicist, professor at Arizona State University and popular author

Highest hope: Finding evidence of life beyond Earth. The decisive test is whether we can find a second sample of life, that is, a form of life that has emerged from scratch via a second genesis, independently of the genesis of life as we know it. It could come from the discovery of life on another planet, from the creation of life in the laboratory, or, in my view more likely, the discovery of a shadow biosphere on Earth, populated by microbes that are a form of life sufficiently different from our own that it must have arisen independently. Scientists have just begun searching for a shadow biosphere. If they find one, it will be the greatest discovery in biology since Darwin's and will strongly suggest that life is widespread in the universe.

Worst fear: As the techniques of synthetic biology become more sophisticated and cheaper, someone will deliberately or inadvertently create a new micro-organism that has unforeseen and devastating consequences for the regular biosphere.

Best bet: I think there is a good chance that we will discover evidence for multiple origins of life on Earth, and perhaps we will also obtain evidence for life on Mars.

the discovery of deep time – was a prediction of principle, not of any particular timescale (see Chapter 2). Now we have a much better idea of the timescales that might be relevant. If there is no big crunch, they are almost immeasurably long – much, much longer than the length of time since the Big Bang, currently reckoned at around fourteen billion years.

Such timescales certainly extend further than the life of stars – around ten billion years for an average-sized star like our Sun. It is far greater for the smaller, slower-burning red dwarfs, which can last for perhaps ten trillion years. But the universe will go on far beyond the life of any stars. At the moment, new stars are being formed from coalescing clouds of hot gas, but eventually, in perhaps ten red-dwarf lifetimes, star formation ceases. The remaining solar furnaces then burn themselves out and the universe enters an era in which the only big concentrations of matter are the remnants of dead stars – black holes, super-dense neutron stars or white dwarfs, depending on the mass of the original star.

This dark prospect, a hundred trillion years or so in the future, is not the end of the slow decay predicted for an expanding universe. Not only does everything carry on moving apart, but the larger elementary particles begin to decay. Although utterly stable for all normal purposes, physics predicts that the proton, one of the two kinds of elementary particles in the nuclei of atoms, eventually breaks down. It has a half- life (the time for half of a given sample to decay) of somewhere around 10^{35} years, so you would need a lot of patience to spot one decaying. But in time, theorists say, they will all break down into lighter subatomic particles.

The other nuclear particle, the neutron, disappears too. Neutrons inside atoms are stable, but free neutrons decay rather quickly into a proton and an electron. The result: after enough proton half-lives have passed there is essentially nothing left to make atoms with. There follow yet further stages, involving accretion of the remaining particles by black holes, which themselves gradually evaporate. The terminal state is an endless vista of cold and dark, with an incredibly thin population of photons. And that's it. This might not mean the end of all universes, as some cosmologies speculate as to the appearance of new big bangs from quantum fluctuations of the vacuum (yes, really). But it would be the silent fate of this one.

LIFE EVERLASTING?

So much for the fate of the universe – which appears uninspiring to say the least. Does that mean that life is also bound to end? A cosmos devoid of stars, planets or even atoms certainly makes it sound like it. But some heroically speculative thinkers have tried to imagine ways in which some kind of life might endure. In practice, they're concerned with the life of the mind, which they interpret, in twenty-first-century fashion, as something that can be reduced to information processing. This kind of argument about the far future follows on from the ideas about computing devices of some kind being the most basic foundation for thinking.

The first detailed consideration of the really long-term prospects here was offered by Freeman Dyson in the now classic paper *Time Without End: Physics and Biology in an Open Universe* (1979). He described the same ideas a bit more accessibly in one of the lectures published in his book *Infinite in All Directions* a decade later. As these titles suggest, Dyson was postulating a universe with no big crunch, and looking at ways in which life could go on indefinitely. As a physicist, Dyson ignores all the details of how life, or intelligence, actually operates and focuses on what he regards as the essentials. In his view, life can be viewed abstractly, ignoring its molecular structure, as a system for processing information that has a measurable complexity. He argues that life, in this abstract sense, can adapt to any environment whatsoever, as long as it contains both matter and energy.

For life, or information processing, to continue, however, there must be differences in energy between one point and another – this is what allows work (again in the physicist's technical sense) or information processing to be done. At first glance it looks as if the ageing universe would rule this out, because temperature differences even out to become immeasurably small. This is what heat death means. However, Dyson, using plausible approximations, reckoned that the energy needs of complex minds are actually pretty small. He estimated, for example, that:

> "For a society of the same complexity as the present human society on Earth, starting from the present time and continuing forever, the total reserve of energy required is about equal to the energy now radiated by the Sun in eight hours."

The second crucial fact about this universal, abstract life is that in an infinitely long future, life can be infinitely patient. In fact, if there is not enough energy immediately on offer to enable it to formulate its next thought, it can shut down for a while – or for billions of years if necessary. In that time, further cooling will have created new temperature differences, and the engine of thought can be started up again. Dyson's infinite future thus involves extremely lengthy bouts of cosmic hibernation punctuated by fleeting instants of consciousness.

Never mind, there is all the time in the universe for these to add up to future thoughts of arbitrary complexity. His conclusion is unequivocal: the ultimate state of the universe is permeated by mind, although incredibly tenuously by our standards, and lasts forever. As he puts it, more positively:

> "No matter how far we go into the future, there will always be new things happening, new information coming in, new worlds to explore, a constantly expanding domain of life, consciousness and memory."

Others pointed out a possible flaw in Dyson's scheme. Some signal is needed to wake the universal mind from hibernation. And it would need to work an infinite number of times – this is one cosmic alarm clock. Could such a device exist that did not itself need an inexhaustible supply of energy? Dyson rose to the challenge, but it's not clear that his design would be durable. According to fellow physicist Laurence Krauss, it would have a limited life because of quantum-mechanical effects.

The fact that this debate got into that kind of detail suggests there is nothing fundamentally wrong with Dyson's approach that is easy to spot. However, there is also general agreement that Dyson's calculations, originally done before the discovery of dark energy, only work in a universe that continues to expand slowly. In the one we see now, in which the expansion is accelerating, the number of thoughts it can support is finite, and there is no way for life to carry on beyond that limit, however many trillions of years ahead it turns out to be.

HEAVEN AT THE OMEGA POINT

Dyson's infinite outlook is intriguing, but unsatisfying to some. What use is endless thought if it just gets slower and slower? One cosmologist who confronted that question directly is Tulane University professor Frank Tipler. His speculations assume a big crunch, not an open universe, and – seemingly paradoxically – allow for eternal life in this finite cosmos. In fact, he argues, this is not just possible but inevitable. Tipler's remarkable book *The Physics of Immortality* (1994) was far out in every possible sense of the term. There, Tipler out-Dysoned Dyson by depicting a future universe in which everyone who has ever been alive is recreated, to enjoy an endless new life that is infinitely more pleasant than the one they had first time round. "Scientist Predicts Heaven", in short.

How will this happen? The argument is long, and depends on a host of assumptions that have been questioned by other physicists, but it is scientific… sort of. Perhaps inspired by Dyson's declaration that "it is impossible to calculate in detail the long-range future of the universe without including the effects of life and intelligence", Tipler tells us that life will manage the collapse of the cosmos to its own advantage. As before, the key is maintaining differences in temperature in a situation in which they might disappear – in a big crunch, everything tends to get unimaginably hot.

Tipler points out that some regions would still stay hotter than others if the great collapse was asymmetric – that is, if things fall in on themselves faster in one direction than others. Left to itself, the universe is unlikely to behave this way, as matter is fairly evenly distributed on the largest scales. But it will not be left to itself, he argues, because intelligence will take on the job of moving it around to get the result it wants.

That's right: Tipler's scheme calls for an intelligence that has colonized every galaxy to engineer the universe. The reward would be an immense supply of energy. And that could be used not to live forever, because this universe would be heading for a terminus, but to create the experience of living forever. The idea is that the vast temperature differences will create a source of energy for unlimited information processing, which can also be arbitrarily fast. The less time is left, the faster it will get, so the amount of "thinking" that is possible during the collapse is infinite.

Tipler calls this the "Omega Point", a term he took from the twentieth-century Catholic mystic Teilhard de Chardin. At the Omega Point, intelligence has the essential attributes of God. It's all-knowing and all-powerful. Among other things, it can recover information from the past about all the beings who ever existed, and recreate them in perfect simulation. This is a perfect, secular version of resurrection. It requires no sinner's repentance or redemption: it is guaranteed. Tipler says that the fact that the Omega Point can exist, according to his equations, means that it must exist. And it is the mission of intelligent life to ensure that it happens.

This extravagant vision, laid out over four hundred pages with a hundred-page mathematical appendix, was described by one scientist in the journal *Nature* as a "masterpiece of pseudoscience", which seems fair enough. For the conventionally religious, it must be a bit bemusing. Theologically, as astrophysicist and Christian Donald York put it with commendable restraint, "there is abundant material here for intellectual offense". It is interesting, though, how much Tipler and the rather more sober Dyson have in common as they try to outline a scientific eschatology. They are both assuming that:

Scientist predicts heaven: How different is cosmologist Frank Tipler's vision of resurrection from Gustave Doré's nineteenth-century depictions for Dante's *Paradiso*?

- Life is about structure, not about the matter that happens to contain that structure.
- The structure that is important is information which can be captured by computation.
- Computers can become conscious, and computer consciousness can eventually be installed in, say, a cloud of galactic dust.

▶ Intelligence defined in this abstract way can colonize the whole universe, by sending information across interstellar, and ultimately intergalactic, space.

These are all very modern assumptions, which are embodied in the idea of von Neumann probes being a potential solution to the problems of life escaping from the solar system. It will be interesting to see how they hold up in the next few decades. In judging them now, bear in mind two things. First, they go along with arguments that computer intelligence will at some time – perhaps some time soon – advance into realms beyond the capabilities of human brains. Such an advanced intelligence might or might not share our interest in the possibility of occupying as much of the universe as possible for as long as possible.

On the other hand, there is a Darwinian twist to the argument for computerized intelligence taking over the cosmos. If the evolution of life on Earth did not begin with a unique event, then there are probably a huge number of planets with intelligent beings. (It seems a bit daft to maintain, for example, that if there turns out to be more than one, there might be only two or three.) If so, then a cosmic version of natural selection will apply. Most intelligences beyond our ken may elect to spend the rest of eternity contemplating their navels, because that is the most fulfilling way they can conceive of spending their time. But if just one gets busy building von Neumann probes and establishes a wave of colonization, the universe still eventually ends up pervaded by their descendants, informationally speaking.

In the end, it's best not to take any of these speculations too seriously. The sentence in Dyson's original paper that is definitely incorrect is surely the one that says of whether he is right, "one day, before long, we shall know". We, or perhaps some being, might know, but it will be a very, very long time before the verdict is in.

Meanwhile, spinning these ideas offers testament to the strength of the impulse to project an indefinite future for consciousness. And it underlines this curious fact about humanity: that once we acquired the ability to project beyond the present, to imagine the course of future events, there is no limit to how far that imagining can reach.

FURTHER EXPLORATION

www.spacefuture.com A highly optimistic view aimed at "everyone who wants to go into space".

www.nasa.gov NASA, ever hopeful, maintains an elaborate website on reasons and means for space settlement.

Damien Broderick (ed) Year Million: Science at the Far Edge of Knowledge (2008) A dazzling collection of essays on life, the universe and everything, in which Year Million is only the beginning of the fun.

Paul Davies The Last Three Minutes (1997) A little less up to date than Gribbin's book (below) on the latest speculations, this is still a clear account of possibilities for the ultimate fate of the universe.

John Gribbin The Universe: A Biography (2007) There are many popular accounts of cosmology – maybe too many – but this is a good one to start with.

Michael Hanlon Eternity: Our Next Billion Years (2009) A knockabout look at earthly futures, in the period defined by the subtitle, rather than the title.

Chris Impey How It Ends: From You to the Universe (2010) A cosmologist's take on the fate of all things, and how we might postpone our own end as a species as long as possible by venturing into space.

picture credits

Front cover © Chad Baker/Getty Images; back cover Space tourism on Mars © Detlev van Ravenswaay/Science Photo Library, Micro-syringe machine with red blood cells in the body (conceptual computer artwork) © Coneyl Jay/Science Photo; inside front cover *Painting of Futuristic Buildings and City* by Anton Brzezinski © Forrest J. Ackerman Collection/Corbis; p.1 Star birth in the Carina Nebula © STScI/NASA/Corbis; p.6 © Danny Lehman/Corbis; p.11 © Rolfe Horn, courtesy of The Long Now Foundation; p.13 © Forrest J. Ackerman Collection/Corbis; p.16 © Bettmann/Corbis; p.25 © Bettmann/Corbis; p.27 © Corbis; p.31 *The Time Machine*, Moviestore © Loew's Incorporated and Galaxy Films, Inc, MGM/George Pal Productions; p.34 © Corbis; p.36 *Parallel Universes* © Mehau Kulyk/Science Photo Library/Corbis; p.39 © Hulton-Deutsch Collection/Corbis; p.49 © Bettmann/Corbis; p.55 © Bettmann/Corbis; p.59 Futurama 1939 by Norman Bel Geddes © Corbis; p.63 © C.P. Cushing/ClassicStock/Corbis; p.65 © Bettmann/Corbis; p.66 © Bettmann/Corbis; p.69 © Martial Trezzini/epa/Corbis; p.75 © Robert Landau/Corbis; p.78 silicon chip © Charles O'Rear/Corbis; p.80 © Larry Downing/Reuters/Corbis; p.83 Mauro Fermariella/Science Photo Library; p.86 Friedrich Saurer/Science Photo Library; p.91 © Charles O'Rear/Corbis; p.95 Downtown Tokyo, Japan, at dusk © James Leynse/Corbis; p.104 © Karen Kasmauski/Science Faction/Corbis; p.113 © Frédéric Soltan/Corbis; p.116 Floating pack ice in Lady Franklin Fjord, Svalbard, Norway © Paul Souders/Corbis; p.122 © Mike McQueen/Corbis; p.127 © Oswaldo Rivas/Reuters/Corbis; p.130 © George W. Wright/Corbis; p.134 The LUZ Solar thermal electric generator, Kramer Junction, California © Roger Ressmeyer/Corbis; p.139 © Gavin Hellier/Robert Harding World Imagery/Corbis; p.142 © Hank Morgan Rainbow/Science Faction/Corbis; p.145 © Joby Energy; p.150 © Institution of Mechanical Engineers (IMechE); p.157 Aral Sea, Kazakhstan © Gerd Ludwig/Corbis; p.161 © George Steinmetz/Corbis; p.165 © LifeStraw, Vestergaard Frandsen; p.169 © George Steinmetz/Corbis; p.171 Soybean harvest, Mato Grosso, Brazil © Paulo Fridman/Corbis; p.175 © Diego Azubel/epa/Corbis; p.181 © Andy Aitchison/In Pictures/Corbis; p.189 © NASA/courtesy of nasaimages.org; p.190 © Oliver Foster, Queensland University of Technology, courtesy of www.verticalfarm.com; p.194 Diver off Bermuda © Julian Calverley/Corbis; p.198 © Anup Shah/Corbis; p.200 © Thomas Marent/Visuals Unlimited/Corbis; p.207 © Jeff Barnett, Barnett Apiaries, Valdosta, Georgia; p.211 © Paulo Fridman/Corbis; p.214 Stem cells © MedicalRF.com/Corbis; p.218 © Bobby Yip/Pool/epa/Corbis; p.226 © Bettmann/Corbis; p.229 © SRI International; p.233 © Mark Thiessen/National Geographic Society/Corbis; p.238 Fire Scout Firing Rocket © US Navy - digital version copy/Science Faction/Corbis; p.241 © Jacques Langevin/Sygma/Corbis; p.246 © Reuters/Corbis; p.248 © Bettmann/Corbis; p.251 © Alaa Al-Shemaree/epa/Corbis; p.257 Giant asteroid collides with Earth © Stocktrek Images/Tobias Roetsch/Stocktrek Images/Corbis; p.265 © (C) Guo Jian She/Redlink/Redlink/Corbis; p.270 Deep Impact, Moviestore Collection, Paramount Pictures/DreamWorks SKG/Zanuck/Brown Productions/Manhattan Project; p.275 Richard Garriott's Tabula Rasa, published by NCsoft, developed by Destination Games; p.277 Le Palais Bulles, Theoule-sur-Mer, France by Pierre Cardin and Antti Lovag © Eric Robert/Corbis SYGMA; p.279 © Bettmann/Corbis; p.286 © Ashley Cooper/Corbis; p.295 © Valentin Flauraud/Reuters/Corbis; p.296 Computer centre at CERN, Geneva, Switzerland © Martial Trezzini/epa/Corbis; p.300 © c,mm,n 2010; p.305 © www.associationofvirtualworlds.com; p.309 *The Truman Show*, Moviestore Collection © Paramount Pictures Corporation/Scott Rudin Productions; p.312 © Bettmann/Corbis; p.315 Technological human brain © Image Source/Corbis; p.319 © Handout/Reuters/Corbis; p.325 *Young Frankenstein*, CinemaPhoto/Corbis, © Twentieth Century-Fox Film Corporation, Gruskoff/Venture Films/Crossbow Productions / Jouer Limited; p.330 © Michael Macor/San Francisco Chronicle/Corbis; p.333 Supernova explosion (computer artwork) © Mehau Kulyk/Science Photo Library/Corbis; p.338 © NASA/courtesy of nasaimages.org; p.340 NASA/courtesy of nasaimages.org; p.344 © NASA/JPL/Cornell University/courtesy of nasaimages.org; p.353 © Bettmann/Corbis

index

A

Abrahamic religions 17
Accelerando (2005) 329
acidification of oceans 129, 149, 201, 261, 336
Adamchak, Raoul 184
Adams, Douglas 7
Addis, Donna 5
Advancement of Learning, The (1605) 19
Afghanistan 108, 242, 244
Africa 92, 109, 110, 163, 185, 244
ageing 91, 101, 223, 231, 235
agriculture 120, 159, 170, 176, 177, 204
AIDS 105, 109, 216, 228, 262
airships 54, 63
air travel 138
alcohol 220, 234
Alcor 331
algae 146, 191
Algebraist, The (2004) 291
alien life 342
Alkon, Paul 19
Allen, Tony 167
Al-Qaeda 254
Alzheimer's disease 234
Amazing Stories magazine 248
Amazon rainforest 209
American Council on Science and Health 186
amphibians 200
Andreadis, Athena 50
Andrews, Arlan 254
Angell, Roger 149
Another Bloody Century: Future Warfare (2005) 239
anthrax 255
antibiotics 219
Anticipations (1901) 32
Apocalyptic AI (2010) 291
Appeal to Reason, An (2008) 131
aquifers 161, 166
Arctic National Wildlife Refuge 124
Arkin, Ron 251
Armageddon (1998) 271
Arnold, General Hap 49
artificial intelligence (AI) 83, 89, 291, 317, 320, 322, 324, 326 *see also* robots and computing
artilects 327
Artilect Wars, The (2005) 327
Association of Virtual Worlds 305
asteroids 258, 267
Atlantic Ocean 335
atomic power 64, 131, 141, 143
atomic weapons 33, 252, 271
Atwood, Margaret 2
augmented reality 307
automobiles 300
aviation 138
Aymara Amerindians 12

B

Bacon, Francis 19
bacteria 119, 219, 336
Badgley, Catherine 181
Balmer, Edwin 269
Bangladesh cyclone 264
Banks, Iain 291
Barnaby, Wendy 167
Barrett, Peter 114
Baxter, Stephen 125
Bear, Greg 255
Beddington Zero Energy Development (BedZed) 286
bees 207
Belkin, Danny 317
Benford, Gregory 156
Berger, Peter 291
Berkowitz, Bruce 247
Bernal, J.D. 62, 317
Berry, Adrian 339
Berube, David 88
Beyond the Limits (1992) 67
Bible, the 7
Big Bang 43, 291, 348
big crunch 348, 352
Bigelow, Robert 341
Bigelow Aerospace 341
Bill and Melinda Gates Foundation 186
biodiversity 194–213
biofuels 140, 146, 152, 174, 178, 179
biological disasters 261
biological weapons 252
biomass 120, 122
bionic limbs 233

biophilia 205
biosphere, the 118, 128, 196
biotechnology 80, 166, 180, 245, 318, 320
birth rate 96, 99
Black Death, the 217
black holes 350
Blade Runner (1982) 208
Blakemore, Colin 225
"Blueprint" scenario 152
Bodin, Félix 21
Book of Dave, The (2006) 125
Bostrom, Nick 70, 261, 275
botulism 253
Boyle, T.C. 125
Bradbury, Robert 347
brain-enhancing drugs 232, 234
brain science 89, 313, 318, 320, 324
  see also neuroscience
Brand, Stewart 112
Branson, Richard 341
Brave New World (1932) 62
Brazil, deforestation in 210
Brazilian Agricultural Research Corporation 180
Brin, David 289
Broderick, Damien 322, 345
Broecker, Wallace 148
Brown, Lester 165, 175, 184
Bruno, Giordano 19
Bryson, Bill 64
Butler, Samuel 30

C

California 138, 142
California water wars 167
Calvin, William 168
cancer 216, 235
capital investment 126
car accidents 216
cars 137, 300
carbon dioxide 118, 128, 131, 135, 144, 152, 335
  carbon capture 144, 147, 149
  carbon markets 144
  carbon pricing 141
  carbon sinks 137
  emission targets 137
  productivity 137
Carr, Nicholas 313
Carrico, Dale 328
carrying capacity of the Earth 22, 70, 98, 114
Carson, Rachel 42, 203
Cascio, Jamais 76
Catholic Church 19
Cat's Cradle (1963) 291
CCTV cameras 311
Center for Strategic and International Studies 101
Centers for Disease Control 220
Centre for Research on Epidemiology of Disasters (CRED) 259, 261
Centre for Responsible Nanotechnology (CRN) 52
CERN (European Organization for Nuclear Research) 262
CFCs 118
chaos theory 39
Chardin, Teilhard de 353
Chatham House 192
Chicago World's Columbian Exhibition (1893) 26, 65
Chicxulub crater (Mexico) 267
child mortality 216
China 92, 105, 139, 163, 175, 176, 184, 243, 259, 265, 288, 293
Chown, Marcus 349
Christian, David 9
Christianity 290
cities 64, 110, 112, 167, 204
Citizen Cyborg (2004) 328
civil society 284
civil war 241
Clark, Andy 321
Clarke, Arthur C. 268
Clarke, I.F. 16, 21
climate change 110, 116–133, 162, 176, 210, 220, 261
Clock of the Long Now 11
CO_2 118, 128, 131, 135, 144, 152, 335
coal 120, 122, 144
colds 226
collectivism 303
Coming Convergence, The (2008) 90
Commission on the Prevention of Weapons of Mass Destruction Proliferation and Terrorism 253
commodity prices 193
common goods 284
computer games 304
computing power 82
computronium 345
Comte, Auguste 42
Condorcet, Marquis de 22, 42
convergence 90, 318, 320
Cool It (2007) 131
Copenhagen Climate Change conference 132, 210
coral reefs 199
Cornucopians 68, 75
cosmic hibernation 352
cosmology 348
Crichton, Michael 125, 199
Criswell, David 147
critical information infrastructure 249

crop hybrids 205
crop yields 129
cross-impact analysis 48
Crowdsourcing (2008) 303
Crutzen, Paul 148
cryonics 330
Crystal, David 293
Cuban Missile Crisis 49
culture 284, 296–314
cyanobacteria 119
cyborgs 90, 320
cyclones 264
cystic fibrosis 228

D

dam collapse, China (1975) 259
dark energy 349, 352
Darwin, Charles 24, 29
Darwin mission 342
data vault 282
Dator, Jim 70
Davies, Paul 350
Deep Green software project 247
Deep Impact (1998) 271
deep time 8, 28
Defense Advanced Research Projects Agency (DARPA) 229, 244
deforestation 209, 210
Delphi, Oracle of 15
Delphi technique 46, 72, 239
Demeny, Paul 102
democracy 43, 289, 304
depression 216
Depression, The Great 63
desalination 166
Desertec project 142
detergent 151
diabetes 221, 235
Diaspora (1997) 329
"Digital Gaia" 326
dimethyl sulphoxide 128
dinosaurs 267
disaster planning 258
disasters 257–276
Dix, Alan 299
Doogie mice 318
DNA 199, 227, 318
DNA sequencing 80
Doctorow, Cory 329
Don Quijote space mission 271
Down and Out in the Magic Kingdom (2003) 329
Dozois, Gardner 329
Drexler, Eric 87, 88
drought 110, 129, 162, 261
drugs 219, 232, 234
Dubailand 206
Dunster, Bill 286
dynamo 24
Dyson, Freeman 213, 240, 346, 351
Dyson sphere 346

E

earthquakes 264
Eco, Umberto 54
eco-homes 112, 285
economic growth 70
ecosystem services 206
Eco-Tourism World 206
Edgerton, David 93
Edison, Thomas 26
education 73, 284, 287, 313
Egan, Greg 329
Ehrlich, Paul 66, 69, 208
Einstein, Albert 348
elderly, the 91, 101, 223, 231, 235
electricity 120, 123, 141, 142, 143
Encyclopedia of Life project 197
End of Food, The (2008) 186
End of History and the Last Man, The (1992) 43, 289
End of Science, The (1996) 326
endocrine disruptors 203
energy 29, 62, 71, 74, 116–133, 134–156, 346
  alternative 126, 131, 134–156
  conservation 140
  efficiency 137
  prices 141
  security 126
  supply 141
  use 120, 137
Energy at the Crossroads (2005) 124
Engines of Creation (1986 & 2006) 87
Enlightenment, the 17, 21, 23
entropy 29, 118
epigenetics 225
Epigone, histoire du siècle futur (1659) 19
Erewhon (1872) 30
Erhlich, Paul 174
Eskow, Richard 310
Essay on the Principle of Population, An (1798) 23
Ettinger, Robert 330, 331
Europe, population of 101
European Organization for Nuclear Research (CERN) 262
European Space Agency 271, 342
Evans, Alex 140
Evans, Robert 141
evolution 28, 29, 32, 195, 198, 235, 317, 327, 345, 354
extinction of species 29, 195, 197, 202
extinction, voluntary 263

F

Failed States Index 244
families 100, 103, 281, 287
farms, vertical 191
Feather, Judith Light 81
feedback loop 128
Feeding the World (2000) 184
Fermi, Enrico 342
Ferris wheel 26
fertility 99, 102
fertilizer 179, 182
Fifty Degrees Below (2007) 125
fish farming 175
Fisk, David 67
Fixing Climate (2008) 148
flash floods 162
Flesh and The Devil, The (1929) 62
Flood (2008) 125
Flood, The (2004) 125
floods 260
flu 217, 219, 262
Flynn, James Robert 313
Fogel, Robert 231
food 23, 74, 171–193
  prices 173
  production 109, 140
  security 168
Food and Agriculture Organization (FAO) 173
Ford, Henry 25
forecasting 38, 40, 47, 56, 68
Foreign Policy magazine 244
Foresight 46
forest fires 127
forests 199, 210
Forster, E.M. 62
Forty Signs of Rain (2004) 125
fossil fuels 117, 137, 124, 141
Freitas, Robert 230
Fresco, Louise 182
Freud, Sigmund 7
Friend of the Earth, A (2000) 125
Fuelling Our Future (2007) 141
Fukuyama, Francis 43, 289
Futurama exhibit 64, 65
Futurama 75
Future Combat Systems project 249
Future of the Internet and How to Stop It, The (2008) 303
Future Shock (1970) 280
Futurecast 2020 (2008) 103
Futurelab 279
Futures of the Wild project 212
futurology 36–58

G

Gaia theory 128, 149
Galactic Suite Space Resort 341
galaxy M31 337
Galileo 348
gamma radiation 274
Garis, Hugo de 327
gas 120, 122, 152
Gee, Maggie 125
Gelernter, David 65, 298, 301, 306
General Motors 64
gene therapy 227
genetic engineering 81, 212
genetic information 227
genetic modification 183, 184, 185, 187, 188, 228 *see also* biotechnology
geoengineering 148, 155, 336
geology 8
geophysical disasters 261
Geraci, Robert 291
Gibson, William 306
Gilbert, Daniel 3
gizmos 84
Global Biodiversity Outlook (2010) 196, 209
Global Canopy Programme 210
Global Catastrophic Risks (2008) 262
Global Convention on Biodiversity 196
Global Earthquake Model 266
global governance 155, 282
global heating 336
Global Hunger Index 173
globalization 154, 283, 287, 293
Global Viral Forecasting Initiative 217
global warming 117, 135, 153, 219, 261
  see also climate change
GM crops 185
God 353
Going Global (2008) 287
Good, I.J. 322
Goodstein, David 124
Google Earth 187, 306
Google 218, 298, 313
Gould, Stephen Jay 7
Governing Lethal Behavior in Autonomous Robots (2009) 251
GPS 301
Grainger, Alan 200
Grangemouth Oil Refinery 122
Gray, Colin 239, 250
Great Cyber Games 304
Greatest Benefit to Mankind, The (1998) 216
Great Exhibition (Britain, 1851) 26
Greeks, Ancient 15
Greeley, Hank 233
Greenfield, Susan 312
greenhouse effect 117, 335
greenhouse gas 127, 129, 135, 137, 148, 155, 178 *see also* carbon dioxide

"green new deal" 293
Grey, Aubrey de 232, 236
Grown Up Digital (2008) 314
Guttin, Jacques 19

H

Haber-Bosch process 179
habitat loss 202
Haiti earthquake 265
Haldane, J.B.S. 61, 80
Hammer of God, The (1973) 268
Hammes, Colonel Thomas 242
Hanson, Robin 326
Harris, Steven 345
health 214–237
Health and Safety Executive (HSE) 51
health care costs 217, 231
heart disease 216, 220
heaven 352
Heat (2006) 138
Heilbroner, Robert 15, 75
Heinlein, Robert 230
Heisenberg, Werner 41
Helmholtz, Hermann von 29
Here Comes Everybody (2008) 303
Herren, Hans 173
Hesketh, Therese 107
Hillis, Danny 10
Hinduism 290
Hindenburg 55, 63
Hiroshima 63, 64
History of Science Fiction, The (2006) 17
history of the future 6, 13–34, 42, 57, 299
HIV/AIDS 105, 109, 216, 228, 262
Hoban, Russell 125
Holland, John 304
Hollander, Jack 174
Holy Fire (1996) 237
Hood, Leroy 225
Horgan, John 326
horizon-scanning 47
hormones 203
housing 56, 112, 149, 285
Howe, Jeff 303
Howe, Leo 56
Hubbert, M. King 124
Hubble, Edwin 348
Hudson Institute 50
Hughes, James 327
Human Development Report, UN 162
human footprint 203
Human Genome Project 81, 225, 345
Human Security Report 240
Hume, David 54
Hurricane Katrina 261, 264
Hutton, James 7, 8
Huxley, Aldous 62
Huxley, Thomas 32
hydrogen 119, 141
hydro-meteorological disasters 261
hydroponics 190

I

iBrain: Surviving the Technological Alteration of the Modern Mind (2008) 313
ice age 335
icecaps 128
I Ching 15
IEEE Spectrum 325
Inayatullah, Sohail 282
India 99, 107, 163, 288
individualism 281
industrial revolution 19, 71, 137
inequality 71, 176, 292
Infinite in All Directions (1989) 351
inflation 132
influenza 217, 219, 262
information 296–314
information beam 298
Institute for Ethics and Emerging Technologies 85
Institute for Soldier Nanotechnologies 246
Institute for Systems Biology 225
Institute for the Future 305, 308
insulin 223
insurance industry 263
intelligence amplification (IA) 326
International Census of Marine Life project 201
International Convention on Biodiversity 209
International Energy Agency 120
International Food Policy Research Institute 168, 173, 177
International Monetary Fund (IMF) 66
International Organization for Migration 110
International Space Station 341
International Union for the Conservation of Nature 199
International Water Management Institute 164, 168
Internet-enabled trends 308
Internet of Things 84, 298, 307
Interpretation of Radium, The (1909) 60
invertebrates 198
In Vitro Meat Consortium 191
IPCC (Intergovernmental Panel on Climate Change) 126, 129, 135, 162, 177
IQ 313

Iran, population of 99
Islam 290, 292
Istanbul, earthquake in 264
ITER project 147
IVF (in vitro fertilization) 228

J

Japan 101, 104, 123
Jenkins, Martin 197
Jenner, Edward 21
jet pack 247, 248
Joachim, Mitchell 285
Joby Energy 145
Jones, Jonathan 153
Joy, Bill 75, 93
Jurassic Park (1990) 199

K

Kahn, Herman 49, 68
Kaku, Micho 79
Kardashev, Nikolai 346
Kedzierski, Peter 248
Kelly, Kevin 85
Kelvin, Lord 28
Khanna, Parag 249
Kirkwood, Tom 236
knowledge explosion 302
Kornbluth, Cyril 191
Krauss, Laurence 352
Krebs, John 148
Kunstler, James Howard 124, 245
Kunzig, Robert 148
Kurzweil, Ray 88, 318, 323, 324, 331, 345

L

L'An 2440 (1771) 21
Land Institute 188
Lang, Tim 178
Laplace, Pierre-Simon 39
Large Hadron Collider 69, 262
Last and First Men (1930) 61
Last Man, The (1826) 21, 217
Law of Accelerating Returns 318
Lawson, Nigel 131
League of Nations 66
Left Behind novel series 18
Le Guin, Ursula K. 2
Leitenberg, Milton 254
Le roman de l'avenir (1834) 21
Levy, David 295
Life and Times of the Thunderbolt Kid, The (2007) 64
life expectancy 99, 101, 216, 235
life extension 236
life in the future 277–295
LifeStraw 165
lighting, street 137
Limits to Growth, The (1972) 48, 66, 117
literacy 73, 313
Living Planet Index 197
Lombardo, Tom 33
Lomborg, Bjørn 54, 131
Long Emergency, The (2005) 124
Long Now Foundation 10, 61, 69
Looking Backward (1888) 27
Los Angeles earthquake 264
Lost World of the Fair, The (1995) 65
Lovelock, James 130, 149
Lovink, Gert 308
Lovins, Amory 140
Lucifer's Hammer (1977) 268
Lyell, Charles 7
Lynch, Zack 90

M

M31 galaxy 337
Machine Stops, The (1909) 62
MacKay, David 143, 148
MacManus, Richard 298
maglev train 138
magnetic fields 272
maize 179, 185
malaria 205, 215, 220
Malawi, agriculture in 185
malnutrition 173, 178
Malthus, Thomas 22, 37, 67, 70, 96
Malthusians 23, 68, 75, 114
Man into Superman (1972) 331
Manuel, Frank 42
many worlds theory 273
Maps of Time: An Introduction to Big History (2004) 9
marine species 201
Marooned in Realtime (1986) 329
marriage 100
Mars 337, 339, 342
Martin, James 92, 158
Marx, Karl 3, 43, 294
Masdar City 112
Massachusetts Institute of Technology (MIT) 67
matrioshka brain 347
May, Robert 198, 208
Mayan calendar 6
McCarthy, Cormac 125
McDonald, Ian 125
McDonald's 288
McGuire, Bill 146
McKibben, Bill 131, 204
McKinsey Global Institute 136
McKinsey report 151
McPhee, John 8

megacities 111
megatsunamis 266
membrane technology 165
Mercier, Louis-Sébastien 21
Merkle, Ralph 330
Metaverse Roadmap 307
Meteorological Office (UK) 127
methane 128, 178
Methuselah Foundation 236
mice, experiments on 223, 236, 318, 320
micro-solar power 109
Middle East 163
migration 110
military research and development 229, 244
military robotics 250
Millennium Development Goals 164
Millennium Project 46, 53, 72, 304
mindware upgrades 321
Mitra, Saibal 273
molecular assembler 52, 86
Monbiot, George 138
moon 269, 337, 339
Moore, Gordon 82
Moore's law 57, 82, 93, 299
Moravec, John 306
Morlocks 32, 62
Moynagh, Michael 287
Mulgan, Geoff 294
Muller-Landau, Helene 201
Mumbai 112
Myers, Norman 208, 211

N

Nano! (1995) 88
nanobots 86, 230
nanoengineering 85
Nanohype (2006) 88
nanomedicine 230, 232
nanotechnology 85, 165, 230, 330, 343
NASA 268, 274
National Bureau of Economic Research (US) 231
National Institute for Science and Technology Policy (Japan) 46
National Institute on Aging (US) 96
nationality 283
National Science Foundation (US) 91
Natural-Born Cyborgs (2003) 321
nature 353
NBIC technology 90
near earth objects (NEOs) 268
Neuromancer (1984) 306
neuroscience 89, 313, 320
neuroseparation 331
Never Ending Days of Being Dead, The (2007) 349
New Atlantis (1627) 19, 25
New Face of War, The (2003) 247
New Green History of the World, A (2007) 195
Newton, Sir Isaac 18, 38, 348
New York 112
New York World's Fair (1939) 62, 65
New York World's Fair (1964) 64, 66
Next 200 Years, The (1976) 68
Next 500 Years, The (1995) 339
nexting 3
NHS (National Health Service, UK) 231
Nicholls, Henry 199
Nielsen, Anne Skare 5
Nigeria, population of 107
Niven, Larry 268
Nostradamus 16
nuclear fusion 147
nuclear power 64, 131, 141, 143
nuclear weapons 33, 252, 271
Núñez, Rafael 12
Nutt, David 234

O

O'Neill cylinder 338
O'Neill, Gerard K. 338
O'Rourke, P.J. 75
Obama, President Barack 138
obesity 176, 221
oceans 149, 158, 261
see also acidification
Odell, Peter 124
oil 117, 120, 122, 152
oil shock 140
Omega Point 353
one-child policy, China's 99
OPEC (Organization of Petroleum Exporting Countries) 140
Optimum Population Trust 114
Oracle of Delphi 15
Order of Cosmic Engineers 328
organic farming 180
Organization for Economic Cooperation and Development (OECD) 66, 105
Origins of Futuristic Fiction (1987) 19
Out of Gas (2004) 124
Oxburgh, Ron 119
ozone 118
ozone layer 155, 335

P

Pamlin, Dennis 202
pandemics 262, 272
Pangaea 335

Pardee Center for the Study of the Longer-Range Future 291
Paris Exposition Universelle (1889) 26
Parkinson's disease 222, 234
Pattern of Expectation, The (1979) 16
Pauly, Daniel 179
peak oil 123, 126
Pelton, Joe 103
People for the Ethical Treatment of Animals (PETA) 191
Pergams, Oliver 206
permafrost 128, 159
Pew Research Center 151, 307
Phaedrus 313
pharmacogenetics 227
phase transition 276
Philosophical Essay on Probabilities (1825) 39
photosynthesis 118, 146
Physics of Immortality, The (1994) 352
phytoplankton 128
Pike, John 250
Pinker, Stephen 240
Pinkerton, James 217
Plan B 3.0: Mobilizing to Save Civilization (2008) 175
plankton 201
Plato 313
Pluto 343
Pohl, Fred 191
polar ice 128
pollution 66, 165, 179, 203
Ponting, Clive 195
population 23, 41, 43, 47, 66, 67, 71, 73, 95–115, 135, 160, 172, 178, 204, 215, 260
Population Bomb, The (1968) 66, 208
post-human 317, 320, 321, 335
Postsingular (2007) 329
Pournelle, Jerry 268
poverty 71, 74, 131, 173
precision farming 187
Predicting the Future (1993) 56
prediction 3, 14, 37, 41, 44
Pretty, Jules 196
preventive medicine 225
Prisco, Giulio 323
privacy 298, 309
probabilities, assessing 40
productionism 178, 183
Prophecies, The (c. 1555) 16
prophecy 14, 16
Prophets of Paris, The (1962) 42
Proxima Centauri 343

Q

Quadrennial Defense Review Report (US) 243
Quantico (2006) 255
quantum physics 273
Quorn 191

R

radioactive waste 143, 154
rainforests 210
RAND Corporation 46, 49, 68, 239
Rapture 18
Rapture Index 18, 73, 275
Real Environmental Crisis, The (2003) 174
red dwarfs 350
Reducing Emissions from Deforestation and Degradation) (REDD) 210
Rees, Martin 253, 334
Regis, Ed 88
religion 290
Renaissance, the 17
Rendezvous with Rama (1972) 268
reputation 282, 284, 329
research and development (R&D) 91, 244
Research 2.0 307
Revenge of Gaia, The (2006) 131
rice 182, 188
Riddley Walker (1980) 125
River of the Gods (2004) 125
Road, The (2007) 125
Roberts, Adam 17
Roberts, Paul 186
Robinson, Kim Stanley 125, 339
robots 47, 86, 105, 228, 245, 249, 295, 328, 343
robot oxygen carrier 230
robot surgery 228
rocket belt 248
Rocky Mountain Institute 140
Romm, Joseph 128
Ronald, Pamela 184
Royal Institute of International Affairs (UK) 185
Rucker, Rudy 329
Russell, Bertrand 29
Russia 105, 221, 259

S

Saffo, Paul 57
Saint-Simon, Henri de 42
San Francisco Exposition (1915) 26
Sanderson, Eric 203
sanitation 160, 164, 177
Sardar, Ziauddin 292
SARS 217, 272
scenarios 49, 50, 152, 239
Schachter, Daniel 5
Schank, Roger 313

Schmidt, Stanley 90
schools 313
Schrock, Richard 180
Schroder, Klaus-Peter 337
Schrödinger, Erwin 273
Schumpeter, Joseph 294
science fiction 53, 329
scientology 291
Scramble scenario 152
S-curve 43, 57
Second Life 306, 328
Self, Will 125
SENS Foundation 236
Shaping Things (2005) 84
Shapiro, Robert 103
Shell 50, 152
Shelley, Mary 21, 217
Shirky, Clay 302
Shock of the Old, The (2006) 93
Siberia 128
Silent Spring (1962) 42, 203
SimCity 304
Simon, Julian 69
Simonyi, Charles 341
simulations 304
Singer, Peter 249
singularity, the 322, 329, 345
Singularity is Near, The (2005) 88, 318, 345
Sixty Days and Counting (2007) 125
Skeptical Environmentalist, The (2007) 55
Sketch for a Historical Picture of the Progress of the Human Mind (1793-94) 22
Skoll Foundation 218
Sling and the Stone: On Warfare in the Twenty-first Century, The (2004) 242
Small, Gary 313
Smil, Vaclav 124, 184
Smith, Robert C. 337
Smithsonian Tropical Research Institute 201
Snow Crash (1992) 306
Soddy, Frederick 60
soil 189, 205
solar energy 142, 143, 147
solar radiation 149, 335
solar storms 271
South Pole Research Station 190
Space Adventures 341
space colonies 337
space elevator 339
Spaceguard 268
Space Merchants (1952) 191
SpaceShipOne 341
space tourism 339, 340, 341
Spaceward Foundation 340
space warfare 252
Spanish flu 217
"Speculations Concerning the First Ultraintelligent Machine" (1965) 322
Spike, The (2001) 322
spimes 84
Spore 304
stabilization wedges 143
Stapledon, Olaf 61, 317
Star Maker (1937) 61
Starship Troopers (1957) 230
State of Fear (2004) 125
State of the Future 46, 72
State of the Future Index (SOFI) 72
stem cells 192, 222
Stephenson, Neal 11, 306
Sterling, Bruce 26, 84
Stern, Lord Nicholas 131
Stern Report, The 132, 178, 210
storms 263
storytelling 41, 52
Strategic Defense Initiative 246
Strategies for Engineered Negligible Senescence (SENS) 236
Stross, Charles 329
Sun, death of the 337
superior intellect 39, 273, 353
Supermen: Tales of the Posthuman Future (2002) 329
superstring theory 349
Superstruct 305
supervolcanoes 267
Surowieki, James 303
sustainable development 178
Sustainable Development Commission 294
Suzman, Richard 96
Swan, Melanie 62
swine flu 218
synthetic biology 82

T

Tabula Rasa 275
Tapscott, Don 303, 314
taxes 284
teeth 224
Terminator, The (1984) 327
terrorism 253
Theory of the Earth (1795) 7
theranostics 230
thermodynamics 28
Thomson, William 28
Thunderball (1965) 248
Time Machine, The (1895) 31, 61
time 1–12
Tipler, Frank 352

tipping points 209
Toffler, Alvin 280, 302, 318
Tomorrow Project 290
Tomorrow's Table (2008) 184
Torino scale 269
Townsend, Anthony 308
trains, high-speed 138
Trans Time Inc 330
transcendence 323
transhumanism 317, 320, 327, 335
transmission grid 142
Transparency International 168
transport 64, 120, 137, 138
trauma pod 229
Treder, Mike 85
Tropical Diseases Research Program (WHO) 220
tropical forests 199, 210
Truman Show, The (1998) 309
Tsien, Joe 318
tsunamis 259, 266, 268
Turse, Nick 244
Tweed, Alistair 223

U

Uchegbu, Ijeoma F. 159
uncertainty principle 41
Undiscovered Mind, The (1999) 326
UNESCO 91
UNICEF 215
Union of Concerned Scientists 210
United Nations (UN) 66, 72, 98, 110, 132, 167
United Nations Environment Programme 183, 197
urbanization 204
USA, population of 108
utopia 70, 75

V

Vaas, Rüdiger 349
Venter, Craig 81, 146
Venus 337
Verne, Jules 25, 28
vertical farms 191
videophilia 206
Vinge, Vernor 325, 329
Virgin Galactic 341
viruses 218, 226
Visions (1998) 79
Visions of the Future (1995) 15
Vita-More, Natasha 332
vitrification 331
Voluntary Human Extinction Movement 263
Vonnegut, Kurt 2, 291
von Neumann, John 343
von Neumann probe 343

W

Wake, David 200
Wallace, Alfred Russell 24
Wallerstein, Immanuel 294
Walmart 151
war 33, 229, 238–256
warfare, fourth generation 242
Warwick, Kevin 320
water 157–170, 177
- drinking 160
- ground 161
- stress 162
- wars 166

Waterworld (1995) 125
weather forecasting 39
Wein, Lawrence 253
Weisman, Alan 213, 263
Wells, H.G. 2, 30, 38, 155, 317
What is Intelligence? (2009) 313
When Worlds Collide (1933) 269
white dwarfs 350
Who Will Feed China? (1995) 184
whole brain emulations (WBEs) 326
Whole Earth Discipline (2010) 112
Why Carbon Fuels Will Dominate the 21st Century's Energy Markets (2004) 124
Wikinomics (2008) 303
Wikipedia 303
Williams, Anthony 303
Williams, Austin 56
Wilson, Edward 211
Wilson, E.O. 205
wind power 145, 149
wind turbines, airborne and tethered 145
Wired for War (2009) 249
Wired magazine 85, 93
Wisdom of Crowds, The (2004) 303
Wisner, Ben 267
Wolf, Aaron T. 166
Wolfe, Nathan 218
work 283, 284
World Bank 66, 167, 176, 261, 292
World Conservation Monitoring Centre 197
World Health Organization (WHO) 215, 219
world modelling 48
World Resources Institute 159, 176
World Set Free, The (1913) 61
world's fairs 26, 63, 65
World Transhumanist Association 327
World Wide Fund for Nature (WWF) 164, 197
World Without Us, The (2007) 263
Worsley, Richard 287
Wright, Joseph 201

WWF 164, 197
Wylie, Philip 269

X

x-rays 90

Y

Year 2000, The (1967) 68
Year Million (2008) 345
Yellowstone Volcano Observatory 267
York, Christian Donald 353

Z

Zittrain, Jonathan 303